Lecture Notes in Computer Science 8307

Commenced Publication in 1973
Founding and Former Series Editors:
Gerhard Goos, Juris Hartmanis, and Jan van Leeuwen

Editorial Board

David Hutchison
Lancaster University, UK

Takeo Kanade
Carnegie Mellon University, Pittsburgh, PA, USA

Josef Kittler
University of Surrey, Guildford, UK

Jon M. Kleinberg
Cornell University, Ithaca, NY, USA

Alfred Kobsa
University of California, Irvine, CA, USA

Friedemann Mattern
ETH Zurich, Switzerland

John C. Mitchell
Stanford University, CA, USA

Moni Naor
Weizmann Institute of Science, Rehovot, Israel

Oscar Nierstrasz
University of Bern, Switzerland

C. Pandu Rangan
Indian Institute of Technology, Madras, India

Bernhard Steffen
TU Dortmund University, Germany

Madhu Sudan
Microsoft Research, Cambridge, MA, USA

Demetri Terzopoulos
University of California, Los Angeles, CA, USA

Doug Tygar
University of California, Berkeley, CA, USA

Gerhard Weikum
Max Planck Institute for Informatics, Saarbruecken, Germany

T0212386

Georges Gonthier   Michael Norrish (Eds.)

Certified Programs and Proofs

Third International Conference, CPP 2013
Melbourne, VIC, Australia, December 11-13, 2013
Proceedings

Springer

Volume Editors

Georges Gonthier
Microsoft Research Cambridge
21 Station Road
Cambridge CB1 2FB, UK
E-mail: gonthier@microsoft.com

Michael Norrish
Canberra Research Lab., NICTA
PO Box 8001
Canberra, ACT 2601, Australia
E-mail: michael.norrish@nicta.com.au

ISSN 0302-9743   e-ISSN 1611-3349
ISBN 978-3-319-03544-4   e-ISBN 978-3-319-03545-1
DOI 10.1007/978-3-319-03545-1
Springer Cham Heidelberg New York Dordrecht London

Library of Congress Control Number: 2013952466

CR Subject Classification (1998): F.3, D.2.4, I.2.3, F.4, D.3, I.1, D.2

LNCS Sublibrary: SL 1 – Theoretical Computer Science and General Issues

© Springer International Publishing Switzerland 2013

This work is subject to copyright. All rights are reserved by the Publisher, whether the whole or part of the material is concerned, specifically the rights of translation, reprinting, reuse of illustrations, recitation, broadcasting, reproduction on microfilms or in any other physical way, and transmission or information storage and retrieval, electronic adaptation, computer software, or by similar or dissimilar methodology now known or hereafter developed. Exempted from this legal reservation are brief excerpts in connection with reviews or scholarly analysis or material supplied specifically for the purpose of being entered and executed on a computer system, for exclusive use by the purchaser of the work. Duplication of this publication or parts thereof is permitted only under the provisions of the Copyright Law of the Publisher's location, in ist current version, and permission for use must always be obtained from Springer. Permissions for use may be obtained through RightsLink at the Copyright Clearance Center. Violations are liable to prosecution under the respective Copyright Law.

The use of general descriptive names, registered names, trademarks, service marks, etc. in this publication does not imply, even in the absence of a specific statement, that such names are exempt from the relevant protective laws and regulations and therefore free for general use.

While the advice and information in this book are believed to be true and accurate at the date of publication, neither the authors nor the editors nor the publisher can accept any legal responsibility for any errors or omissions that may be made. The publisher makes no warranty, express or implied, with respect to the material contained herein.

Typesetting: Camera-ready by author, data conversion by Scientific Publishing Services, Chennai, India

Printed on acid-free paper

Springer is part of Springer Science+Business Media (www.springer.com)

Preface

This volume contains the papers presented at CPP 2013, the Third International Conference on Certified Proofs and Programs, held during December 11–13, 2013, in Melbourne.

The CPP series of meetings aims to cover those topics in computer science and mathematics in which certification via formal techniques is crucial. This year's edition of CPP was co-located with APLAS 2013 (Asian Symposium on Programming Languages and Systems), similarly to CPP and APLAS 2012 in Japan, and CPP and APLAS 2011 in Taiwan. The next CPP will, however, be colocated with POPL 2015 in India, and the plan is to eventually locate CPP in Europe and North America as well as in Asia. A manifesto for CPP, written by Jean-Pierre Jouannaud and Zhong Shao, appears in the proceedings of CPP 2011 (LNCS 7086).

We were pleased that Dan Licata and Carroll Morgan accepted our invitation to be invited speakers for CPP 2013 and that Nick Benton also agreed to be a keynote speaker addressing both APLAS 2013 and CPP 2013.

The Program Committee for CPP 2013 was composed of 20 researchers from 12 countries. We received a total of 39 submissions and accepted 18 papers. Every submission was reviewed by at least four Program Committee members and their selected reviewers.

We wish to thank the Program Committee members and their reviewers for their efforts in helping to evaluate the submissions: it was a privilege to work with them. The EasyChair conference management system helped us to deal with all aspects of putting together our program. It was a pleasure to work with Peter Schachte the general chair for CPP and APLAS 2013, and with Chung-chieh Shan, the Program Committee chair for APLAS 2013. We also wish to thank the invited speakers, the authors of submitted papers, and the reviewers for their interest and strong support of this new conference series.

October 2013

Georges Gonthier
Michael Norrish

Organization

General chair

Peter Schachte   University of Melbourne, Australia

Steering Committee

Andrew Appel   Princeton University, USA
Nikolaj Bjørner   Microsoft Research
Georges Gonthier   Microsoft Research
John Harrison   Intel Corporation, USA
Jean-Pierre Jouannaud   Université Paris-Sud, France and Tsinghua
  (Co-chair)   University, China
Xavier Leroy   Inria, France
Gerwin Klein   NICTA, Australia
Tobias Nipkow   Technische Universität München, Germany
Zhong Shao (Co-chair)   Yale University, USA

Program Committee

Derek Dreyer   MPI-SWS, Germany
William Farmer   McMaster University, Canada
Jean-Christophe Filliâtre   CNRS, France
Cédric Fournet   Microsoft Research
Georges Gonthier   Microsoft Research
Benjamin Grégoire   Inria, France
Aquinas Hobor   National University of Singapore, Singapore
Reiner Hähnle   TU Darmstadt, Germany
Gyesik Lee   Hankyong National University, South Korea
Toby Murray   NICTA, Australia
Cesar Muñoz   NASA, USA
Gopalan Nadathur   University of Minnesotta, USA
Michael Norrish   NICTA, Australia
Claudio Sacerdoti Coen   University of Bologna, Italy
Peter Sewell   University of Cambridge, UK
Bas Spitters   Radboud University of Nijmegen,
                                    The Netherlands
Gang Tan   Lehigh University, USA
Alwen Tiu   Nanyang Technical University, Singapore
Yih-Kuen Tsay   National Taiwan University
Lihong Zhi   Academia Sinica, Taiwan

Additional Reviewers

Aldini, Alessandro
Aransay-Azofra, Jesus
Asperti, Andrea
Avigad, Jeremy
Barthe, Gilles
Betarte, Gustavo
Boldo, Sylvie
Bormer, Thorsten
Braibant, Thomas
Brotherston, James
Cha, Reeseo
Clouston, Ranald
Conchon, Sylvain
Dawson, Jeremy
Dupressoir, Francois
Gacek, Andrew
Harrison, John
Hasan, Osman
Hritcu, Catalin
Hur, Chung-Kil
Ilik, Danko
Im, Hyeonseung
Kim, Ik-Soon
Li, Guodong

Marché, Claude
Melquiond, Guillaume
Mu, Shin-Cheng
Nahas, Mike
Nanevski, Aleks
Narkawicz, Anthony
Neis, Georg
Palmskog, Karl
Paskevich, Andrei
Paulin, Christine
Paulin-Mohring, Christine
Pientka, Brigitte
Rayadurgam, Sanjai
Ruemmer, Philipp
Sozeau, Matthieu
Tassi, Enrico
Tsai, Ming-Hsien
Urban, Christian
Wang, Bow-Yaw
Wang, Shuling
Zavattaro, Gianluigi
Zhang, Lijun
Ziliani, Beta
Zuliani, Paolo

The "Probabilistic Information-Order for Noninterference" Competition: Do we have a winner?

Carroll Morgan

University of New South Wales, Sydney, Australia

Many information measures compete for our attention when we want to decide whether one program leaks no more information than another: an early favourite was Shannon Entropy, of course. But more recent entries are Bayes Vulnerability, Marginal Guesswork, Guessing Entropy... and the field has become rather full. It's a bit of a zoo. All of the measures have their own advantages and disadvantages, their adherents and detractors. But do any of them have mathematical properties relevant to specification and refinement? Do they admit a program algebra? Do they allow compositional reasoning?

Individually, they don't; but taken together, perhaps they do. There is new information order, recently discovered by two groups independently [1,2], which generalises Landauer and Redmond's Lattice of Information and seems to generalise much of the zoo: furthermore, it has the following two important properties. First (soundness), if a specification is refined by an implementation in this order, then the implementation leaks no more than the specification for any measure in the zoo and in any context (as determined by a pGCL-like programming language [3] equipped with hidden state). Second (completeness), if there is a (pGCL) context in which the implementation has more Bayes Vulnerability than did the specification, then the (purported) implementation cannot actually have been a refinement of the specification in this order.

I will describe the new order, and where it comes from, and will explain the properties above and why they are so important.

The practical implications are that if one wants to implement a noninterference-based secure system in a compositional way, i.e. from a specification via stepwise refinement, then there is a very strong case for using this new order to determine the proof obligations, no matter which of the other popular (and more specialised) orders your application ultimately requires.

References

1. McIver, A.K., Meinicke, L.A., Morgan, C.C. Compositional closure for Bayes risk in probabilistic noninterference. *Proc. ICALP 2010*.
2. Alvim, M.S., Chatzikokolakis, K., Palamidessi, C., Smith, G. Measuring information leakage using generalized gain functions. *Proc. CSF 2012*.
3. McIver, A.K. and Morgan, C.C. *Abstraction, Refinement and Proof for Probabilistic Systems*. Springer Monographs in Computer Science, 2005.

Table of Contents

Session 4: Mathematics

Session 5: Certified Transformations

Session 6: Security

$\pi_n(S^n)$ in Homotopy Type Theory

Daniel R. Licata[1] and Guillaume Brunerie[2]

[1] Wesleyan University
dlicata@wesleyan.edu
[2] Université de Nice Sophia Antipolis
brunerie@unice.fr

1 Introduction

Homotopy type theory [Awodey and Warren, 2009; Voevodsky, 2011] is an extension of
Martin-Löf's intensional type theory [Martin-Löf, 1975; Nordström et al., 1990] with
new principles such as Voevodsky's univalence axiom and higher-dimensional inductive types [Lumsdaine and Shulman, 2013]. These extensions are interesting both from
a computer science perspective, where they imbue the equality apparatus of type theory
with new computational meaning, and from a mathematical perspective, where they allow higher-dimensional mathematics to be expressed cleanly and elegantly in type theory. One example of higher-dimensional mathematics is the subject of homotopy theory,
a branch of algebraic topology. In homotopy theory, one studies topological spaces by
way of their points, paths (between points), homotopies (paths between paths), homotopies between homotopies (paths between paths between paths), and so on. This infinite tower of concepts—spaces, points, paths, homotopies, and so on—is modeled in
type theory by types, elements of types, proofs of equality of elements, proofs of equality of proofs of equality, and so on. A space corresponds to a type A. Points of a space
correspond to elements $a, b : A$. Paths in a space are modeled by elements of the identity
type (propositional equality), which we notate $p : a =_A b$. Homotopies between paths p
and q correspond to elements of the iterated identity type $p =_{a=_A b} q$. The rules for the
propositional equality type allow one to define the operations on paths that are considered in homotopy theory. These include identity paths $\mathsf{id} : a = a$ (reflexivity of equality),
inverse paths $!\, p : b = a$ when $p : a = b$ (symmetry of equality), and composition of paths
$q \circ p : a = c$ when $p : a = b$ and $q : b = c$ (transitivity of equality), as well as homotopies
relating these operations (for example, $\mathsf{id} \circ p = p$), and homotopies relating these homotopies, etc. This equips each type with the structure of a (weak) ∞-*groupoid*, as studied
in higher category theory [Lumsdaine, 2009; van den Berg and Garner, 2011]. In category theoretic terminology, the elements of a type correspond to objects (or 0-cells), the
proofs of equality of elements to morphisms (1-cells), the proofs of equality of proofs
of equality to 2-morphisms (2-cells), and so on.

One basic question in algebraic topology is calculating the *homotopy groups* of a
space. Given a space X with a distinguished point x_0, the *fundamental group of X at
the point* x_0 (denoted $\pi_1(X, x_0)$) is the group of loops at x_0 up to homotopy, with composition as the group operation. This fundamental group is the first in a sequence of
homotopy groups, which provide higher-dimensional information about a space: the

G. Gonthier and M. Norrish (Eds.): CPP 2013, LNCS 8307, pp. 1–16, 2013.
© Springer International Publishing Switzerland 2013

homotopy groups $\pi_n(X,x_0)$ "count" the n-dimensional loops in that space up to homotopy. $\pi_2(X,x_0)$ is the group of homotopies between id_{x_0} and itself, $\pi_3(X,x_0)$ is the group of homotopies between $id_{id_{x_0}}$ and itself, and so on. *Calculating a homotopy group* $\pi_n(X,x_0)$ is to construct a group isomorphism between $\pi_n(X,x_0)$ and some explicit description of a group, such as $\mathbb{Z}$ or $\mathbb{Z}_k$ ($\mathbb{Z}$ mod k).

The homotopy groups of a space can be difficult to calculate. This is true even for spaces as simple as the n-dimensional spheres (the circle, the sphere, ...)—some homotopy groups of spheres are currently unknown. A category-theoretic explanation for this fact is that the spheres can be presented as *free ∞-groupoids* constructed by certain generators, and it can be difficult to relate a presentation of a space as a free ∞-groupoid to an explicit description of its homotopy groups. For example, the circle is the free ∞-groupoid generated by one point and one loop:

base is a point (object) on the circle, and loop is a path (morphism) from base to itself. That the circle is the free ∞-groupoid on these generators means that all the points, paths, homotopies, etc. on the circle are constructed by applying the ∞-groupoid operations to these generators in a free way. The generator loop represents "going around the circle once counter-clockwise." Using the groupoid operations, one can construct additional paths, such as ! loop (going around the circle once clockwise) and loop ∘ loop (going around the circle twice counter-clockwise). Moreover, there are homotopies between paths, such as loop∘! loop = id (going clockwise and then counter-clockwise is the same, up to homotopy, as staying still). In this case, one can prove that, up to homotopy, every loop on the circle is either id or (loop ∘ loop ... ∘ loop) (n times, for any n) or (! loop∘! loop ... ∘! loop) (n times, for any n), and thus that the loops on the circle are in bijective correspondence with the integers. Moreover, under this bijection, concatenation of paths corresponds to addition of integers. Thus, the fundamental group of the circle is $\mathbb{Z}$.

However, in general, it can be quite difficult to relate a presentation of a space as a free ∞-groupoid to an explicit description of its homotopy groups, in part because of *action across levels*. For example, the sphere can be presented as the free ∞-groupoid generated by one point (0-cell) base and one homotopy (2-cell) $loop_2$ between id_{base} (the path that stands still at base) and itself—think of $loop_2$ as "going around the the surface of the sphere." An ∞-groupoid has group operations at each level, so just as we have identity, inverse, and composition operations on paths (1-cells), we have identity, inverse, and composition operations on homotopies (2-cells). Thus, we can form homotopies such as $loop_2 \circ loop_2$ (going around the surface of the sphere twice) and $loop_2∘! loop_2$ (going around the surface of the sphere once in one direction, and then in the opposite direction)—and, analogously to above, there is a homotopy-between-homotopies relating the latter path to the constant homotopy ($loop_2∘! loop_2 = id_{id_{base}}$). Thus, one would expect that the homotopies (2-cells) on the sphere have the same structure as the paths (1-cells) on the circle, and this is indeed the case: $\pi_2(S^2)$ is also $\mathbb{Z}$. However, ∞-groupoids have more structure than just the group operations at each level—for example, lower-dimensional generators can construct higher-dimensional paths. An example of this is that $\pi_3(S^2)$, the group of homotopies between homotopies (3-cells) on the sphere, is also $\mathbb{Z}$, *despite the fact that there are no generators for 3-cells in the presentation of the sphere!* The paths arise from the applying the algebraic operations of an ∞-groupoid to the 2-cell generator loop_2—and this action across levels is one reason that homotopy groups are so difficult to calculate.

One enticing idea is to use homotopy type theory to calculate homotopy groups: by doing so, we can give computer-checked proofs of these calculations, and we can potentially exploit constructivity and the type-theoretic perspective ∞-groupoids to attack these difficult problems in algebraic topology. To pose the problem of calculating a homotopy group in homotopy type theory, we use two ingredients.

First, we describe basic spaces using *higher inductive types*, which generalize ordinary inductive types by allowing constructors not only for elements of the type, but for paths (proofs of equality) in the type. For example, the circle is represented by a higher inductive type with two constructors

$$\mathsf{base} : S^1$$
$$\mathsf{loop} : \mathsf{base} =_{S^1} \mathsf{base}$$

This says that base is a point on the circle, while loop is a path from base to base. In type theory, we express that S^1 is the *free* type with these generators by an elimination rule: to define a function from S^1 into any other type C, it suffices to give a point in $c : C$, which is the image of base, and a loop $p : c =_C c$, which is the image of loop:

$$\frac{c : C \quad p : c =_C c}{S^1 - \mathsf{rec}_C(c, p) : S^1 \to C}$$

That is, to define a function $S^1 \to C$, it suffices to find a "circle" in C, which gives the image of the generators.

The computation rules for this elimination rule are as follows:

$$S^1 - \mathsf{rec}_C(c, p)\mathsf{base} :\equiv c$$
$$\mathsf{ap}(S^1 - \mathsf{rec}_C(c, p))\mathsf{loop} := p$$

ap ("**action on paths**") applies a function $f : A \to B$ to a path $p : a =_A a'$ to produce a path $fa =_B fa'$. Note that the first computation rule is a definitional equality, while the second is a propositional equality/path—while future versions of homotopy type theory might take this to be a definitional equality, the known semantics of higher inductive types justifies only a propositional equality [Lumsdaine and Shulman, 2013]. To express freeness, we also need to know that $S^1 - \mathsf{rec}_C(c, p)$ is the unique such map, up to homotopy. This can be expressed either by generalizing the simple elimination rule to a dependent elimination rule, or by adding an η-equality axiom (see [Awodey et al., 2012] for a discussion of these alternatives for ordinary inductive types).

The second ingredient is to define the homotopy groups of a type. One might think that we could define the homotopy groups by iterating the identity type:

$$\pi_1(X, x_0) := x_0 =_X x_0$$
$$\pi_2(X, x_0) := \mathsf{id}_{x_0} =_{(x_0 =_X x_0)} \mathsf{id}_{x_0}$$
$$\pi_3(X, x_0) := \mathsf{id}_{\mathsf{id}_{x_0}} =_{(\mathsf{id}_{x_0} =_{x_0 =_X x_0} \mathsf{id}_{x_0})} \mathsf{id}_{\mathsf{id}_{x_0}}$$

and so on. However, these iterated identity types may still have non-trivial higher-dimensional structure. The nth homotopy group considers only the structure up to level n, so we need to "kill" the higher-dimensional structure of these types. Thus, we first define the nth *loop space* $\Omega^n(X,x_0)$ so that

$$\Omega^1(X,x_0) := x_0 =_X x_0$$
$$\Omega^2(X,x_0) := id_{x_0} =_{(x_0=_X x_0)} id_{x_0}$$
$$\Omega^3(X,x_0) := id_{id_{x_0}} =_{(id_{x_0}=_{x_0=_X x_0} id_{x_0})} id_{id_{x_0}}$$

and so on. We write $\Omega(X,x_0)$ for $\Omega^1(X,x_0)$.

Then we can define

$$\pi_n(X,x_0) := ||\Omega^n(X,x_0)||_0$$

where $||A||_0$, the *0-truncation of A*, is a *set* (a type with no higher structure—any two paths are homotopic) constructed by "killing" the higher-dimensional structure of A—i.e. equating any two paths between the same two points. For a more leisurely introduction to these definitions, we refer the reader to previous work [Licata and Shulman, 2013; The Univalent Foundations Program, 2013].

Using these two ingredients, we can use homotopy type theory to calculate homotopy groups of spaces: define the space X as a higher inductive type, and give a group isomorphism between the type $\pi_n(X,x_0)$ (with path composition as the group structure) and an explicit description of a group like $\mathbb{Z}$. In this note, we give a calculation of the fact that $\pi_n(S^n) = \mathbb{Z}$. That is, we describe a proof in homotopy type theory that the nth homotopy group of the n-dimensional sphere is isomorphic to $\mathbb{Z}$. This proof is interesting for several reasons:

– Calculating $\pi_n(S^n)$ is a fairly easy theorem in algebraic topology (e.g. it would be covered in a first- or second-year graduate algebraic topology course), but it is more complex than many of the results that had previously been proved in homotopy type theory. For example, it was one of the first results about an infinite family of spaces, of variable dimension, to be proved in homotopy type theory.
– When doing homotopy theory in a constructive/synthetic style in homotopy type theory, there is always the possibility that classical results will not be provable—the logical axioms for spaces might not be strong enough to prove certain classical theorems about them. Our proof shows that the characterization of $\pi_n(S^n)$ does follow from a higher-inductive description of the spheres, in the presence of univalence, which provides evidence for the usefulness of these definitions and methods.
  Moreover, while we do not yet have a full computational interpretation of univalence, one can see, in the proof, a computational process that transforms n-dimensional loops on the n-sphere into integers. This is one of the first examples of computation with arbitrary-dimensional structures that has been considered in homotopy type theory.
– The proof is not a transcription of a textbook homotopy theoretic proof, but mixes classical ideas with type-theoretic ones. The type-theoretic techniques used here have been applied in other proofs. For example, the proof described here led to a simpler proof of a more general theorem, the Freudenthal Suspension Theorem (Lumsdaine's proof is described in the HoTT

book [The Univalent Foundations Program, 2013]), which gives a shorter calculation of $\pi_n(S^n)$ (also described in the HoTT book). This, in turn, led to a proof of an even more general theorem, the Blakers-Massey theorem [Finster et al., 2013].

- We give a direct higher-inductive definition of the n-dimensional sphere S^n as the free type with a base point base and a loop in $\Omega^n(S^n)$. This definition does not fall into the collection of higher inductive types that has been formally justified by a semantics, because it involves a path constructor at a variable level (i.e. in Ω^n, where n is an internal natural number variable). However, our result shows that it is a useful characterization of the spheres to work with, and it has prompted some work on generalizing schemas for higher inductive types to account for these sorts of definitions.
- The proof we present here includes an investigation of some of the type-theoretic structure of loop spaces. We will characterize $\Omega^n(A)$ for various types A, which explains how concepts such as function extensionality and univalence induce paths at higher levels. This characterization could potentially inform investigations into the computational interpretation of univalence, a major open problem.
- The proof has been formalized in Agda [Norell, 2007], and is available on GitHub in the repository github.com/dlicata335/hott-agda (tag pinsn-cpp-paper). The proof includes a library of lemmas about iterated loop spaces that is independent of the particular application to n-dimensional spheres.

In the remainder of this paper, we give an informal overview of some of the interesting aspects of the proof, and discuss the Agda formalization. From this point forward, we assume that the reader is familiar with the calculation of $\pi_1(S^1)$ [Licata and Shulman, 2013] and with Part I and Chapter 8 of the HoTT book [The Univalent Foundations Program, 2013].

2   Overview of the Proof

2.1   Definition of the Spheres

We define the n-dimensional sphere S^n (for $n \geq 1$) as the higher inductive type generated by one point base and one point in the nth loop space of S^n at base:

$$\text{base}_n : S^n$$
$$\text{loop}_n : \Omega^n(S^n, \text{base})$$

The corresponding elimination rule, *sphere recursion*, says that to define a function $S^n \to C$, it suffices to give a point $c : C$ and a loop in $\Omega^n(C, c)$:

$$\frac{C : \text{Type} \quad c : C \quad p : \Omega^n(C, c)}{S^n - \text{rec}_C(c, p) : S^n \to C}$$

The computation rules for this elimination rule are as follows:

$$S^n - \text{rec}_C(c, p)\text{base}_n :\equiv c$$
$$\text{ap}^n(S^n - \text{rec}_C(c, p))\text{loop}_n := p$$

where ap^n applies a function $f : A \to B$ to an n-dimensional loop in $\Omega^n(A,a)$ to get an n-dimensional loop in $\Omega^n(B, fa)$. We discuss the definition of $\Omega^n(X, x_0)$ and ap^n in Section 2.4 below.

We also require a dependent elimination rule, *sphere induction*:

$$\frac{C : S^n \to \mathsf{Type} \quad c : C(\mathsf{base}_n) \quad p : \Omega^n_{\mathsf{loop}_n}(C, c)}{S^n\text{--elim}_C(c, p) : \Pi x{:}S^n.\, C(x)}$$

Here, the type $\Omega^n_p(C, c)$ represents an "n-dimensional loop-over-a-loop"; it is well-formed when $C : A \to \mathsf{Type}$ and $p : \Omega^n(A, a)$ and $c : C(a)$ (for some A and a). Topologically, it represents an n-dimensional path at c in the total space of C that projects down to p. We discuss the definition of $\Omega^n_p(C, c)$ in Section 2.4 below. The computation rules for sphere induction are similar to those for sphere recursion.

Note that the constructor loop_n is a path whose level depends on n: for $n = 1$, it is a path, for $n = 2$, it is a path between paths, and so on. Because of this, the above definition of S^n does not fall into any of the schemas for higher inductive types that have been formally studied. However, it seems like a sensible notion, because for any fixed n, it expands to a type a higher inductive constructor would be permitted to have: For $n = 1$, it is $\mathsf{base}_1 =_{S^1} \mathsf{base}_1$, for $n = 2$, it is $\mathsf{id} =_{\mathsf{base}_2 =_{S^2} \mathsf{base}_2} \mathsf{id}$, and so on. All of these are iterated identity types in S^n, which is the type being defined. We leave it to future work to justify this kind of definition semantically. Another possible justification would be to take the above rules not as a specification of a higher-inductive type, but as an interface, and implement it by the definition of S^n by iterated suspension [The Univalent Foundations Program, 2013, Section 6.5].

2.2 Calculation of $\pi_n(S^n)$

We now describe the calculation that $\pi_n(S^n) = \mathbb{Z}$ for $n \geq 1$.[1] Formally, this statement means that there is a group isomorphism between the group $\pi_n(S^n)$ (with composition as the group operation) and the additive group $\mathbb{Z}$. In what follows, we will discuss the proof that the *type* $\pi_n(S^n)$ is equivalent (and hence equal, by univalence) to the type $\mathbb{Z}$, and omit the proof that this equivalence sends composition to addition.

The first step is an induction on n. In the base case, we use the homotopy type theory proof of $\pi_1(S^1) = \mathbb{Z}$ described in previous work [Licata and Shulman, 2013]. In the inductive step, the key lemma is that $\pi_{n+1}(S^{n+1}) = \pi_n(S^n)$, which, combined with the inductive hypothesis gives the result.

To show that $\pi_{n+1}(S^{n+1}) = \pi_n(S^n)$, we calculate as follows:[2]

---

[1] We write $\pi_n(X)$ for $\pi_n(X, x_0)$ (and similarly for Ω) when the base point is clear from context. For the spheres, if we elide the base point, it is the constructor base_n.

[2] We use the convention of eliding the base point heavily here. The base point of $\Omega(A)$ is id_a, when a is the base point of A. The base point of $||A||_k$ is $|a|$, where a is the base point of A, and $|-|$ is the constructor for the n-truncation type $||A||_n$ [The Univalent Foundations Program, 2013, Section 7.3].

$$
\begin{aligned}
\pi_{n+1}(S^{n+1}) &= ||\Omega^{n+1}(S^{n+1})||_0 & \text{definition} \\
&= ||\Omega^n(\Omega(S^{n+1}))||_0 & \text{unfold } \Omega^{n+1} \\
&= \Omega^n(||\Omega(S^{n+1})||_n) & \text{swap truncation and loop space} \\
&= \Omega^n(||S^n||_n) & \text{main lemma} \\
&= ||\Omega^n(S^n)||_0 & \text{swap truncation and loop space} \\
&= \pi_n(S^n) & \text{definition}
\end{aligned}
$$

Several of these steps are relatively easy lemmas. For example, we can unfold $\Omega^{n+1}(X)$ as $\Omega^n(\Omega(X))$—one might take this as the definition of Ω^{n+1}, but for the definition below it is a lemma, which we will call LoopPath. Additionally, there is a rule for swapping a truncation with a loop space, incrementing the index:

$$||\Omega(X)||_n = \Omega(||X||_{n+1})$$

Intuitively, $||\Omega(X)||_n$ is a type built from $\Omega(X)$ by equating all $(n+1)$-cells in $\Omega(X)$. However, $\Omega(X)$ is the space of 1-cells (paths) in X, so the $(n+1)$-cells in $\Omega(X)$ are the $(n+2)$-cells in X. Thus, it is equivalent to equate all $(n+2)$-cells in X, and then take the loop space. Iterating this reasoning gives the equation used above, that

$$||\Omega^n(X)||_0 = \Omega^n(||X||_n)$$

This reasoning reduces the problem to proving the main lemma, that

$$||\Omega(S^{n+1})||_n = ||S^n||_n$$

That is, the loop space on the $(n+1)$-sphere is equivalent to the n-sphere, when appropriately truncated. $\Omega(S^{n+1})$ is, by definition, the type $\text{base}_{n+1} =_{S^{n+1}} \text{base}_{n+1}$, so this lemma is characterizing (the truncation of) a path space of a higher-inductive type. A general template for doing such characterizations is the *encode-decode method* [The Univalent Foundations Program, 2013, Section 8.9], which we apply here.

2.3   The Encode-Decode Argument

The bulk of the proof consists of proving that $||\Omega(S^{n+1})||_n = ||S^n||_n$. To build intuition, consider the case of $||\Omega(S^2)||_1 = ||S^1||_1$. Setting aside the truncations for the moment, this means we are comparing points on S^1 (the circle) with loops on S^2 (the sphere). The idea is to set up a correspondence where the base point of the circle (base_1) corresponds to the constant path at the base point of the sphere ($\text{id}_{\text{base}_2}$), and going n times around the loop of the circle (loop_1^n) corresponds to going n times around the surface of the sphere (loop_2^n). In classical topology, it is clear that this correspondence induces a bijection between the set of points on the circle and the set of loops on the sphere (both considered up to homotopy): the circle has one connected component, and any loop on the sphere can be contracted to the constant loop. Moreover, it induces a bijection between the set of loops on the circle and the set of 2-loops on the sphere (again considered up to homotopy), because *every* loop on the circle is loop_1^n for some n, and *every* 2-loop on the sphere is loop_2^n for some n. While proving these facts is essentially what we are doing in this section, it should at least be intuitively plausible that the points and loops on the circle are the same as the loops and 2-dimensional loops on the sphere. But the loops and 2-dimensional loops on the sphere are the points and loops of the loop space of the sphere, $\Omega(S^2)$, so the points and loops of S^1 are the same as the points and loops of $\Omega(S^2)$.

However, it is not the case that S^1 and $\Omega(S^2)$ are equivalent types, because S^2 has non-trivial 3-dimensional paths, while S^1 has only trivial 2-dimensional paths. Thus, the situation is that the points and paths of S^1 are in correspondence with the points and paths of $\Omega(S^2)$, but the correspondence does not extend to higher dimensions. The role of the truncation is to account for this difference. Comparing $||S^1||_1$ with $||\Omega(S^2)||_1$ considers only points and paths, not any higher-dimensional cells, and restricted to points and paths the above correspondence is in fact an equivalence.

To prove the lemma, we proceed as follows. First, we define a map $S^n \to \Omega(S^{n+1})$ by sphere recursion, where

$$\text{promote} : S^n \to \Omega(S^{n+1})$$
$$\text{promote}(\text{base}_n) :\equiv \text{id}_{\text{base}_{n+1}}$$
$$\text{ap}^n \text{ promote}(\text{loop}_n) := \text{loop}_{n+1}$$

In the third line, we omit an application of LoopPath, which coerces $\Omega^{n+1}(S^{n+1})$, the type of loop_{n+1}, to the required type $\Omega^n(\Omega(S^{n+1}))$—$\text{loop}_{n+1}$ can be seen as an n-dimensional loop in $\Omega(S^{n+1})$. promote is one direction of the correspondence described above; for example, it sends loops on the circle to 2-dimensional loops on the sphere. Because functions are functors, we specify only the action on the generators base_n and loop_n; the function automatically preserves identity, composition, etc., and thus, for example, takes the n-fold composition loop_1^n to the n-fold composition loop_2^n, as desired. Thinking of S^{n+1} as the suspension of S^n, this is the meridian map of the suspension, which embeds X into $\Omega(\Sigma X)$—i.e. it's the unit of the suspension/loop space adjunction.

Because truncation is functorial, promote extends to a map $||S^n||_n \to ||\Omega(S^{n+1})||_n$:

$$\text{decode}' : ||S^n||_n \to ||\Omega(S^{n+1})||_n$$
$$\text{decode}'(|x|) = |\text{promote}(x)|$$

(where peeling off the truncation is permitted by truncation-elimination, because $|| - ||_n$ is an n-type).

We would like to show that this map is an equivalence. To define the inverse map $||\Omega(S^{n+1})||_n \to ||S^n||_n$, we need to define a map out of (the truncation of) a path space. One central tool for doing this is univalence: we define a map from S^{n+1} into the universe, so that the base point of S^{n+1} is sent to $||S^n||_n$ and paths in S^{n+1} are sent to equivalences, and then we apply the equivalence determined by a loop to the base point of $||S^n||_n$. In this case, we define a fibration Codes by sphere recursion

$$\text{Codes} : S^{n+1} \to \text{Type}$$
$$\text{Codes}(\text{base}_{n+1}) :\equiv ||S^n||_n$$
$$\text{ap}^n \text{ Codes}(\text{loop}_{n+1}) := (\ldots : \Omega^{n+1}(\text{Type}, ||S^n||_n))$$

The fiber over the base point is $||S^n||_n$, so Codes will send an element of $\Omega(S^{n+1})$ (which, recall, is notation for $\text{base}_{n+1} =_{S^{n+1}} \text{base}_{n+1}$) to an equivalence $||S^n||_n \simeq ||S^n||_n$.

Thus, we can define a function $\text{encode}' : ||\Omega(S^{n+1})||_n \to ||S^n||_n$ by applying Codes to the given path (after peeling off the truncation brackets, which is allowed because the result is an n-type), and then applying the resulting equivalence to $|\text{base}_n|$:

$$\text{encode}' : ||\Omega(S^{n+1})||_n \to ||S^n||_n$$
$$\text{encode}'(|p|) = (\text{ap}(\text{Codes})p)|\text{base}_n|$$

Eliding truncations for a moment, the term $\text{ap}^n(\text{encode}')$ is a function from $\Omega^n(\Omega(S^{n+1}))$ to $\Omega^n(S^n)$, so it determines (by LoopPath) a function from $\Omega^{n+1}(S^{n+1})$ to $\Omega^n(S^n)$. Because we would like encode' to be inverse to decode', we need to fill in the ... in the definition of Codes so that $\text{ap}^n \text{encode}'$ sends loop_{n+1} to loop_n (modulo truncations and LoopPath). Thus, we need an element of $\Omega^{n+1}(\text{Type}, ||S^n||_n)$ that is somehow determined by loop_n, so that we get loop_n back out when we apply encode'.

The key maneuver is to apply an equivalence between $\Omega^{n+1}(\text{Type}, A)$ and $\Pi x{:}A.\,\Omega^n(A, x)$ (discussed below). That is, an $n+1$-dimensional loop in the space of types with base point A is the same as a family of n-dimensional loops in A, given for each point in A. This reduces the problem to giving an element of type

$$\Pi x{:}||S^n||_n.\,\Omega^n(||S^n||_n, x)$$

which (modulo some truncation manipulation) is defined by sphere-elimination, sending base_n to loop_n, and proving that this choice respects loop_n. For $n = 1$, a small calculation is needed to prove this final condition, and for any greater n it is trivial by truncation reasons. This "packages up" loop_n in the Codes fibration in such a way that encode' extracts it.

This defines a fibration Codes over S^{n+1}, where the fiber over the base point is $||S^n||_n$, and the lifting of the $n+1$-loop is an n-dimensional homotopy given by "going around loop_n once". For $n = 1$, this is a fibration over S^2, where the fiber over the base is S^1 (S_1 is already a 1-type, so the truncation cancels), and the lifting of loop_2 goes around the circle—i.e., it is the Hopf fibration (though we do not need to calculate the total space to prove our theorem).

Now that we have defined encode' and decode', the task is to show that they are mutually inverse. The remaining steps required to do so are as follows:

– First, we do a calculation to show that $\text{encode}'(\text{decode}'c) = c$. The proof uses sphere-elimination, and calculations with the loop space library—this is where we prove that encode' takes loop_{n+1} to loop_n.
– The definition of encode' given above in fact has a more general type: it works not only for loops, but for paths to any endpoint x:

$$\text{encode} : \Pi x{:}S^{n+1}.\,(\text{base}_{n+1} =_{S^{n+1}} x) \to \text{Codes}(x)$$
$$\text{encode}(|p|) = (\text{ap}(\text{Codes})p)|\text{base}_n|$$

– Through a somewhat involved calculation with the loop space library, we can show that decode' extends to a function

$$\text{decode} : \Pi x{:}S^{n+1}.\,\text{Codes}(x) \to (\text{base}_{n+1} =_{S^{n+1}} x)$$

This function is defined by sphere elimination; when x is base_{n+1}, the function is decode'; then we have to prove that this choice respects the loop.

– Now that we have generalized to encode and decode, it is easy to show that

$$\Pi x : S^{n+1}, p : \|\mathsf{base}_{n+1} = x\|.\mathsf{decode}_x(\mathsf{encode}_x(p)) = p$$

by path induction, because on the identity path, it is true by definition.

The details of these steps are somewhat intricate, so we refer the reader to the Agda proof.

2.4 Loop Space Library

Next, we give a brief overview of some of the key lemmas in the loop space library.

Definitions, groupoid and functor structure First, we define

$$\Omega(X : \mathsf{Type}, x_0 : X) : \mathsf{Type}$$
$$\Omega(X, x_0) :\equiv (x_0 =_X x_0)$$

$$\Omega^{n \geq 1}(X : \mathsf{Type}, x_0 : X) : \mathsf{Type}$$
$$\Omega^1(X, x_0) :\equiv \Omega(X, x_0)$$
$$\Omega^{1+n}(X, x_0) :\equiv \Omega(\Omega^n(X, x_0), \mathsf{id}^n)$$

$$\mathsf{id}^n : \Omega(X, x_0)$$
$$\mathsf{id}^1 :\equiv \mathsf{id}_{x_0}$$
$$\mathsf{id}^{1+n} :\equiv \mathsf{id}_{\mathsf{id}^n}$$

That is, we define Ω^n mutually with a point id^n of it. This definition unfolds as $\Omega^{1+n}(X) = \Omega(\Omega^n(X))$. An alternative is to define

$$\Omega^1_0(X, x_0) :\equiv \Omega(X, x_0)$$
$$\Omega^{1+n}_0(X, x_0) :\equiv \Omega^n_0(\Omega(X), \mathsf{id}_{x_0})$$

i.e. $\Omega^{1+n}_0(X) = \Omega^n_0(\Omega(X))$. By the LoopPath lemma, these two definitions are equivalent, so we can unfold Ω^{1+n} in both ways. The current loop space library takes Ω as the main definition and uses Ω_0 as an auxiliary notion. However, since the different definitions have different definitional behaviors, it would be interesting to try revising the library based on taking Ω_0 as the main notion, to see if it is simpler or not.

In addition to id^n, there are also inverse and composition operations on each loop space:

$$!^n(l : \Omega^n(X, x_0)) : \Omega^n(X, x_0)$$
$$(l_1 : \Omega^n(X, x_0)) \circ^n (l_2 : \Omega^n(X, x_0)) : \Omega^n(X, x_0)$$

We prove various groupoid laws for these operations (unit, involution).

Many loop space operations are defined by induction on n. For example, consider applying a function to a loop. Intuitively, the idea is that $\mathsf{ap}^n f\, l$ iterates ap to apply f

at the appropriate level. For example, $\mathrm{ap}^2 f$ should be $\mathrm{ap}(\mathrm{ap}\, f)$, while $\mathrm{ap}^3 f$ should be $\mathrm{ap}(\mathrm{ap}\,(\mathrm{ap}\, f))$. In general, it is defined as follows:

$$\mathrm{ap}^n(f : X \to Y)(l : \Omega^n(X, x_0)) : \Omega^n(Y, f x_0)$$
$$\mathrm{ap}^1 f\, l :\equiv \mathrm{ap} f l$$
$$\mathrm{ap}^{1+n} f\, l :\equiv \mathrm{ap}^n\text{-id} \circ \mathrm{ap}(\mathrm{ap}^n f) l \circ (!\mathrm{ap}^n\text{-id})$$

In the $1 + n$ case, $l : \Omega(\Omega^n(X))$. The recursive call $\mathrm{ap}^n f$ has type $\Omega^n(X, x_0) \to \Omega^n(Y, f(x_0))$. Thus, using ap to apply this to the path l gives an element of

$$\mathrm{ap}^n f\, \mathrm{id}^n = \mathrm{ap}^n f\, \mathrm{id}^n$$

We require an element of $\mathrm{id}^n = \mathrm{id}^n$, so we compose on both sides with a proof ap^n-id that ap^n preserves identities. Many definitions on Ω^n follow this template: an induction on n, defined mutually with a lemma stating preservation of identities. ap^n also preserves inverses and composition, and is functorial in the function position:

$$\mathrm{ap}^n(\lambda x.x)l = l$$
$$\mathrm{ap}^n(g \circ f)l = \mathrm{ap}^n g(\mathrm{ap}^n f\, l)$$

Another key lemma is that ap^{1+n} can be unfolded in the other assocativity: above, we essentially defined $\mathrm{ap}^{1+n} = \mathrm{ap}(\mathrm{ap}^n f)$. We can prove that it is also equal to $\mathrm{ap}^n(\mathrm{ap} f)$ (with the appropriate LoopPath coercions inserted).

Some properties hold only for $n \geq 2$. For example,

$$\mathrm{ap}^n\, !\, l = !^n l$$

This follows from the Eckmann-Hilton argument, which shows that the higher homotopy groups are abelian, and that the two different ways of composing higher-dimensional loops are homotopic.

Loops in Types. For many types A, one can give a straightforward characterization of the paths in A. For example:

- Paths $f =_{A \to B} g$ are equivalent to paths $\Pi x{:}A.fx =_B gx$, by function extensionality.
- Paths $A =_{\mathrm{Type}} B$ are equivalent to equivalences between A and B, by univalence.
- Paths $e_1 =_{A \simeq B} e_2$ between equivalences are equivalent to paths $e_1 =_{A \to B} e_2$ (where we implicitly cast e_i from an equivalence to a function), because being an equivalence is an hprop.

One important piece of the loop space library is an investigation of how these characterizations extend to higher-dimensional loop spaces.

Functions. First, we characterize $\Omega^n(\Pi x{:}A.B, f)$. For $n = 1$, it is equal to $\Pi x{:}A.\Omega^n (B, fx)$ by function extensionality. For $n = 2$, the question is to characterize the type

$$\mathrm{id}_f =_{f =_{\Pi x{:}A.B} f} \mathrm{id}_f$$

But by applying function extensionality (and using its action on id_f), this type is equivalent to

$$(\lambda x.\mathrm{id}_{fx}) =_{\Pi x:A.\,fx=_{B(x)}fx} (\lambda x.\mathrm{id}_{fx})$$

Using function extensionality again, this type is equivalent to

$$\Pi x{:}A.\,\mathit{id}_{fx} =_{fx=_{B(x)}fx} \mathrm{id}_{fx}$$

which is $\Pi x{:}A.\,\Omega^2(B(x),fx)$.

Indeed, in general, we prove

$$\Omega^n(\Pi x{:}A.B, f) = \Pi x{:}A.\,\Omega^n(B(x),fx)$$

That is, a loop in a function space is a family of loops.

Paths between types. Second, we characterize $\Omega^n(A =_{\mathsf{Type}} A, p)$, where p is path in the universe from A to A. Consider $n = 1$: by univalence, we know that $\Omega(A = A, p)$ is equivalent to $\Omega(A \simeq A, p^*)$, where p^* is the equivalence induced by the path p. But a path between equivalences is equivalent to a path between the underlying functions: an equivalence between A and B is a pair (f, i) where $f : A \to B$ and $i : \mathsf{IsEquiv}(f)$, and being an equivalence is an hprop, so the second components of such pairs are always equal. Thus,

$$\Omega(A = A, p) = \Omega(A \to A, \mathsf{coe}(p))$$

where $\mathsf{coe}(p : A = B) : A \to B$—i.e. coe ("coerce") can be thought of as turning the path into an equivalence, and then selecting the "forward" direction. We can prove by induction that this rule extends to higher dimensions, so that

$$\Omega^n(A = A, p) = \Omega^n(A \to A, \mathsf{coe}(p))$$

Loop spaces. The LoopPath lemma mentioned above also fits this pattern: it characterizes an n-dimensional loop in a loop space as a $(1 + n)$-dimensional loop in the underlying space:

$$\Omega^n(\Omega(A), \mathsf{id}_a) = \Omega^{1+n}(A, a)$$

Putting it all together. Combining the previous three lemmas, we have that

$$\begin{aligned}
\Omega^{1+n}(\mathsf{Type}, A) &= \Omega^n(\Omega(\mathsf{Type}, A), \mathsf{id}_A) && \text{loop in loop space} \\
&= \Omega^n(A =_{\mathsf{Type}} A, \mathsf{id}_A) && \text{definition} \\
&= \Omega^n(A \to A, \lambda x.x) && \text{loop in path between types} \\
&= \Pi x{:}A.\,\Omega^n(A, x) && \text{loop in function type}
\end{aligned}$$

This equivalence is the "key maneuver" that we used to define the Codes fibration in Section 2.3 above.

Loops over a Loop. The notion of a "path over a path" is a key ingredient of the dependent elimination rule for higher inductive types. Given $p : a_1 =_A a_2$, and a fibration $B : A \rightarrow \text{Type}$, a path over p relates $b_1 : B(a_1)$ to $b_2 : B(a_2)$. This is a kind of heterogeneous equality [McBride, 2000] between elements of different instances of B. However, the notion of a path over a path need not be taken as primitive, since it can be represented by a homogeneous path $\text{transport}_B \, p \, b_1 =_{B(a_2)} b_2$.

In the library, we define an n-dimensional loop over a loop $\Omega_p^n(B, b_0)$ (where $p : \Omega^n(A, a_0)$ and $b_0 : B(a_0)$) by induction on n, using transport at each level to account for the heterogeneity. For small n, this definition gives the expected dependent elimination rule for, e.g., S^1 and S^2 and S^3, the specific spheres whose higher inductive elimination rule had previously been written out. For example, for S^2, sphere elimination with $B : S^2 \rightarrow \text{Type}$ and $b : B(\text{base})$ requires a proof of

$$(\text{transport}(\lambda x.(\text{transport} \, B \, x \, b) = b) \, \text{loop}_2 \, \text{id}) = \text{id}$$

A key lemma relates this definition of loop-over-a-loop to an alternate characterization, which coalesces all of the transports into a single use of ap^n combined with the equivalence defined above between $\Omega^{1+n}(\text{Type}, A)$ and $\Pi x{:}A.\, \Omega^n(A, x)$. The reason this lemma is important is that we then give rules for $\text{ap}^n A$ driven by the structure of A, such as when A is $\lambda x.A_1(x) \rightarrow A_2(x)$ or when A is $\lambda x.f(x) = g(x)$ or when A is $\lambda x.\|A_1(x)\|_k$. These rules are higher-dimensional analogues of the computation rules for transport_A that are driven by the structure of A.

3 Formalization

The calculation of $\pi_n(S^n)$ described in Sections 2.2 and 2.3 is in homotopy/PiNSN.agda; it is about 250 lines of code. The loop space library described in Section 2.4 is in lib/loopspace/; it is about 1500 lines of code. The proof of $\pi_n(S^n)$, including specifying the lemmas it uses from the loop space library, took a few days. The loop space library then took a couple of weeks to complete, working for perhaps 4 hours per day.

The formalization has one cheat, which is that it takes place in an Agda homotopy type theory library that uses type:type as a terser form of universe polymorphism than Agda's. More recent homotopy type theory libraries use Agda's universe polymorphism instead of type:type, and we believe that the proof could be ported to such a library.

Agda does not provide very much proof automation for this proof: the bulk of the proof is manual equational calculations with the loop space operations. However, with an improved computational understanding of homotopy type theory, some of these calculation steps might be definitional equalities.

That said, Agda was quite useful as a proof checker, and for telling us what we needed to prove next. The terms involved in the calculations get quite long, so it would be difficult to do these calculations, or to have confidence that they were done correctly, without the use of a proof checker.

It is worth describing one new device that we developed for this proof, which is a combination of a mathematical and a engineering insight. Often in this proof, we are manipulating paths that have the form $p \circ q \circ \, ! \, p$ (q conjugated by p), where q is thought of as "the actual path of interest" and p is some "coercion" or "type cast" that shows that it has some desired type. Early in the development of the proof, we got stuck, because manipulating these coercions explicitly gets quite cumbersome.

The engineering insight is that, if we define a function adj p q that returns p ∘ q ∘ !p, but make it *abstract* (i.e. hide its definition), then Agda can fill in in terms of the form adj _ q in the middle of an equational calculation by unification. By stating the coercions at the beginning and end of the proof, and using lemmas that propagate this information without explicitly mentioning it, we need not state the coercions at each step of the proof. Though adj p q is abstract, we export a propositional equality equating it to p ∘ q ∘ !p, so that we can use this technique in the intermediate steps of a calculation; the overall theorem is the same.

The mathematical insight is that, for an element p of a doubly-iterated identity type (i.e. when p is at least a path between paths), for any coercions q and q' of the same type, $(q \circ p \circ ! q) = (q' \circ p \circ !q')$. This is a consequence of the higher homotopy groups being abelian.

Combining these two insights, we can let Agda infer the coercions as we proceed through the steps of the proof, and then, at the end, when we need the inferred coercion to turn out to be a specific one, we simply apply the lemma. As a practical matter, this technique for managing these coercions was essential to our being able to complete this proof.

For example, here is a snippet of an equational deduction without applying the technique:

```
adjust (ap^-id n (λ f → f x) {f}) (ap (λ f → apl n f x) (adjust (λl-id n) (ap (λl n) (λ≃ a)))) ≃⟨ ... ⟩
adjust (ap^-id n (λ f → f x) {f}) (ap (λ f → apl n f x) (adj _ (ap (λl n) (λ≃ a)))) ≃⟨ .. ⟩
adj (ap^-id n (λ f₁ → f₁ x)) (ap (λ f → apl n f x) (adj _ (ap (λl n) (λ≃ a)))) ≃⟨ ... ⟩
adj (ap^-id n (λ f₁ → f₁ x) ∘ ap (λ f' → apl n f' x) (λl-id n)) (ap (λ f → apl n f x) (ap (λl n) (λ≃ a))) ≃⟨ ... ⟩
adj (ap^-id n (λ f₁ → f₁ x) ∘ ap (λ f' → apl n f' x) (λl-id n)) (ap (λ f → apl n (λl n f) x) (λ≃ a)) ≃⟨ ... ⟩
adj ((ap^-id n (λ f₁ → f₁ x) ∘ ap (λ f' → apl n f' x) (λl-id n)) ∘ ap≃ (! (β n (λ x₁ → id^ n)))) (ap (λ f → f x) (λ≃ a)) ≃⟨ ... ⟩
adj ((ap^-id n (λ f₁ → f₁ x) ∘ ap (λ f' → apl n f' x) (λl-id n)) ∘ ap≃ (! (β n (λ x₁ → id^ n)))) (a x) ≃⟨ ... ⟩
adj id (a x) ≃⟨ ! (adj-id _) ⟩
a x ∎
```

Here is the same snippet, where we use the technique and replace the first argument to adj with _:

```
adjust (ap^-id n (λ f → f x) {f}) (ap (λ f → apl n f x) (adjust (λl-id n) (ap (λl n) (λ≃ a)))) ≃⟨ ... ⟩
adjust (ap^-id n (λ f → f x) {f}) (ap (λ f → apl n f x) (adj _ (ap (λl n) (λ≃ a)))) ≃⟨ ... ⟩
adj _ (ap (λ f → apl n f x) (adj _ (ap (λl n) (λ≃ a)))) ≃⟨ ... ⟩
adj _ (ap (λ f → apl n f x) (ap (λl n) (λ≃ a))) ≃⟨ ... ⟩
adj _ (ap (λ f → apl n (λl n f) x) (λ≃ a)) ≃⟨ ... ⟩
adj _ (ap (λ f → f x) (λ≃ a)) ≃⟨ ... ⟩
adj _ (a x) ≃⟨ ... ⟩
adj id (a x) ≃⟨ ... ⟩
a x ∎
```

Moreover, if we did not appeal to the fact that any two coercions give equal results, it is unclear how we would even prove, between the second-to-last and third-to-last lines, that

$$((ap^\wedge\text{-id } n\ (\lambda\ f_1 \to f_1\ x) \circ ap\ (\lambda\ f' \to apl\ n\ f'\ x)\ (\lambda l\text{-id } n)) \circ ap\simeq (!\ (\beta\ n\ (\lambda\ x_1 \to id^\wedge\ n)))) = id$$

The inferred coercion (the left-hand-side of this equation) uses several loop space lemmas that are defined by induction on n, and it is unclear how to prove that they cancel each other.

4   Conclusion

In this paper, we have described a computer-checked calculation of $\pi_n(S^n)$ in homotopy type theory. One important direction for future work is to develop a computational interpretation of homotopy type theory; our proof would be a good test case for such an interpretation. Given a number k, how does the proof compute the path loop_n^k? Or, more interestingly, given a path on S^n, how does the proof compute a number? Another direction would be to investigate the relationship between this proof and proofs of $\pi_n(S^n)$ in classical homotopy theory. The proof we have described here has since been generalized to a proof of the Freudenthal Suspension Theorem [The Univalent Foundations Program, 2013], which is one way that $\pi_n(S^n)$ is proved in classical homotopy theory. However, it would be interesting to see whether the more specific proof presented here has been (or can be) phrased in classical terms.

Acknowledgments. We thank the participants of the Institute for Advanced Study's special year on homotopy type theory for uncountably many helpful conversations.

This material is based in part upon work supported by the National Science Foundation under grants CCF-1116703 and DMS-1128155 and by the Institute for Advanced Study's Oswald Veblen fund. Any opinions, findings, and conclusions or recommendations expressed in this material are those of the author(s) and do not necessarily reflect the views of the National Science Foundation or any other sponsoring entity.

References

Awodey, S., Warren, M.: Homotopy theoretic models of identity types. In: Mathematical Proceedings of the Cambridge Philosophical Society (2009)

Awodey, S., Gambino, N., Sojakova, K.: Inductive types in homotopy type theory. In: IEEE Symposium on Logic in Computer Science (2012)

Finster, E., Licata, D.R., Lumsdaine, P.L.: The Blakers-Massey theorem in ∞-topoi and homotopy type theory (in preparation, 2013)

Licata, D.R., Shulman, M.: Calculating the fundamental group of the cirlce in homotopy type theory. In: IEEE Symposium on Logic in Computer Science (2013)

Lumsdaine, P.L.: Weak ω-categories from intensional type theory. In: Curien, P.-L. (ed.) TLCA 2009. LNCS, vol. 5608, pp. 172–187. Springer, Heidelberg (2009)

Lumsdaine, P.L., Shulman, M.: Higher inductive types (in preparation, 2013)

Martin-Löf, P.: An intuitionistic theory of types: Predicative part. In: Rose, H., Shepherdson, J (eds.) Proceedings of the Logic Colloquium, Logic Colloquium 1973. Studies in Logic and the Foundations of Mathematics, vol. 80, pp. 73–118. Elsevier (1975)

McBride, C.: Dependently Typed Functional Programs and Their Proofs. PhD thesis, University of Edinburgh (2000)

Nordström, B., Peterson, K., Smith, J.: Programming in Martin-Löf's Type Theory, an Introduction. Clarendon Press (1990)

Norell, U.: Towards a practical programming language based on dependent type theory. PhD thesis, Chalmers University of Technology (2007)

The Univalent Foundations Program. Homotopy Type Theory: Univalent Foundations of Mathematics. Institute for Advanced Study (2013), http://homotopytypetheory.org/book

van den Berg, B., Garner, R.: Types are weak ω-groupoids. Proceedings of the London Mathematical Society 102(2), 370–394 (2011)

Voevodsky, V.: Univalent foundations of mathematics. In: Beklemishev, L.D., de Queiroz, R. (eds.) WoLLIC 2011. LNCS, vol. 6642, p. 4. Springer, Heidelberg (2011)

Mostly Sound Type System Improves
a Foundational Program Verifier

Josiah Dodds and Andrew W. Appel

Princeton University

Abstract. We integrate a verified typechecker with a verified program
logic for the C language, proved sound with respect to the operational
semantics of the CompCert verified optimizing C compiler. The C language is known to not be type-safe but we show the value of a provably
mostly sound type system: integrating the typechecker with the program
logic makes the logic significantly more usable. The computational nature of our typechecker (within Coq) makes program proof much more
efficient. We structure the system so that symbolic execution—even tactical (nonreflective) symbolic execution—can keep the type context and
typechecking always in reified form, to avoid expensive re-reification.

1 Introduction

Hoare logics [15] and separation logics are valuable tools for program understanding. These logics can be straightforward to design when they target small,
type-safe programming languages. Program logics for less cooperative programming languages often have complex inference rules that are difficult to apply,
requiring extensive proofs even for simple operations.

Despite these difficulties we built a usable program logic for C, proved sound
with respect to CompCert's operational semantics (a thorough description of
this logic will soon be available [4]). Applying the Hoare logic directly to the C
program instead of a lower-level language makes it easy for users to understand
how the program relates to the proof. Other tools such as Frama-C [12] and
VCC [11] utilize intermediate languages, and translate back to C source when
user interaction is required. Our tool permits user interaction at the source level,
the *same* source program the user wrote.

Some features of C are unfriendly to Hoare logic, in particular subexpressions
with side effects, so we use a slightly different language. This factored language,
called C light, is already one of the high-level intermediate languages of the
CompCert compiler. Every C light program is a C program and every C program[1]
can be translated to C light with only automatic local transformations. If the C
program is already in the C light subset, the first phase of CompCert will leave
it unchanged (except for parsing it from ASCII into abstract syntax trees).

Our program logic (a higher-order impredicative concurrent separation logic
[2]) can be used in (at least) two ways: by applying it interactively in a proof

---

[1] We use the same specification of the C language as CompCert [16].

G. Gonthier and M. Norrish (Eds.): CPP 2013, LNCS 8307, pp. 17–32, 2013.
© Springer International Publishing Switzerland 2013

assistant, or by using the program logic to prove the soundness of a fully automatic static analysis. We have previously demonstrated such a foundationally verified shape analysis [3]; in this paper we focus on interactive proof.

We address the problem of numerous and complex verification side conditions that arise when verifying C programs. In idealized presentations of Hoare logics, we write $P[e/x]$ meaning "assertion P with the value of expression e substituted for program variable x." We implicitly assume that e *has* a value, that is, will evaluate deterministically in the current dynamic context without getting stuck; and we implicitly assume that the value will match the type of variable x (pencil-and-paper presentations may even assume a unityped language). Proof rules for a real language (especially one not designed for Hoare logic) will need many side conditions to establish that these assumptions hold. These hypotheses and side conditions become tedious proof obligations for the user. In this paper we show how to use the type system of C to automatically discharge these hypotheses in the majority of cases. To make this process efficient, we design for the use of computational reflection (Section 6). In many cases we are able to make interactive proofs work in the same way as a pencil-and-paper proof would.

One might think this is obvious: *well-typed programs don't go wrong.* But that is only in a language with a sound type system, and C does not have a sound type system: it is *mostly sound.* In this paper we show that a *mostly sound* type system can still be very useful. In a mostly sound type system well-typed programs go wrong only in well defined cases that can be avoided by proving specific obligations.

Contributions. We formalize a mostly sound type system for C, we prove its mostly-soundness, we implement it computationally in Coq, we integrate it into a program logic for C, we prove the integration is entirely sound with respect to the operational semantics of CompCert C, and we demonstrate that our type system integrates into a tactical proof system (written in Ltac) that is convenient and efficient to apply to C programs using semiautomatic forward symbolic execution[2]. In our mostly sound type system, the typechecker does not just succeed or fail, it calculates an appropriate precondition assertion for the safe evaluation of an expression. In practice, this assertion is **True** for many expressions.

We also discuss other design decisions regarding the interface of a program logic to operational semantics of the C language.

2   Example

Consider a naive Hoare assignment rule for the C language.

$$\vdash \{P[e/x]\}\, x := e\, \{P\} \qquad \text{(naive-assignment)}$$

This rule is not sound with respect to the operational semantics of C. We need a proof that e evaluates to a value. It could, for example, be a division

---

[2] Our source code can be found at http://vst.cs.princeton.edu/typecheck/.

by zero, in which case the program would crash and Hoare triples would not hold. This is an example of the mostly-sound property of the C type system. The expression e might typecheck in the C compiler, but can still get stuck in the operational semantics ("crash" during expression evaluation). A better assignment rule requires e to evaluate:

$$\frac{\exists v.e \Downarrow v}{\vdash \{P[v/x]\}\, x := e\, \{P\}}\text{assignment-ex}$$

The proof of this rule is relatively easy, but the rule is inconvenient to apply because we must use the operational semantics to show that v exists. In fact, any time that we wish to talk about the value that results from the evaluation of an expression, we must add an existential quantifier to our assertion. Showing that an expression evaluates can require a number of additional proofs. If our expression is (y / z), we will need to show that our precondition implies: y and z are both initialized, $z \neq 0$, and $\neg(y = \text{int\_min} \wedge z = \text{-1})$. The latter case causes overflow, which is undefined in the C standard. These requirements will become apparent as we apply the semantic rules.

We can remove the existential variables and make the requirements for evaluation easier to discover by creating specialized rules:

$$\frac{\text{initialized}(y) \quad \text{initialized}(z) \quad z \neq 0 \quad \neg(y = \text{int\_min} \wedge z = -1)}{\vdash \{P[(y/z)/x]\}\, x := y/z\, \{P\}}\text{intdiv}$$

This moves the proof of expression evaluation into the rule's soundness proof where it only needs to be done once. Such an approach would lead to an overwhelming number of new rules—and it hardly allows for any nested expressions, requiring substantial program rewrites.

Instead, we build a static analysis to generate simple preconditions that will ensure expression evaluation. We define a function typecheck_expr (Section 7) to tell us when expressions evaluate. Now our assignment rules are,

$$\frac{}{\Delta \vdash \{\text{typecheck\_expr}(e, \Delta) \wedge P[e/x]\}\, x := e\, \{P\}}\text{tc-assignment}$$

$$\frac{}{\Delta \vdash \{\text{typecheck\_expr}(e, \Delta) \wedge P\}\, x := e\, \{\exists v.x = \text{eval}(e[v/x]) \wedge P[v/x]\}}\text{tc-floyd-assignment}$$

The typecheck_expr is not a side condition, as it is not simply a proposition (Prop in Coq) but a separation-logic predicate quantified over an environment. When run on the expression (y/z) it computes to the assertion $z \neq 0 \wedge \neg(y \neq \text{int\_min} \vee z \neq -1)$ where z and y are not the variables, but the values that result when z and y are evaluated in some environment. The assertions initialized(y) and initialized(z) may not be produced as proof obligations if the type-and-initialization context Δ assures that y and z are initialized (Section 5). The calculation of Δ is also part of our type system.

We use the Floyd-style forward assignment rule, instead of the Hoare-style weakest-precondition rule. This is not related to type-checking; separation logic with backward verification-condition generation gives us magic wands which are best avoided when possible [6].

3   Expression Evaluation

CompCert defines expression evaluation by an inductive relation eval_expr; failure to evaluate is denoted by omitting tuples from the relation. This is a standard technique in operational semantics, but it is inconvenient in a program logic: assertions would need existential quantifiers, as in $\exists v. e \Downarrow v \wedge P(v)$, "there exists some value such that e evaluates to v and P holds on v." It is much cleaner to say $P(\text{eval\_expr}(e))$, or "P holds on the evaluation of e". We want eval_expr(e) to be a value, not a value-option, or else we (again) need existential quantifiers. Defining evaluation as a function also makes proofs more computational—more efficient to build and check.

We simplify eval_expr in our program logic—and make it computational—by leveraging the typechecker's guarantee that evaluation will not fail. Our total recursive function eval_expr (e: expr) (rho: environ): in environment ρ, expression e evaluates to the value (eval_expr e ρ). When CompCert.eval_expr fails, our own eval_expr (though it is a total function) can return an arbitrary value. We can do this because the function will be run on a program that typechecks—the failure is unreachable in practice. We then prove the relationship between the two definitions of evaluation on expressions that typecheck (we state the theorem in English and in Coq):

Theorem 1. *For all logical environments ρ that are well typed with respect to a type context Δ, if an expression e typechecks with respect to Δ, the Comp-Cert evaluation relation relates e to the result of the computational expression evaluation of e in ρ.*

Lemma eval_expr_relate :
$\forall \Delta \rho e m ge ve te$, typecheck_environ $\Delta \rho \rightarrow$ mkEnviron ge ve te $= \rho \rightarrow$
    denote_tc_assert (typecheck_expr Δ e) $\rho \rightarrow$
    Clight.eval_expr ge ve te m e (eval_expr e ρ)

Expression evaluation requires an environment, but when writing assertions for a Hoare logic, we actually write assertions that are functions from environments to Prop. So if we wish to say "the expression e evaluates to 5", we write fun $\rho \Rightarrow$ eq (eval_expr e ρ) 5. Because Coq does not match or rewrite under lambda (fun), assertions of this form hinder proof automation. Our solution is to follow Bengtson *et al.* [5] in *lifting* eq over ρ: `eq (eval_expr e) `5. This produces an equivalent assertion, but one that we are able to rewrite and match against. The first backtick lifts eq from val→ val→ Prop to (environ→ val)→ (environ→ val)→ Prop, and the second backtick lifts 5 from val to a constant function in environ→ val.

4   C Light

Our program logic is for C, but the C programming language has features that are unfriendly to Hoare logic: *side effects within subexpressions* make it impossible to simply talk about "the value of e" and *taking the address of a local variable* means that one cannot reason straightforwardly about substituting for a program variable (as there might be aliasing).

The first passes of CompCert translate *CompCert C* (a refined and formalized version of C90 [16]) into *C light*. These passes remove side effects from expressions and distinguish *nonaddressable* local variables from *addressable* locals.[3] We recommend that the user do this in their C code, however, so that the C light translation will exactly match the original program.

C has pointers and permits pointer dereference in subexpressions: d = p→ head+q→ head. Traditional Hoare logic is not well suited for pointer-manipulating programs, so we use a separation logic, with assertions such as $(p \rightarrow$ head $\mapsto x) * (q \rightarrow$ head $\mapsto y)$. Separation logic does not permit pointer-dereference in subexpressions, so to reason about d = p→ head+q→ head the programmer should factor into: t = p→ head; u = q→ head; d=t+u; where dereferences occur only at top-level in assignment commands. Adding these restrictions to C light gives us *Verifiable C*, which is not a different semantics but a proper sublanguage, enforced by our typechecker.

A well typed C program might still go wrong. These are the cases where the typechecker must generate assertions. A few of these cases might be surprising, even to experienced C programmers. The following operations are undefined in the C standard, and *stuck* in CompCert C:

- shifting an integer value by more than the word size,
- dividing the minimum int by −1 (overflows),
- subtracting two pointers with different base addresses (i.e., from different malloc'ed blocks or from different addressable local variables),
- casting a float to an int when the float is out of integer range,
- dereferencing a null pointer, and
- using an uninitialized variable.

Some operations, like overflow on integer addition, are undefined in the C standard but defined in CompCert. The typechecker permits these cases.

5   Type Context

Expression evaluation requires an expression and an environment. An expression will evaluate to different values (or Vundef) depending on the environment. To guarantee that certain expressions will evaluate, we will need to control what values can appear in environments. We use a type context to describe environments where our expressions will evaluate to defined values.

Definition tycontext: Type :−
    (PTree.t (type ∗ bool) ∗ (PTree.t type) ∗ type ∗ (PTree.t global_spec)).

---

[3] Xavier Leroy added the SimplLocals pass to CompCert 1.12 at our request, pulling nonaddressable locals out of memory in C light. Prior to 1.12, source-level reasoning about local variables (represented as memory blocks) was much more difficult.

PTree.t(τ) is CompCert's efficient computational mapping data-structure (from identifiers to τ) implemented and proved correct in Coq. The elements of the type context are

- a mapping from temp-var names to type and initialization information,
- a mapping from local variable names to types,
- a return type for the current function, and
- a mapping from global variable names to types (and Hoare specifications for global functions).

The first, second, and fourth items match exactly with the three parts of an environment (environ), which is made up of temporary variable mappings, local variable mappings, and global variable mappings.

A temporary variable is a (local) variable whose address is not taken anywhere in the procedure. Unlike local and global variables, temporaries do not alias—so we can statically determine when their values are modified. If the typechecker sees that a temporary variable is initialized, it knows that it will stay initialized. If the typechecker is unsure, it can emit an assertion guard for the user to prove initialization. Calculating initialization automatically is a significant convenience for the user; proofs in the previous generation of our program logic were littered with definedness assertions in invariants.

The initialization information is a Boolean that tells us if a temporary variable has certainly been initialized. The rules for this are simple, if a variable is assigned to, that variable will always be initialized in code executed after it. The initialization status on leaving an **if-then-else** is the GLB of the two branches. Loops have similar rules.

typecheck_environ checks an environ with respect to a tycontext. It does not generate assertions as typecheck_expr does, it simply returns a Boolean that if true claims all of the following:

- If the type context contains type information for a temporary variable the temp environment contains a value for that variable. If the variable is claimed to be initialized, that value must belong to the type claimed in the type environment.
- If the type context contains type information for a (addressable) local variable, the local variable environment contains a local variable of matching type.
- If the type context contains type information for a global variable, the global environment contains a global variable of matching type.
- If the type context contains type information for a global variable, either
  - the local variable environment does not have a value for that variable or
  - the type context has type information for that variable as a local variable.

The fourth point is required because local variables shadow global variables.

Initialization information is changed by statements. We only know a variable is initialized once we see that it is assigned to. Our typechecker only needs to operate at the level of expressions, so we can merge maintainence of the type context into the definition of our logic rules. We will now give some of these rules and explain how they work to keep the type context correct.

We provide a function

Definition func_tycontext (func: function) (V: varspecs) (G: funspecs): tycontext

that automatically builds a correct type context (for the beginning of the function body) given the function, local, and global specifications. The resulting context contains every variable used in the function matched with its correct type. We have proved that the environment created by the operational semantics when entering a function body typechecks with respect to the context generated by this function. Once the environment is created, the Hoare rules use the function updatetycon to maintain the type context across statements.

$$\frac{\Delta \vdash \{P\} \, c \, \{Q\} \qquad \Delta' = \mathsf{updatetycon}(\Delta, c) \qquad \Delta' \vdash \{Q\} \, d \, \{R\}}{\Delta \vdash \{P\} \, c; d \, \{R\}} \mathrm{seq}$$

updatetycon tells us that variables are known to be initialized after they are assigned to. It also says that variables are initialized if they were initialized before the execution of any statement, and that a variable is initialized if we knew it was initialized at the end of both branches of a preceding **if** statement. When we say initialized, we mean *unambiguously* initialized, meaning that it will be initialized during all possible executions of the program.

The type context is deeply integrated with the type rules. We write our hoare judgment as $\Delta \vdash \{P\} \, c \, \{Q\}$. We added the type context Δ because instead of quantifying over all environments as a normal Hoare triple does, we quantify only over environments that are well typed with respect to Δ. This has a huge benefit to the users of the rules: they do not need to worry about the contents of Δ, and they do not need to show that the environment typechecks or mention Δ explicitly in preconditions. Our *rule of consequence* illustrates what we always know about Δ:

$$\frac{\mathsf{typecheck\_environ}(\rho, \Delta) \wedge P \vdash P' \qquad \Delta \vdash \{P'\} \, c \, \{R\}}{\Delta \vdash \{P\} \, c \, \{R\}}$$

The conjunct typecheck_environ(ρ, Δ) gives the user more information to work with in proving the goal. Without this, the user would need to explicitly strengthen assertions and loop invariants to keep track of the initialization status of variables and the types of values contained therein.

With func_tycontext and updatetycon the rules can guarantee that the type context is sound at all times. To keep the type context updated, the user must simply apply the normal Hoare rules, with our special Hoare rule for statement sequencing shown above.

6   Keeping it Real

Proof by reflection is a three-step process. A program is *reified* (made real) by translating it from Prop to a data structure that can be reasoned about computationally. Computation is then performed on that data structure and the result is *reflected* back into Prop where it can be used in a proof (see bottom of Fig. 1). Reification is costly, however, so our approach is different. We provide a brief example of standard reflection in order to discuss the differences.

We could use reflection, for example, to remove True and False from propositions containing conjunctions and disjunctions. Chlipala discusses a similar problem in more detail [10]. The first step is to define a syntax that represents the propositions of interest. Our tc_assert syntax has 14 cases—to cover issues described in Section 4—of which we show the first four, followed by a function to *reflect* this syntax into the logic of propositions:

Inductive tc_assert :=
| tc_FF | tc_TT: tc_assert
| tc_andp': tc_assert → tc_assert → tc_assert
| tc_nonzero: expr → tc_assert
| ... **end**.

Definition denote_tc_nonzero (v: val) :=
    match v **with** Vint i ⇒ **if** negb (Int.eq i Int.zero) **then** True **else** False
                  | _ ⇒ False **end**.
⋮

Fixpoint denote_tc_assert (a: tc_assert) : environ → Prop :=
  match a **with**
  | tc_FF ⇒ 'False | tc_TT ⇒ 'True
  | tc_andp' b c ⇒ 'and (denote_tc_assert b) (denote_tc_assert c)
  | tc_nonzero e ⇒ 'denote_tc_nonzero (eval_expr e)
  | ... **end**.

If we were doing standard reflection—which we are not—we would then write a *reification* tactic,

```
Ltac p_reify P :=
  match P with
  | True ⇒ tc_TT        | False ⇒ tc_FF
  | ?P1 ∧ ?P2 ⇒ let t1 := p_reify P1 in let t2 := p_reify P2 in constr:(tc_andp t1 t2) ...
```

Finally, we do write a simplification function that operates by recursion on tc_assert. Comparing the steps, we see that the reflection step, as well as any transformations on our reified data, will be computational. Reification, on the other hand, operates by matching proof terms. The computational steps are efficient because they operate in the same way as any functional program. Ltac is less efficient because it operates by matching on arbitrary proof terms.

To avoid the costly reification step, the typechecker generates *syntax* directly—so we can perform the computation on it immediately, without need for reification. This keeps interactive proofs fast. The typechecker keeps all of its components real, meaning there are no reification tactics associated with it.

We use this design throughout the typechecker. We keep data reified for as long as possible, reflecting it only when it is in a form that the user needs to solve directly. The difference between the two approaches can be seen in Fig. 1.

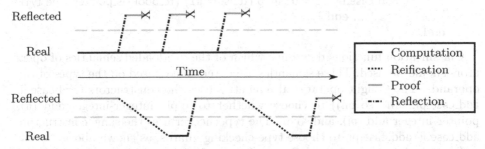

Fig. 1. Our approach (top) vs. standard reflection (bottom)

7   Typechecker

The typechecker produces assertions that, if satisfied, prove that an expression will always evaluate to a value.

In the C light abstract syntax produced by CompCert from C source code, every subexpression is syntactically annotated with a C-language type, accessed by (typeof e). Thus our typing judgment does not need to be of the form $\Delta \vdash e : \tau$, it can be $\Delta \vdash e$, meaning that e typechecks according to its own annotation.

We define a function to typecheck expressions with respect to a type context:

```
Fixpoint typecheck_expr (Δ : tycontext) (e: expr) : tc_assert :=
   let tcr := typecheck_expr Δ in  match e with
     | Econst_int _(Tint __) ⇒ tc_TT
     | Eunop op a ty ⇒ tc_andp
          (tc_bool (isUnOpResultType op a ty) (op_result_type e)) (tcr a)
     | Ebinop op a1 a2 ty ⇒ tc_andp
          (tc_andp (isBinOpResultType op a1 a2 ty) (tcr a1)) (tcr a2)
   ... end.
```

This function traverses expressions emitting conditions that ensure that the expressions will evaluate to a value in a correctly typed environment. The typechecker is actually a mutually recursive function: one function typechecks rvalues and the other typechecks lvalues. For convenience, this paper only discuss rvalues. Although CompCert's operational semantics are written as an inductive Coq type, they also have parts that are computational. For example, when we need to typecheck operation expressions, we use functions from CompCert that *classify* them. The following function is used to typecheck binary operations:

Definition isBinOpResultType op a1 a2 ty : tc_assert :=
match op **with**
  | Oadd ⇒ **match** classify_add (typeof a1) (typeof a2) **with**
    | add_default ⇒ tc_FF
    | add_case_ii _ ⇒ tc_bool (is_int_type ty)
    | add_case_pi _ _⇒ tc_andp (tc_isptr a1) (tc_bool (is_pointer_type ty))
      ... **end**
  ... **end**.

Classification functions determine which of the overloaded semantics of operators should be used. These semantics are determined based on the types of the operands. The C light operational semantics uses the constructors (add_case_ii, add_case_pi, (and so on)) to choose whether to apply integer-integer add (ii), pointer-integer add (pi), and so on. The typechecker uses the same constructors add_case_ii, add_case_pi, to choose type-checking guards, as shown above.

Despite the reuse of CompCert code on operations, the bulk of the typechecker's code checks binary operations. This is because of the operator overloading on almost every operator in C. The typechecker looks at eight types of operations (shifts, boolean operators, and comparisons can be grouped together as they have the exact same semantics with respect to the type returned). Each of these has approximately four behaviors in the semantics giving a total of around thirty cases that need to be handled individually for binary operations.

The code above is a good representation of how the typechecker is implemented. The first step is to match on the syntax. Next, if the expression is an operation, we use CompCert's classify function to decide which overloaded behavior to use. From there, we generate the appropriate assertion.

8   Soundness

The soundness statement for our typechecker is:

Theorem 1. *If the dynamic environment ρ is well typed with respect to the static type context Δ (Section 5), and the expression e typechecks with respect to Δ producing an assertion that in turn is satisfied in ρ, then the value we get from evaluating e in ρ (Section 3) will match the type that e is labeled with.*

typecheck_environ ρ Δ = true → denote_tc_assert (typecheck_expr Δ e) ρ →
typecheck_val (eval_expr e ρ) (typeof e) = true.

This guarantees that an expression will evaluate to the right kind of value: integer, or float, or pointer. As a corollary we guarantee the absence of Vundef, which has no type.

The proof proceeds by induction on the expression. One of the most difficult parts of the soundness proof is the proofs about binary operations. We need to prove that when a binary operation typechecks it evaluates to a value as a case for the main soundness proof. The proof is difficult because of the number of cases. When all integers and floats of different sizes and signedness are taken into account, there are seventeen different CompCert types. This means that there are 289 combinations of two types. A proof needs to be completed for each combination of types for all seventeen C light operators, leading to a total of 4913 cases that need to be proved. Each proof requires a decent amount of work, so the amount of memory taken by the proof becomes a problem. We use lemmas to group some of the cases together to keep the proof time reasonable.

These cases are not all written by hand: we automate using Ltac. Still, the proofs are large, and Coq takes almost 4 minutes to process the file containing the binary operation proofs.

9 A Tactical Proof

In this section, we apply our C light program logic to verify a simple C program interactively in Coq. We will verify the C program:

int assigns (int a) { int b, c; c = a*a; b = c/3; return b; }

We begin by passing our program through the CompCert clightgen tool to create a file that we can read into Coq. The next step is to specify our program. The specification for this program is:

$$\Delta \vdash \{\mathsf{Vint}(v) = \mathsf{eval}\ a\ \rho\}\ \mathsf{assigns}\ldots\ \{\mathsf{retval}\ \rho = \mathsf{Vint}((v * v)/3)\}$$

Barring any unexpected evaluation errors, we expect this specification to hold. The specification states that in an arbitrary initial state, the program will either infinite loop or terminate in a state in which retval = (a*a)/3. For this example we will focus on proving the specification of the function body:

Lemma body_test : semax_body Vprog Gtot f_assigns assign_spec.
Proof. start_function. name a _a. name b _b. name c _c.
    forward. forward. go_lower. normalize. solve_tc. forward. go_lower.
    (* ... prove that the function-body postcondition implies the
    function-specification postcondition ... *) **Qed**.

The function semax_body creates a triple for a function body given a list of global variable specifications (Vprog, the empty list), a list of global function specifications (Gtot, list of this function and main), the pointer to the function (f_assigns, pointer from program .v file), and a specification (assign_spec, the Coq version of the triple shown above). The tactic start_function unfolds semax_body and ensures that the precondition is in a usable form. The relation semax defines the triple we have seen throughout the paper.

The name tactic, and the name hypotheses it generates, relate variable names to value names. For example _a is the name of the variable a in the program. The tactic name a _a tells the tactics that values associated with evaluating _a should be called a.

We will examine the proof state at a few points to highlight the forward and go_lower tactics and show goals generated by the typechecker. We have replaced

C light AST with C-like syntax in the lines marked *(∗pseudocode ∗)*. Assertions
are in a canonical form, separated into PROP (propositions that don't require
state), LOCAL (assertions lifted over the local environment), and SEP (separation
assertions over memory). Empty assertions for any of these mean True.

```
a : name _a
b : name _b
c : name _c
Δ := initialized _c (func_tycontext f_assigns Vprog Gtot) : tycontext
================================
  semax Δ (PROP ()
     LOCAL ('eq (eval_id _c) (eval_expr(_a ∗ _a)); ('eq (eval_id _a) v))  SEP ())
     (_b = _c / 3; return _b;) (∗ pseudocode standing for C-light AST ∗)
     (function_body_ret_assert tint (_a ∗ _a / 3) = retval)
```

Above is the state after we apply forward for the first time. This tactic performs
forward symbolic execution using Coq tactic programs, as various authors have
demonstrated [1,9,5,17]. In effect, forward applies the appropriate Hoare rule for
the next command, using the sequence rule if necessary. The backtick (') is the
"lifting" coercion of Bengtson *et al.* [5]. function_body_ret_assert tells us that our
postcondition talks about the state when the program returns successfully. A
program that does not return successfully will not satisfy this triple.

The forward tactic makes a decision when it sees an assignment. In general,
it uses the Floyd assignment rule that existentially quantifies the "old" value
of the variable being assigned to (in this case c). It needs to do this because
otherwise we would lose information from our precondition by losing the old
value of c. This means the postcondition would end up in the form "∃ old, ...".
If the variable doesn't appear in the precondition, however, the existential can
be removed because it will never be used. The forward tactic checks to see if it
needs to record the old value of the variable or not. In this case, it sees that c is
not in the precondition and does not record its old value.

In the proof so far (after symbolic execution of the command c=a∗a;) we have
not yet seen a typechecking side condition—not because they were automatically
solved, but because they were never generated in the first place. They were
checked computationally, but no assertion about them is given. The condition
that a be initialized immediately evaluates to True and is dispelled trivially.

Finally we notice that Δ has been updated with initialized _c. This was done
by the sequence rule as discussed in Section 5.

Applying forward again gives the following separation-logic side condition:

```
a : name _a
b : name _b
c : name _c
Δ := initialized _b (initialized _c (func_tycontext f_assigns Vprog Gtot) : tycontext
============================
  PROP()  LOCAL(tc_environ Δ; 'eq (eval_id _c) (eval_expr (_a ∗ _a)); ('eq (eval_id _a) v))
  SEP('TT) ⊢ local (tc_expr Δ (_c / 3)) && local (tc_temp_id _b tint Δ)
```

This is an entailment, asking us to prove the right hand side given the left
hand side. We need to show that the expression on the right hand side of the

assignment typechecks, and that the id on the left side typechecks. We would expect to see that: c is initialized, $3 \neq 0$ and $\neg(c = \text{min\_int} \wedge 3 = -1)$.

Why is it useful to have tc_environ Δ? This entailment is *lifted* (and quantified) over an abstract environ ρ; if we were to intro ρ and make it explicit, then we would have conditions about eval_id _c ρ, and so on. To prove these entailments, we need to know that (eval_id _a ρ) and (eval_id _c ρ) are defined and well-typed.

In a paper proof it is convenient to think of an integer *variable* _a as if it were the same as the *value* obtained when looking _a up in environment ρ—we write this (eval_id _a ρ). In general, we can not think this way about C programs because in an arbitrary environment, _a may be of the incorrect type or uninitialized. In an environment ρ that typechecks with respect to some context Δ, however, we can bring this way of thinking back to the user. Our automatic go_lower tactic, after introducing ρ, uses the name hints to replace every use of (eval_id _a ρ) with simply a, *and* it proves a hypothesis that the value a has the expected type. In the case of an int, it does one step more: knowing that the value (eval_id _a ρ) typechecks implies it must be Vint x for some x, so it introduces a as that value x. (Again, the name a is chosen from the hint, a: name _a.) Thus, the user can think about values, not about evaluation, just as in a paper proof. Our go_lower tactic, followed by normalize for simplification converts the entailment into

```
c : int
a : int
H0 : Vint c = Vint (Int.mul a a) (*simplified*)
==============================
denote_tc_nodivover (Vint c) (Vint (Int.repr 3))
```

All we are left with is the case that the division doesn't overflow. The other conditions (c is initialized, $3 \neq 0$) have computed to True and simplified away. We can no longer see the variables _c and _a.

Now we can apply some simple Boolean rewrite rules with solve_tc and solve the goal. Not all typechecker-introduced assertions will be so easy to solve, of course; in place of solve_tc the user might have to do some real work.

The rest of the proof advances through the return statement, then proves that the postcondition after the return matches the postcondition for the specification. In this case it is easy, just a few unfolds and rewrites.

10   Related Work

Frama-C is a framework for verifying C programs [12]. It presents a unified assertion language to enable static analysis cooperation. The assertion language allows users to specify only first-order properties about programs, and does not include separation logic. The Value analysis [8] uses abstract interpretation to determine possible values, giving errors for programs that might have runtime errors. The WP plugin uses weakest precondition calculus to verify triples. WP is only correct when first running Value which may result in some verification conditions that can then be verified by WP along with the function specifications. Frama-C does not seem to have any soundness proof.

VCC is a verifier for concurrent C programs. It works by translating C programs to Boogie, which is a combination of an intermediate language, a VC generator, and an interface to pass VCs off to SMT solvers such as Z3. VCC adds verification conditions that ensure that expressions evaluate.

Greenaway et al. [14] show a verified conversion from C into a high-level specification that is easy for users to read. They do this by representing the high-level specification in a monadic language. They add guards during their translation out of C in order to ensure expression evaluation (this is done by Norrish's C parser [18]). Many of these guards will eventually be removed automatically. Their tool is proved correct with respect to the semantics of an intermediate language, not the semantics of C. The expression evaluation guards are there to ensure that expressions always evaluate in the translated program, because there is no concept of undefined operations in the intermediate language. Without formalized C semantics, however, the insertion of guards must be trusted to actually do this. This differs from our approach where the typechecker is proved sound with respect to a C operational semantics; so we have more certainty that we have found all the necessary side conditions. Another difference is that they produce all the additional guards and then solve most of them automatically, while we avoid creating most such assertions. Expression evaluation is not the main focus of Greenaway's work, however, and the ideas presented for simplifying C programs could be useful in conjunction with our work.

Bengtson et al. [5] provide a framework for verifying correctness of Java-like programs with a higher-order separation logic similar to the one we use. They use a number of Coq tactics to greatly simplify interactive proofs. Chlipala's Bedrock project [9] also aims to decrease the tedium of separation logic proofs in Coq, with a focus on tactical- and reflection-based automation of proofs about low level programs. Bengtson operates on a Java-like language and Chlipala uses a simple but expressive low-level continuation-based language. Earlier versions of our work (Appel [1]) used a number of tactics to automate proofs as well. In this system, the user was left with the burden of completing proofs of expression evaluation.

The proof rules we use in this paper are also used in the implementation of a verified symbolic execution called VeriSmall [3]. VeriSmall does efficient, completely automatic shape analysis.

Tuerk's HolFoot [19] is a tactical system in HOL for separation logic proofs in an imperative language. Tuerk uses an idealized language "designed to resemble C," so he did not have to address many of the issues that our typechecker resolves.

One of Tuerk's significant claims for HolFoot is that his tactics solve purely shape-analysis proofs without any user assistance, and as a program specification is strengthened from shape properties to correctness properties, the system smoothly "degrades" leaving more proof obligations for the interactive user. This is a good thing. As we improve our typechecker to include more static analysis, we hope to achieve the same property, with the important improvement that the static analysis will run much faster (as it is fully reified), and only the user's proof goals will need the tactic system.

Our implementation of computational evaluation is similar to work on executable C semantics by Ellison and Rosu [13] or Campbell [7]. Their goals are different, however. Campbell, for example, used his implementation to find bugs in the specification of the CompCert semantics. We, on the other hand, are accepting the CompCert semantics as the specification of the language we are operating on. Ellison and Rosu have the goal of showing program correctness, which is a similar goal to ours. They show program correctness by using their semantics as a debugger or an engine for symbolic execution.

11 Conclusion

By integrating a typechecker with a program logic, we improve the usability of the logic. Our system maintains type and initialization information through a Hoare-logic proof in a continuously reified form, leading to efficiency and improved automation. Automatic maintenance of the well-typedness of the local-variable environment (tc_environ) makes it easy to discharge "trivial" (but otherwise annoying) subgoals. We have a proof of soundness of the whole system (w.r.t. the CompCert C operational semantics) even though C does not actually have a sound type system.

We have used the tool to prove full functional correctness of programs such as list sum, list reverse, imperative thread queue, and object-oriented message passing. We are currently working on safety and correctness of an implementation of the SHA-256 hash function.

Our style of integrating a static analysis with a program logic should not be limited to a typechecker. Many of the ideas presented in this paper could be used to integrate other static analyses with program logics. Symbolic execution, abstract interpretation, and more (or less) sound typecheckers could all be integrated in a similar fashion.

Acknowledgments. This material is based on research sponsored by the Air Force Office of Scientific Research under agreement FA9550-09-1-0138 and by DARPA under agreement number FA8750-12-2-0293. The U.S. Government is authorized to reproduce and distribute reprints for Governmental purposes not withstanding any copyright notation thereon. The views and conclusions contained herein are those of the authors and should not be interpreted as necessarily representing the official policies or endorsements, either expressed or implied, of DARPA of the U.S. Government.

References

1. Appel, A.W.: Tactics for separation logic (2006)
2. Appel, A.W.: Verified Software Toolchain. In: Barthe, G. (ed.) ESOP 2011. LNCS, vol. 6602, pp. 1–17. Springer, Heidelberg (2011)

3. Appel, A.W.: VeriSmall: Verified Smallfoot shape analysis. In: Jouannaud, J.-P., Shao, Z. (eds.) CPP 2011. LNCS, vol. 7086, pp. 231–246. Springer, Heidelberg (2011)
4. Appel, A.W., Dockins, R., Hobor, A., Beringer, L., Dodds, J., Stewart, G., Blazy, S., Leroy, X.: Program Logics for Certified Compilers. Cambridge (to appear, 2014)
5. Bengtson, J., Jensen, J.B., Birkedal, L.: Charge! A framework for higher-order separation logic in Coq. In: Beringer, L., Felty, A. (eds.) ITP 2012. LNCS, vol. 7406, pp. 315–331. Springer, Heidelberg (2012)
6. Berdine, J., Calcagno, C., O'Hearn, P.W.: Symbolic execution with separation logic. In: Yi, K. (ed.) APLAS 2005. LNCS, vol. 3780, pp. 52–68. Springer, Heidelberg (2005)
7. Campbell, B.: An executable semantics for compCert C. In: Hawblitzel, C., Miller, D. (eds.) CPP 2012. LNCS, vol. 7679, pp. 60–75. Springer, Heidelberg (2012)
8. Canet, G., Cuoq, P., Monate, B.: A value analysis for C programs. In: Ninth Source Code Analysis and Manipulation, pp. 123–124. IEEE (2009)
9. Chlipala, A.: Mostly-automated verification of low-level programs in computational separation logic. In: PLDI 2011, pp. 234–245 (2011)
10. Chlipala, A.: Reflection. In: Certified Programming With Dependent Types. MIT Press (2013)
11. Cohen, E., Dahlweid, M., Hillebrand, M., Leinenbach, D., Moskal, M., Santen, T., Schulte, W., Tobies, S.: VCC: A practical system for verifying concurrent C. In: Berghofer, S., Nipkow, T., Urban, C., Wenzel, M. (eds.) TPHOLs 2009. LNCS, vol. 5674, pp. 23–42. Springer, Heidelberg (2009)
12. Cuoq, P., Kirchner, F., Kosmatov, N., Prevosto, V., Signoles, J., Yakobowski, B.: Frama-C. In: Eleftherakis, G., Hinchey, M., Holcombe, M. (eds.) SEFM 2012. LNCS, vol. 7504, pp. 233–247. Springer, Heidelberg (2012)
13. Ellison, C., Roşu, G.: An executable formal semantics of C with applications. In: Proceedings of the 39th Symposium on Principles of Programming Languages (POPL 2012), pp. 533–544. ACM (2012)
14. Greenaway, D., Andronick, J., Klein, G.: Bridging the gap: Automatic verified abstraction of C. In: Beringer, L., Felty, A. (eds.) ITP 2012. LNCS, vol. 7406, pp. 99–115. Springer, Heidelberg (2012)
15. Hoare, C.A.R.: An axiomatic basis for computer programming. Communications of the ACM 12(10), 578–580 (1969)
16. Leroy, X.: The CompCert verified compiler, software and commented proof (June 2013), http://compcert.inria.fr
17. McCreight, A.: Practical tactics for separation logic. In: Berghofer, S., Nipkow, T., Urban, C., Wenzel, M. (eds.) TPHOLs 2009. LNCS, vol. 5674, pp. 343–358. Springer, Heidelberg (2009)
18. Norrish, M.: C-to-isabel parser (2013), http://www.ssrg.nicta.com.au/software/TS/c-parser/
19. Tuerk, T.: A formalisation of Smallfoot in HOL. In: Berghofer, S., Nipkow, T., Urban, C., Wenzel, M. (eds.) TPHOLs 2009. LNCS, vol. 5674, pp. 469–484. Springer, Heidelberg (2009)

Computational Verification
of Network Programs in Coq

Gordon Stewart*

Princeton University

Abstract. We report on the design of the first fully automatic, machine-checked tool suite for verification of high-level network programs. The tool suite targets programs written in NetCore, a new declarative network programming language. Our work builds on a recent effort by Guha, Reitblatt, and Foster to build a machine-verified compiler from NetCore to OpenFlow, a new protocol for software-defined networking.

1 Introduction

The past few years have witnessed a groundswell of interest in *software-defined networks (SDNs)*, as evidenced by the popularity of new standards for programmable networking such as OpenFlow [9]. In an SDN, rules for packet processing still live on network switches with dedicated hardware, as in traditional networks (data plane). But unlike in traditional networks, SDNs allow decisions about when and how to update network policies in response to network events (control plane) to be handled by a dedicated *controller* program running on one or more general-purpose computers. The controller machine(s) and switches interoperate via an open standard, such as OpenFlow, that allows on-the-fly switch re-programming via special *configuration* or *control* messages.

Several recent research efforts have capitalized on the modular structure of SDNs to build new high-level programming languages for networks, such as Nettle [13], Frenetic [2], and NetCore [10]. These high-level languages, which are typically compiled to low-level OpenFlow forwarding rules, are characterized by a focus on *declarative* and *modular* programming of network policies: The programmer defines *what* a particular policy is, not *how* it is implemented; and programs are constructed by composing small, reusable components. These two features of the new breed of network languages make them an ideal target for program verification. Yet there have been few, if any, efforts to-date to build verification tools for network programs written in these languages.

As an initial foray, this paper presents the first machine-certified toolset for verifying network programs written in a high-level network programming language. We build on recent work by Monsanto *et al.* [10], which defined the syntax and semantics of the network programming language NetCore, and on work by Guha, Reitblatt, and Foster [3], which presented a Coq formalization of Net-Core and of a lightweight version of the OpenFlow protocol called Featherweight

* Supported in part by the National Science Foundation under grant CNS-0910448.

G. Gonthier and M. Norrish (Eds.): CPP 2013, LNCS 8307, pp. 33–49, 2013.
© Springer International Publishing Switzerland 2013

OpenFlow. Drawing on the NetCore compilation algorithm of Monsanto *et al.*, Guha *et al.* formalized these models in Coq in order to verify the correctness of a compiler and runtime for NetCore targeting Featherweight OpenFlow. The result of their work was a fully machine-verified network programming platform that targeted actual switch hardware.

In this paper we start where Guha *et al.* left off, by building a suite of tools for verifying properties of the NetCore programs that are the input to their verified compiler. For many concrete network specifications—for example, reachability or security specifications targeting a particular network topology with known port identifiers—our verification tools are fully automatic: In order to prove the specification $\{P\}$ pg $\{Q\}$ of NetCore program pg, we calculate the weakest precondition $wp(pg, Q)$ of Q given pg. Then we verify that P implies $wp(pg, Q)$ by checking the implication $P \implies wp(pg, Q)$ in a special-purpose resolution theorem prover coded in Gallina, the functional programming language embedded within Coq. Because all of our tools are proved sound in Coq with respect to an extension of the NetCore semantics presented by Guha *et al.* [3], we end up with a fully automatic verification toolset that when connected to Guha *et al.*'s verified compiler will provide strong guarantees on the correctness of generated OpenFlow programs. To demonstrate our tool suite, we use it to verify a multiplexing network address translation module (§5).

Contributions. The novel contributions of this paper are the following.

1. We develop the first suite of machine-checked tools for verifying correctness and security properties of network programs written in a high-level programming language (NetCore) targeting an open SDN platform (OpenFlow).
2. To fully automate proofs of NetCore specifications within our system, we develop (§3) two related weakest precondition calculi for NetCore, and build a special-purpose resolution theorem prover, in Coq's Gallina language, for checking entailments of NetCore program specifications. We prove the resolution prover sound in Coq, and the weakest precondition calculi both sound and complete, with respect to the NetCore semantics.
3. Because our tool suite targets an extension of the NetCore semantics of Guha *et al.* [3], it can be connected to Guha *et al.*'s verified NetCore compiler to provide strong guarantees on generated OpenFlow programs. However, we have not yet fully integrated the Coq proofs of our NetCore verifier with those of Guha *et al.*'s NetCore compiler due to engineering concerns.
4. We use the tool suite to verify correctness and security properties of a network address translation module (§5). Section 8 describes additional applications.

An alternative to checking entailments within Gallina using a custom theorem prover is to send verification conditions to an external first-order prover or SMT solver, and then to check proof certificates *post hoc*. We prefer the Gallina approach for two reasons. First, implementing the entailment checker in Gallina means we can prove it sound once and for all. In the certificate checking approach, the potential for soundness bugs in an external tool means that program verification may fail unnecessarily when a bad certificate is detected. Second, building a custom entailment checker means that we can apply domain-specific reasoning in ways that may not be directly exploited by an external tool. The final few paragraphs of Section 3 provide one such example, in which we exploit Coq's theory of inductive datatypes in order to reason constructively by inversion, without explicit first-order inversion laws. There are of course disadvantages to using a special-purpose entailment checker as well: our tool is necessarily much less sophisticated than state-of-the-art provers such as Z3 [1]. However, it is still sufficient to discharge the verification conditions that arise when verifying a range of network programs, as later sections of this paper will demonstrate.

The Coq Development. This paper is closely tied to a mechanized proof development in Coq, which can be downloaded from the address below.[1] In code listings, line numbers refer to the corresponding files in the mechanized development.

2 Software-defined Networking

In a traditional network, control logic is distributed among a number of physically distinct routers and switches, each with its own *flow table*. The flow tables define how packets are processed and forwarded through the router or switch, and are typically implemented using dedicated hardware such as ternary content addressable memories (TCAMs), in order to process packets at line rate.

Software-defined networks differ from traditional networks by splitting the control plane from the data plane, providing logically centralized control in the form of a general-purpose *controller*. The controller, which is connected to the switches over a secure, special-purpose link, decides when and how to update the flow rules installed on the switches in response to network packets and other network events.

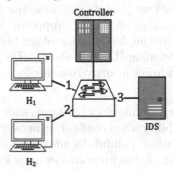

Fig. 1. A software-defined network with a single switch (center), two endhosts (H_1, H_2), a middlebox that performs intrusion detection (*IDS*), and a general-purpose controller machine

High-level network programming languages such as NetCore build another layer of abstraction above software-defined networking platforms such as OpenFlow. Network programs in NetCore are built from small, reusable packet-processing actions that are composed both in sequence—in order to apply multiple modifications, in order, to a single packet—and in parallel—in order to apply multiple forwarding or modification rules to multiple copies of a single packet. In order to scope the applicability of a network program constructed in

---

[1] http://www.cs.princeton.edu/~jsseven/papers/netcorewp

this manner, NetCore programs can be restricted by *predicates* on the location and header fields of network packets. For example, it is good style in NetCore to write, *e.g.*, a forwarding policy for HTTP traffic first as if it applied to all packets, and then to restrict the resulting policy just to those TCP packets that have port **tcpDst** = 80.

Example. To make this concrete, consider the simple network topology depicted in Figure 1. The network, which is adapted from Guha *et al.* [3], comprises a single switch S_1 connected to two endhosts H_1 and H_2, a middlebox *IDS* that performs intrusion detection, and a controller. The endhosts are connected to the switch on ports 1 and 2 respectively, while *IDS* is connected on port 3. The controller machine is connected to the switch via a special-purpose link.

Now imagine we want to impose the following traffic policy, also adapted from Guha *et al.* [3]: HTTP traffic on TCP port 80 gets forwarded to *IDS* (as well as to its original destination), SSH traffic is dropped, and all other traffic gets forwarded to the appropriate endhost. In an SDN platform like OpenFlow, this policy would be defined as a set of *flow tables*, or packet forwarding rules—one set of rules for each switch. For example, the following rule expressed in Guha *et al.*'s notation causes SSH traffic to be dropped: **Add** 10 {**tcpDst** = 22} {| |}. The 10 after **Add** is the rule's priority. On the switch, any rules with lower priority will be applied only if this rule fails to fire. The expression **tcpDst** = 22 is a *pattern*: it limits the applicability of the rule to packets with TCP destination port equal to 22. The expression {| |} denotes the empty multiset of ports. It specifies that packets matching the pattern should be forwarded along *no* ports (that is, they should be dropped). After configuring this rule on the switch, an OpenFlow implementation of the high-level policy would add lower-priority rules for forwarding HTTP traffic to *IDS* and to the appropriate endhost, in addition to even lower priority rules for forwarding the remaining traffic.

NetCore. While SDN platforms such as OpenFlow give programmers a great deal of flexibility when configuring networks, writing network controllers in OpenFlow is still quite painful. In addition to the actual forwarding logic of the network application, the programmer must keep track of numerous low-level details such as dependencies between rule priorities. He also must determine (manually) on which switch to install each forwarding rule. In essence, the programmer is writing a low-level distributed program by hand. This can become quite a difficult task, especially as networks scale to tens or hundreds of switches.

High-level SDN programming languages such as NetCore mitigate many of these challenges by providing a programming model that is at once more declarative and more modular than those provided by traditional SDN platforms. NetCore, for example, provides as one of its key abstractions the notion of *whole-network programmability*: instead of defining forwarding rules for particular switches, a NetCore program defines the behavior of the entire network all at once. The NetCore compiler and runtime system determine on which switch to install each rule, and with what priority.

```
103 (* basic actions *)
104 Inductive action :=
105 | Id: action
106 | UpdIpSrc: Word32.t → action | UpdTcpSrc: Word16.t → action
107 | UpdIpDst: Word32.t → action | UpdTcpDst: Word16.t → action
108 | Fwd: Word16.t → action | Drop: action.
116 (* selected packet patterns *)
117 Inductive ppat :=
124 | DlSrc: Word48.t → ppat (*MAC src*)
125 | DlDst: Word48.t → ppat (*MAC dst*)
133 | TpSrc: Word16.t → ppat (*TCP src*)
134 | TpDst: Word16.t → ppat (*TCP dst*)
140 (* atomic predicates *)
141 Inductive atom :=
142 | Wild: atom | Location: lpat → atom | Packet: ppat → atom.
146 (* Boolean predicates *)
147 Inductive pred :=
148 | Atom: atom → pred
149 | And: pred → pred → pred | Or: pred → pred → pred
150 | Not: pred → pred.
154 (* NetCore programs *)
155 Inductive prog :=
156 | Act: pred → action → prog | Restrict: prog → pred → prog
157 | Par: prog → prog → prog | Seq: prog → prog → prog.
```

Listing 1. Excerpts from the syntax of NetCore (`src/NetCoreSyntax.v`)

To illustrate, consider the following implementation in NetCore, the syntax of which is given in Listing 1, of the high-level policy for the network in Figure 1.[2] First, we define the rules that establish point-to-point connectivity in the network.

```
18 Definition pg1 := DLDST=H1 ⇒ FWD 1.
```

For example, program pg1 defines a basic guarded command in NetCore that forwards to port 1 (FWD 1) any packets satisfying the predicate DLDST=H1, that is, with destination MAC address equal to H1. Here DLDST=H1 is syntactic sugar for the atomic predicate Atom (Packet (DlDst (Val H1))) (*cf.* Listing 1).

In NetCore, we can compose this first program with a second program that defines the routing policy for host 2 as follows.

```
21 Definition pg2 := pg1 'PAR' DLDST=H2 ⇒ FWD 2.
```

The combinator 'PAR', which is infix notation for the Par constructor of Listing 1, defines the parallel composition of two NetCore programs. Semantically, it duplicates its input packets, applying the program on the left (pg1) to one of the duplicated input packets and the program on the right (DLDST=H2 ⇒ FWD 2)

---

[2] The code that follows can be found in file `src/examples/Guha.v` in the code distribution that accompanies this paper.

to the other. The result is a set of packets that will be transferred across links and further processed by other switches in the network, if any.

The resulting program pg2 can be further composed with the routing policy for the intrusion detection system, resulting in the following program.

```
24| Definition pg3 := pg2 'PAR' TPDST=80 ⇒ FWD 3.
```

Program pg3 forwards HTTP packets (TPDST=80) to the IDS middlebox on port 3, packets destined for MAC address H1 to host 1, and those destined for MAC address H2 to host 2.

Finally, in order to satisfy the high-level policy described above we need to ensure that SSH traffic on port 22 is dropped. In NetCore, this is accomplished by restricting program pg3 by a predicate that scopes the resulting program. Packets that do not satisfy the predicate are implicitly dropped.

```
28| Definition routing := RESTRICT pg3 BY (NOT (TPDST=22)).
```

Here, RESTRICT pg3 BY (NOT (TPDST=22)) is syntactic sugar for an application of the Restrict constructor of Listing 1. This has the effect of applying pg3 to any packet satisfying the predicate NOT (TPDST=22) (that is, with TCP destination port not equal to 22) and dropping all other packets (*i.e.*, those on port 22), which is the behavior we intended.

3 Verification

Now that we have defined the routing policy for the network topology in Figure 1, we can begin proving properties of the resulting network. For example, imagine we would like to prove that the routing network defined in Section 2 *actually does* drop all SSH traffic. For this particular network, the property is of course trivial: the network program is guarded by a RESTRICT that filters packets satisfying exactly this predicate! However, for more complicated network programs, security properties such as this one can be significantly less obvious. In any case, it will be instructive to present our verification methodology in the context of this simple example; we consider more interesting networks and verification problems in Sections 5 and 8.

In order to state the theorem described informally above, we first briefly describe the semantics of NetCore programs. In our Coq development, NetCore programs are interpreted as inductively defined relations on *located packets*, where a located packet is a pair of a packet, including its header fields and payload, and a location, which is a pair of a switch identifier and a port number.[3]

```
72| Inductive progInterp: prog → lp → lp → Prop :=
73| (* ··· *)
80| | InterpUpdSrcIp: ∀x x' ip cond,
81|     (predInterp cond x)=true →
82|     upd_ip_src x ip = Some x' →
```

---

[3] The semantics of NetCore is defined in file src/NetCoreSemantics.v.

```
83|    progInterp (Act cond (UpdIpSrc ip)) x x'
84| (* ... *)
```

For example, the InterpUpdSrcIp constructor of the relation states that packet x is related to packet x' by program Act cond (UpdIpSrc ip)), which in sugared form is cond => UpdIpSrc ip, if (1) the predicate cond is satisfied by x (predInterp cond x=true) and (2) updating the IP address of packet x to ip succeeds, resulting in x' (our semantics must handle situations in which x is not a valid IP packet, in which case the upd_ip_src operation will fail).

A bit more formally now, the security property we would like to prove is: for all packets x, if (predInterp (TPDST=22) x)=true then progInterp routing x x' is false. That is, no input packet with TCP destination port equal to 22 is ever routed as output packet x'. We could state (and prove) this theorem directly, but instead we will encapsulate the general kind of specification as a Hoare triple, with the following definition.

```
120| Definition triple (P: pred) (pg: prog) (Q: pred) :=
121|   ∀x y, (predInterp P x)=true →
122|   progInterp pg x y →
123|   (predInterp Q y)=true.
```

That is, a program pg satisfies triple P pg Q when it takes packets x satisfying precondition P (predInterp P x=true) to packets satisfying postcondition Q (predInterp Q y=true).

Now, with the help of some syntactic sugar for triple P pg Q, we can restate the theorem as follows. Using our NetCore tool suite, the proof is a single line.

```
32| Lemma ssh_traffic_dropped: |- [TPDST=22] routing [NOT WILD].
33| Proof. Time checker. (*0. secs (0.0156001u,0.s)*) Qed.
```

Here NOT WILD is the representation of False in the NetCore predicate language. Thus |- [TPDST=22] routing [NOT WILD] states that packets satisfying TPDST =22 are never routed (*i.e.*, they are always dropped).

To prove this theorem, one could reason from the definitions of the triple |- [P] pg [Q] and of the interpretation relation progInterp, perhaps proving a few general-purpose Hoare rules along the way. Indeed, this would be the conventional way to proceed in an interactive proof assistant such as Coq. However, we would like to automate this proof, and others like it. In general, we will avoid making use of the semantic meaning of the Hoare triple defined above whenever possible, instead relying on the computational verification procedure given in Listing 2.[4]

The function check takes as arguments a bound n on the number of iterations of the procedure, a background theory th, the program pg to be verified, and its specification spec. The main steps of the procedure are the following.

---

[4] The code that follows is found in file **src/Checker.v** in our source distribution.

1. Calculate wp pg Q, the weakest precondition of the postcondition Q with respect to program pg. By soundness of the weakest precondition calculus, |- [wp pg Q] pg [Q]. Thus by the rule of consequence for Hoare triples, |- [P] pg [Q] if P $\implies$ wp(pg, Q).

```
1138 Definition check (n: nat) (th: pred) (pg: prog) (spec: pred*pred) :=
1139   let P := fst spec in
1140   let Q := snd spec in
1141   let vc := th 'AND' P 'AND' (NOT (wp pg Q)) in
1142     go n nil (preprocess (clausify (normalize n vc) nil nil) nil).
```

Listing 2. Top-level Verification Procedure

2. Prove that P $\implies$ wp pg Q. This entails: Encoding the *negation*[5] of the implication P $\implies$ wp pg Q as a formula in clausal normal form; Simplifying the resulting formula by removing tautological and subsumed conjuncts; and Proving that the resulting simplified formula is unsatisfiable. In the code in Listing 2, these steps correspond to the calculation of vc and the call to go, the top-level loop of the resolution prover. In the definition of vc, the negation of P $\implies$ wp pg Q is implicitly simplified to And P (Not (wp pg Q)).

Weakest Preconditions. Of these two steps, the calculation of the weakest precondition of Q with respect to program pg is the most straightforward. Because NetCore contains no looping constructs, and therefore no loop invariants are required, we can calculate wp pg Q using the recursive function defined in Listing 3.

```
 87 Fixpoint wp (pg: prog) (R: pred): pred :=
 88   match pg with
 89   | Act cond Id => cond ==> R
 90   | Act cond (Fwd pt) => cond ==> subst_port pt R
 91   | Act cond (UpdIpSrc ip) =>
 92       cond ==> Atom (Packet IsIp) ==> subst_ip_src ip R
 93   (* ... *)
102   | Act cond Drop => Atom Wild
103   | Restrict pg' cond => cond ==> wp pg' R
104   | Par pg1 pg2 => wp pg1 R 'AND' wp pg2 R
105   | Seq pg1 pg2 => wp pg1 (wp pg2 R)
106   end.
```

Listing 3. Weakest Precondition Calculus for NetCore (excerpt)

For example, the weakest precondition of postcondition R and the the guarded identity action Act cond Id (in sugared form cond ⇒ Id) is just R, under the assumption that cond evaluates to true (cond $\implies$ R). Likewise, the weakest precondition of the parallel composition of two programs pg1 and pg2 (Par pg1 pg2)

---

[5] Although this procedure follows the usual proof-by-contradiction approach of automated tools for propositional and first-order logic, it can be done without classical axioms in Coq because the language of NetCore predicates is decidable.

is just the conjunction of the weakest preconditions of the component programs (wp pg1 R 'AND' wp pg2 R), while the weakest precondition of the sequential composition of pg1 with pg2 is the weakest precondition of pg1 given postcondition wp pg2 R.

The weakest preconditions of commands that update packet headers or locations (Fwd, UpdIpSrc, UpdTcpSrc) are calculated as one would calculate the weakest precondition of an assignment statement in a typical imperative language. That is, the weakest precondition of $x := e$ for R is $R[e/x]$. However, instead of substituting an expression e for variable x in R, we substitute *true* for occurrences of location or packet predicates that are consistent with a packet modification, and *false* for any such atomic predicates that are inconsistent. For example, the following code excerpt (file `src/WP.v`) performs the substitution that is required for forwarding actions.

```
11| Fixpoint subst_port (x: Word16.t) (p: pred) {struct p} :=
12|    match p with
13|      | Atom Wild => Atom Wild
14|      | Atom (Location (Switch _)) => p
15|      | Atom (Location (Port y)) =>
16|          if Word16.eq x y then WILD else NOT WILD
17|      | Atom (Packet _) => p
23|      (* ⋯ *)
24|    end.
```

In our Coq development, we have proved that wp as defined above is both sound and complete.

```
257| Lemma wp_sound: ∀pg R, |- [wp pg R] pg [R].
```

```
445| Lemma wp_complete:
446|    ∀P pg Q,
447|    |- [P] pg [Q] →
448|    ∀I, (predInterp P I)=true → (predInterp (wp pg Q) I)=true.
```

The proof of soundness is straightforward by induction on the program pg. Completeness requires that the language of predicates be full-featured enough to express equality on located packets. That is, we must define a predicate Eq x such that predInterp (Eq x) x' if, and only if, x=x'. The equality predicate is used in the Seq case of the proof to constrain intermediate packets.

Resolution. After we have calculated the weakest precondition wp pg Q of postcondition Q and program pg, we next must check that P entails wp pg Q (in fact, this implication *must* be provable in order for |- [P] pg [Q] to hold since wp is proved complete). Here we resort to *resolution* [12], a standard method from automated theorem proving, in order to check the implication automatically within Gallina.

To do so, we first encode the negation of the implication as a set of *clauses*, or disjunctions of logical literals. Literals are, in turn, either positive or negative (*i.e.*, negated) atomic predicates.

```
127│Inductive lit := Neg: atom → lit | Pos: atom → lit.
```

Clauses are defined in the code as lists of literals, and are interpreted as the following disjunction of their elements.

```
141│Definition clauseInterp (cl: clause) (I: Ip) :=
142│  foldInterp (fun p => litInterp p I) orb false cl.
```

Here the function foldInterp folds an interpretation function (litInterp) and a combinator (orb) over the constituent elements of the list (cl), with unit false.

Encoding a formula as a set of clauses entails: (1) Converting the formula to *negation normal form (NNF)*, by moving negations inwards using De Morgan equalities; (2) Distributing disjunctions over conjunctions; and (3) Rewriting the resulting formula as a set of (implicitly conjoined) clauses. Once we have encoded the negation of the initial implication P $\implies$ wp pg Q, together with the background theory th, as a set of clauses (its so-called *clausal normal form*) we simplify the resulting clause set to remove tautologous and otherwise redundant clauses, then begin searching for a contradiction by iterating the following procedure.

```
967│Definition step (act pas: list clause): result :=
968│  match pas with
969│  | nil => if invert act then Unsat else Sat act
970│  | nil :: pas' => Unsat
971│  | given :: pas' =>
972│      let act' := given::act in
973│      let resolvents := map condense (resolve given act' nil) in
974│      let resolvents' := filter (negb ∘ subsumedBy pas) resolvents in
975│        Later act' (pas' ++ resolvents')
976│  end.
```

The step function implements a variation of what is known as the *given clause algorithm* for saturating a search space by resolution, which was popularized by the OTTER theorem prover [8]. It operates on two sets of clauses: act, the set of active or usable clauses, and pas, the set of passive clauses that have not yet taken part in resolution inferences. Initially, all clauses are in pas.

At each iteration of step, we do a case analysis on pas, resulting in a three-way branch: Either (1) pas is empty, in which case the search space is saturated (traditional first-order provers would return Sat at this point); or (2) the head of pas is the "always false" clause nil[6] (pas = nil :: pas'); or (3) the given clause at the head of pas (pas = given :: pas') contains at least one literal.

Case (3) is the most interesting. Here we add given to act, resulting in the new clause set act', then attempt to resolve given with each clause in act' (resolve given act' nil), including itself. The resulting set of resolvents is then *condensed* to remove unnecessary duplicate literals (map condense $\cdots$). Finally, newly resolved clauses that are subsumed by clauses already in pas are filtered away as

---

[6] Recall that clauses are interpreted as the disjunction of their component literals, with unit false; thus the nil clause is never satisfiable.

redundant (resolvents' := ···) and the resulting set is appended to pas'. The Later constructor is used to communicate the updated clause set to the main loop of the prover, which is not shown in the code above.

In case (2), pas contains the nil, or always false, clause. Thus the prover immediately returns Unsat: nil is unsatisfiable.

In case (1), a traditional resolution prover would return Sat: Because resolution is refutation complete, the procedure is guaranteed to derive the nil clause when given an unsatisfiable initial clause set. But since all clauses have been processed (pas is empty) without the nil clause being discovered, it must be the case that the initial input clause set has a model.

We could stop here. Indeed, standard resolution provers would stop at this point. Instead, we use the fact that we are constructing a custom prover to build in an additional level of inference by inversion on inductive types (invert act).

To see why this is useful, consider a clause set act that contains a pair of singleton clauses asserting TPSRC=22 and TPSRC=80 respectively. Both of these assertions cannot be true simultaneously. Yet a traditional first-order prover would not be able to derive a contradiction at this point; the standard inversion principles we get when reasoning about the inductive packet and nat types in Coq must first be explicitly added to the prover's background theory. This can be done. For example, in this particular case, we can safely add the clause

```
1| Neg (TPSRC=22) :: Neg (TPSRC=80) :: nil
```

which asserts that TPSRC=22 and TPSRC=80 cannot be true simultaneously. However, a set of more general inversion principles would clutter the search space with many (usually unnecessary) clauses. It is quite convenient, instead, to be able to do a domain-specific check, at the point at which all other first-order inferences have been exhausted, for clauses that are incompatible by inversion.

4   Reachability

In Section 3, we described a general methodology for proving theorems of the form

```
1| Lemma ssh_traffic_dropped: |- [TPDST=22] routing [NOT WILD].
```

in which the triple |- [P] pg [Q] made a claim about *all* packets that may result after routing packets satisfying P but did not ensure that at least one such packet *existed*. This form of Hoare triple was useful for writing specifications of security properties such as "all packets with TCP destination port equal to 22 are dropped." We will see in Section 5 that this triple is also useful for specifying the security properties of a network address translation module.

However, it is also quite useful when describing high-level properties of a network to be able to specify *reachability*, in addition to security properties. That is, we would like to be able to prove that, given a packet x satisfying some predicate P, *there exists* a second packet y such that y satisfies the predicate Q. Furthermore, it should be the case that progInterp pg x y for the NetCore program pg in question, *i.e.*, x is actually routed to y by pg. For example, if P is specialized to PORT=1 and Q is specialized PORT=2, then a reachability specification for P and Q states that a host located on port 2 is reachable by a host on port 1.

In order to specify and prove reachability queries of this form, we have adapted the weakest precondition calculus of Section 3 to the following variation of the Hoare triple of that section.

```
130  Definition triple' (P: pred) (pg: prog) (Q: pred) :=
131    ∀x, (predInterp P x)=true →
132    ∃y, progInterp pg x y ∧ (predInterp Q y)=true.
```

This Hoare triple states that there *exists* a y for which progInterp pg x y holds, and such that predInterp Q y evaluates to true. In what follows, we will use the notation |-r [P] pg [Q] to denote reachability specifications of this form.

Adapting the weakest precondition calculus of Section 3 to reachability specifications is reasonably straightforward. For example, here are the weakest precondition rules for Restrict, Par, and Seq.

```
112  Fixpoint wp' (pg: prog) (R: pred) :=
113    match pg with
114    (* ··· *)
124    | Restrict pg' cond => cond 'AND' wp' pg' R
125    | Par pg1 pg2 => wp' pg1 R 'OR' wp' pg2 R
126    | Seq pg1 pg2 => wp' pg1 (wp' pg2 R)
127    end.
```

They are essentially dual to those given by wp of Section 3.

With these definitions in place, we can easily adapt the verification procedures of Section 3 to prove reachability theorems such as the following.

```
37  Lemma http_reaches_ids: |-r [TPDST=80] routing [PORT=3].
38  Proof. Time checker'. (*0. secs (0.u,0.s)*) Qed.
```

This theorem states that all packets with TCP destination port equal to 80 are forwarded to port 3, the network location of the intrusion detection middlebox.

5   Network Address Translation

In *Network Address Translation (NAT)*, IP packet headers are modified on the fly as packets are routed through a network, typically to implement IP sharing. For example, in private networks, the source IP addresses of packets routed from internal hosts to hosts outside the private network will be modified to the IP address of an externally visible router. The result is that only the router need be assigned a globally unique IP; internal hosts are not directly visible to the outside world. The technique can be extended to handle multiple internal hosts by storing information about the sender in an auxiliary field of the packet header. For example, for TCP traffic, the source port of the sender might be stored as the TCP source port.

As a concrete example, consider the network topology depicted in Figure 2. It consists of a single switch (center), three endhosts (H_1-H_3), and a general-purpose controller. The endhosts are connected on ports 1 through 3, while port 4 maintains connectivity with the Internet.

The network policy we would like to implement is: HTTP packets on port 80 destined for external hosts are forwarded to port 4, but only after their source IP address has been overwritten to Externallp, the IP address of the gateway switch. In order to correctly multiplex HTTP packets, the network program must also update the source TCP port of outgoing HTTP packets to equal the switch location of the sending host. Accordingly, incoming external HTTP packets should be sent to the switch port given by the packet's TCP destination, but only after the packet's TCP and IP fields have been restored. We implement this policy by combining a number of small NetCore programs, as follows.

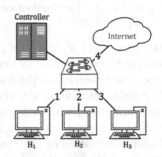

Fig. 2. *Network Address Translation.* HTTP packets sent from internal endhosts to external hosts (*Internet*) are multiplexed over the single external IP address assigned to a gateway switch (center)

First, we define[7] a NetCore program fragment that overwrites a packet's source IP address, then forwards the packet to port 4.

```
26 Definition pg1 :=
27   WILD ⇒ UPD_IP_SRC Externallp 'SEQ' WILD ⇒ FWD 4.
```

Program pg1 can be combined with a parameterized NetCore program that updates a packet's TCP source field to equal the sender's switch port, as follows.

```
36 Definition pg2 (n: Z) :=
37   PORT=n ⇒ UPD_TCP_SRC n 'SEQ' pg1.
```

All together, the rule for outgoing packets is the restriction of pg2, for hosts 1 through 3, to internal packets that are both *not* located on port 4 and have TCP destination port equal to 80.

```
44 Definition outgoing :=
45   RESTRICT (pg2 1 'PAR' pg2 2 'PAR' pg2 3)
46   BY (NOT PORT=4 'AND' TPDST=80).
```

The rule for incoming packets on port 4 is defined in a similar way, by first restoring the packet's IP and TCP destination fields, then forwarding the packet to the internal port given by the packet's initial TCP destination field.

Now that we have defined a NetCore program that implements NAT for the topology given above, we can verify that the program behaves as we expect. For example, the following lemma states that packets sent by hosts H_1 through H_3

---

[7] The code that follows is found in `src/examples/NAT.v`.

are forwarded to port 4, with source IP address modified to equal Externallp and
TCP source port set to the host port number h.

```
95| Lemma nat_ok: ∀h, List.In h hosts →
96|    |- [PORT=h] incoming 'PAR' outgoing
97|        [NWSRC=Externallp 'AND' TPSRC=h 'AND' PORT=4].
98| Proof. Time unfold outgoing; check_all hosts. Qed.
99| (*Finished transaction in 0. secs (0.3125u,0.s)*)
```

Here hosts = [1; 2; 3]. The proof of this theorem relies on an additional automation tactic provided by our Coq library (check_all). The check_all tactic proves
correctness theorems of the form ∀x, List.In x all → |- [P x] pg [Q x], where
all is a finite multiset of x's, and P and Q are predicates on x's. The tactic works
by breaking the theorem down into a finite set of verification conditions, which
are all then proved automatically using the checker tactic of previous sections.

Using a variation of the check_all tactic for reachability queries, we can quite
easily prove that TCP packets destined for external hosts are forwarded to port
4 of the gateway.

```
116| Lemma nat_outgoing: ∀h, List.In h hosts →
117|    |-r [TPDST=80 'AND' PORT=h 'AND' IS_IP 'AND' IS_TCP]
118|        incoming 'PAR' outgoing
119|        [PORT=4].
120| Proof. Time check_all' hosts. Qed.
121| (*Finished transaction in 0. secs (0.40625u,0.s)*)
```

Recall that the outgoing NetCore program updates both the the TCP source
port and the source IP of outgoing packets. Thus the theorem holds only for
packets that are indeed valid TCP/IP packets. Our Coq development proves an
analogous reachability theorem for incoming traffic.

6   Measurements

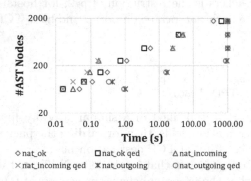

Fig. 3. NAT timings for 2 to 65 endhosts

To better understand the performance profile of our tool suite,
we extended the network address
translation example of the previous section to scale from 2 to
65 endhosts.[8] The plot in Figure 3 presents timing results, on
a log-log scale, for the NAT security and reachability theorems we
stated and proved in the previous
section. The y-axis gives the size
of the verified NAT programs in

---

[8] File src/examples/NATBench.v in our development.

number of AST nodes. We measured both the time to execute proof scripts in Coq, and the time to typecheck proof terms at Qed (x-axis).

While these experiments are still quite preliminary, they seem to indicate that our tool suite is quite suitable for interactive use, at least for moderately sized programs. Verifications of programs of up to approximately 250 AST nodes usually took no more than a second or two. On the other hand, there is still room for improvement. We would like to increase the efficiency of our resolution backend, by using more efficient data structures and also term orderings. We also plan to experiment with certificate-producing backends, in order to better understand the concomitant tradeoffs. For example, it is not immediately clear which matters more: the time to check proof certificates versus the potential gains from using a highly tuned external prover.

7 Related Work

There has been a great deal of work on verification of low-level network configurations in recent years [5–7, 11]. VeriFlow [6] uses an incremental analysis of OpenFlow rule updates to check network invariants such as reachability in real time. Anteater [7], an earlier effort, reduces verification of data plane invariants to SAT. Header space analysis [5] does static analysis of low-level network configurations using a geometric abstract domain. Reitblatt *et al.* [11] apply techniques from model checking to verify invariants of OpenFlow configurations. The VeriFlow paper [6] provides a good summary of additional related work.

The techniques described above all operate directly on switch and router configurations, in the form of unstructured flow tables. They therefore incorporate very little of the high-level structure present in the NetCore programs we analyze in this paper. In addition, our NetCore weakest precondition calculi are *proved complete* (and sound) in Coq. The analyses cited above provide no such formal guarantees. On the other hand, techniques such as header space analysis, which operates on a geometric abstraction of headers as uninterpreted bit vectors, make fewer assumptions about the underlying network protocols, and therefore are more general than the analyses we describe in this paper.

8 Conclusions

We have only scratched the surface of potential applications of the verification techniques we describe in this paper. In our Coq development, we do example verifications[9] of the conditions that arise when proving disjointness of virtual networks, or VLANs, using the network *slice* abstraction recently proposed by Gutz *et al.* [4] We have also begun to explore the use of our tool to detect, and prove the absence of, loops in multi-switch networks, using a static analysis that depends heavily on our weakest precondition calculus for reachability. Although this work is still in progress, completeness of the weakest precondition calculus

---

[9] File src/examples/Slice.v.

should allow us to *prove* the absence of network loops, given a network program and topology, rather than just detect them, as is done using existing techniques such as Header Space Analysis [5].

At the same time, our Coq library is still in its early stages, and is therefore limited in some ways. For example, at the moment we only target static NetCore programs running on concrete network topologies (that is, in which the number of switches, ports, and hosts are all known in advance). We would like to experiment with using the library as a subcomponent of a larger tool suite, in order to do verification of controller programs that generate streams of NetCore programs in response to a stream of network input events. In addition, our specification language currently only targets expressions in the NetCore predicate language. In the future, we plan to extend it, and the accompanying tool support, to enable verification of richer properties. Finally, because our NetCore semantics was defined before Guha *et al.*'s verified compiler was publically available, it differs in some details from the Guha *et al.* implementation. For example, in order to do proof by reflection in Coq, we specify packets using a fixed-width machine integer library that supports computable equality, whereas Guha *et al.* use an axiomatization of machine words. We also support sequential composition, whereas Guha *et al.*'s compiler does not. It is details like these that have so far prevented complete convergence of our mechanized development with theirs.

Acknowledgments. I am indebted to the members of the Princeton programming languages group for reading and commenting on early versions of this paper, and to the anonymous reviewers for their insightful comments.

References

1. de Moura, L., Bjørner, N.: Z3: An efficient SMT solver. In: Ramakrishnan, C.R., Rehof, J. (eds.) TACAS 2008. LNCS, vol. 4963, pp. 337–340. Springer, Heidelberg (2008)
2. Foster, N., Harrison, R., Freedman, M.J., Monsanto, C., Rexford, J., Story, A., Walker, D.: Frenetic: A network programming language. In: ICFP (2011)
3. Guha, A., Reitblatt, M., Foster, N.: Machine-verified network controllers. In: PLDI (2013)
4. Gutz, S., Story, A., Schlesinger, C., Foster, N.: Splendid isolation: A slice abstraction for software-defined networks. In: Hot Topics in SDNs. ACM (2012)
5. Kazemian, P., Varghese, G., McKeown, N.: Header space analysis: Static checking for networks. In: NSDI (2012)
6. Khurshid, A., Zhou, W., Caesar, M., Godfrey, P.: Veriflow: Verifying network-wide invariants in real time. In: Hot Topics in SDNs. ACM (2012)
7. Mai, H., Khurshid, A., Agarwal, R., Caesar, M., Godfrey, P., King, S.T.: Debugging the data plane with Anteater. ACM SIGCOMM CCR 41(4) (2011)
8. McCune, W., Wos, L.: Otter: The CADE-13 competition incarnations. JAR 18, 211–220 (1997)
9. McKeown, N., Anderson, T., Balakrishnan, H., Parulkar, G., Peterson, L., Rexford, J., Shenker, S., Turner, J.: OpenFlow: Enabling innovation in campus networks. ACM SIGCOMM CCR 38(2), 69–74 (2008)

10. Monsanto, C., Foster, N., Harrison, R., Walker, D.: A compiler and run-time system for network programming languages. In: POPL (2012)
11. Reitblatt, M., Foster, N., Rexford, J., Schlesinger, C., Walker, D.: Abstractions for network update. In: SIGCOMM (2012)
12. Robinson, J.A.: A Machine-Oriented Logic Based on the Resolution Principle. Journal of the ACM 12, 23–41 (1965)
13. Voellmy, A., Hudak, P.: Nettle: Taking the sting out of programming network routers. In: Rocha, R., Launchbury, J. (eds.) PADL 2011. LNCS, vol. 6539, pp. 235–249. Springer, Heidelberg (2011)

Aliasing Restrictions of C11 Formalized in Coq

Robbert Krebbers

ICIS, Radboud University Nijmegen, The Netherlands

Abstract. The C11 standard of the C programming language describes dynamic typing restrictions on memory operations to make more effective optimizations based on alias analysis possible. These restrictions are subtle due to the low-level nature of C, and have not been treated in a formal semantics before. We present an executable formal memory model for C that incorporates these restrictions, and at the same time describes required low-level operations.

Our memory model and essential properties of it have been fully formalized using the Coq proof assistant.

1 Introduction

Aliasing is when multiple pointers refer to the same object in memory. Consider:

```
int f(int *p, int *q) { int x = *q; *p = 10; return x; }
```

When f is called with aliased pointers for the arguments p and q, the assignment to *p also affects *q. As a result, a compiler cannot transform the function body of f into *p = 10; return (*q);.

Unlike this example, there are many situations in which pointers *cannot* alias. It is essential for an optimizing compiler to determine when aliasing cannot occur, and use this information to generate faster code. The technique of determining whether pointers are aliased or not is called *alias analysis*.

In *type-based alias analysis*, type information is used to determine whether pointers are aliased or not. Given the following example

```
float g(int *p, float *q) { float x = *q; *p = 10; return x; }
```

a compiler should be able to assume that p and q are not aliased as their types differ. However, the static type system of C is too weak to enforce this restriction because a union type can be used to call g with aliased pointers.

```
union { int x; float y; } u = { .y = 3.14 }; g(&u.x, &u.y);
```

A union is C's version of a *sum* type, but contrary to ordinary sum types, unions are *untagged* instead of *tagged*. This means that their current variant cannot be obtained. Unions destroy the property that each memory area has a unique type that is statically known. The *effective type* [6, 6.5p6-7] of a memory area thus depends on the *run time behavior* of the program.

The *strict-aliasing restrictions* [6, 6.5p6-7] imply that a pointer to a variant of a union type (not to the whole union itself) can only be used for an access (a read

G. Gonthier and M. Norrish (Eds.): CPP 2013, LNCS 8307, pp. 50–65, 2013.
© Springer International Publishing Switzerland 2013

or store) if the union is in that particular variant. Calling g with aliased pointers (as in the example where u is in the y variant, and is accessed through a pointer p to the x variant) thus results in *undefined behavior*, meaning the program may do literally anything. C uses a "garbage in, garbage out" principle for undefined behavior to refrain compilers from having to insert (possibly expensive) checks to handle corner cases. A compiler thus does not have to generate code that tests whether effective types are violated (here: to test whether p and q are aliased), but is allowed to assume no such violations occur.

As widely used compilers (*e.g.* GCC and Clang) perform optimizations based on C's aliasing restrictions, it is essential to capture these in a formal memory model for C. Not doing so, makes it possible to prove certain programs to be correct when they may crash when compiled with an actual C compiler.

Approach. The main challenge of formalizing C's strict-aliasing restrictions is that both *high-level* (by means of typed expressions) and *low-level* (by means of byte-wise manipulation) access to memory is allowed. Hence, an abstract "Java-like" memory model would not be satisfactory as it would disallow most forms of byte-wise manipulation.

Significant existing formal semantics for C (*e.g.* Leroy *et al.* [10], Ellison and Rosu [3]) model the memory using a finite partial function to objects, where each object consist of an array of bytes. Bytes are symbolic to capture indeterminate storage and pointer representations. However, because no information about the variants of unions is stored, this approach cannot capture C's strict-aliasing restrictions. We refine this approach in two ways.

– Instead of using an array of bytes as the contents of each object, we use well-typed trees with arrays of bits that represent base values as leafs.
– We use symbolic bits instead of bytes as the smallest available unit.

The first refinement is to capture strict-aliasing restrictions: effective types are modeled by the state of the trees in the memory model. Our use of trees also captures restrictions on padding bytes[1] simply because these are not represented. The second is to deal with bit fields as part of structs (in future work) where specific bits instead of whole bytes may be indeterminate.

The novelty of our memory model is that it also describes low-level operations such as byte-wise copying of objects and type-punning. As depicted in Figure 1, the model has three layers: (a) *abstract values*: trees with mathematical integers and pointers as leafs, (b) *memory values*: trees with arrays of bits as leafs, and (c) arrays of bits. Memory values are internal to the memory model, and abstract values are used for its external interface. Pointers are represented by a pair of a cell identifier and a path through the corresponding memory value.

In order to enable type-based alias analysis, we have to ensure that only under certain conditions a union can be read using a pointer to another variant than

---

[1] In particular: "When a value is stored in an object of structure or union type, including in a member object, the bytes of the object representation that correspond to any padding bytes take unspecified values" [6, 6.2.6.1p6].

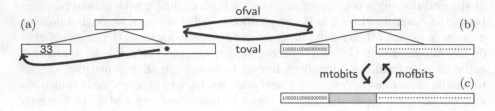

Fig. 1. The representations of struct { short x, *p; } s = { 33; &s.x }

the current one (this is called *type-punning* [6, 6.5.2.3]). Since the C11 standard is unclear about these conditions[2], we follow the GCC documentation [4] on it. It states that "type-punning is allowed, provided the memory is accessed through the union type". This means that the function f has defined behavior[3]:

```
union U { int x; float y; };
int f() { union U t; t.y = 3.0; return t.x; }
```

whereas the function g exhibits undefined behavior:

```
int g() { union U t; int *p = &t.x; t.y = 3.0; return *p; }
```

We formalize the previously described behavior by decorating the formal definition of pointers with annotations. Whenever a pointer to a variant of some union is stored in memory, or used as the argument of a function, the annotations are changed to ensure that type-punning is no longer possible via that pointer.

We tried to follow the C11 standard [6] as closely as possible. Unfortunately, it is often ambiguous due to its use of natural language (see the example above, this message [12] on the standard's committee's mailing list, and Defect Report #260 and #236 [5]). In the case of ambiguities, we tried to err on the side of caution. Generally, this means assigning undefined behavior.

Related work. The first formalization of a significant part of C is due to Norrish [14] using HOL4. He considered C89, in which C's aliasing restrictions were not introduced yet, and thus used a memory model based on just arrays of bytes. Tuch *et al.* [18] also consider a memory model based on just arrays of bytes.

Leroy *et al.* have formalized a large part of a C memory model as part of CompCert; a verified optimizing C compiler in Coq [11,10]. The first version of their memory model [11] uses type-annotated symbolic bytes to represent integer, floating point, and pointer values. This version describes some aliasing restrictions (namely those on the level of base types), but at the cost of prohibiting any kind of "bit twiddling". In the second version of their memory model [10], type information has been removed, and symbolic bytes were only used for pointer values and indeterminate storage. Integers and floating points were represented

---

[2] The term *type-punning* is merely used in a footnote, but for the related *common initial segment* rule, it uses the notion of *visible*, which is not clearly defined either.

[3] Provided size_of(int) ≤ size_of(float) and ints do not have trap values.

using numeric bytes. We adapt their choice of using symbolic representations for indeterminate storage and pointers. Moreover, we adapt their notion of *memory extensions* [11]. As an extension of CompCert, Robert and Leroy have verified an untyped alias analysis [16].

Ellison and Rosu [3] have defined an executable semantics of the C11 standard in the $\mathbb{K}$-framework. Their memory model is based on the CompCert memory model by Leroy *et al.* and does not describe the aliasing restrictions we consider.

The idea of a memory model that uses trees instead of arrays of plain bits, and paths instead of offsets to model pointers, has been used for object oriented languages before. It goes back to at least Rossie and Friedman [17], and has been used by Ramananandro *et al.* [15] for C++. However, we found no evidence in the literature of using trees to define a memory model for C.

Contribution. This work presents an executable mathematically precise version of a large part of the (non-concurrent) C memory model. In particular:

- We give a formal definition of the core of the C type system (Section 2).
- Our formalization is parametrized by an abstract interface to allow implementations that use multiple integer representations (Section 3).
- We define a memory model that describes a large set of subtly interacting features: effective types, byte-level operations, type-punning, indeterminate memory, and pointers "one past the last element" (Sections 4 to 6).
- We demonstrate that our memory model is suitable for formal proofs by verifying essential algebraic laws, abstract version of `memcpy`, and an essential property for aliasing analysis (Section 6).
- All proofs have been formalized using the Coq proof assistant (Section 7).

As this paper describes a large formalization effort, we often just give representative parts of definitions due to space restrictions. The interested reader can find all details online as part of our Coq formalization.

Notations. We let B^{opt} denote the *option type*, which is inductively defined as either $\bot$ or x for some $x \in B$. We often implicitly lift operations to operate on the option type, which is done using the *option monad* in the Coq formalization. A *partial function* $f : A \to B^{\text{opt}}$ is called *finite* if its *domain* dom f is finite. The operation $f[x := y]$ stores the value y at index x.

2   Types

We treat the most relevant C-types: integers, pointers, arrays, structs, unions, and the void type. Floating point and function types are omitted as these are orthogonal to the aliasing restrictions described in this paper. The void type plays a dual role, it is used for functions without a return value, and for pointers to data of an unspecified type.

Definition 2.1. Integer, base, *and* full types *are inductively defined as:*

$$si \in \mathsf{signedness} ::= \mathsf{signed} \mid \mathsf{unsigned}$$
$$\tau_i \in \mathsf{inttype} ::= si\ k$$
$$\tau_b \in \mathsf{basetype} ::= \tau_i \mid \tau*$$
$$\tau \in \mathsf{type} ::= \tau_b \mid \mathsf{void} \mid \tau[n] \mid \mathsf{struct}\ s \mid \mathsf{union}\ u$$

In the above definition, k ranges over *integer ranks* (see Section 3), and $s, u \in$ tag range over struct and union names (called *tags*). *Environments* ($\Gamma \in \mathsf{env}$) are finite partial functions from tags to lists of types representing struct and union fields. Since fields are represented using lists, they are nameless. We allow structs and unions with the same name for simplicity.

The above definition still allows ill-formed types as `void[0]`. Also, we have to ensure that cyclic structs and unions are only allowed when recursion is guarded by a pointer. The type `struct T1 { struct T1 x; };` should thus be prohibited whereas `struct T2 { struct T2 *p; };` should be allowed.

Definition 2.2. *The judgment* $\Gamma \vdash_b \tau_b$ *describes valid base types,* $\Gamma \vdash \tau$ *valid types, and* $\Gamma \vdash_* \tau$ *types to which pointers are allowed.*

$$\frac{}{\Gamma \vdash_b \tau_i} \qquad \frac{\Gamma \vdash_* \tau}{\Gamma \vdash_b \tau*} \qquad \frac{\Gamma \vdash_b \tau_b}{\Gamma \vdash \tau_b} \qquad \frac{\Gamma \vdash \tau \quad 0 < n}{\Gamma \vdash \tau[n]} \qquad \frac{\Gamma\,s = \vec{\tau}}{\Gamma \vdash \mathsf{struct}\ s} \qquad \frac{\Gamma\,u = \vec{\tau}}{\Gamma \vdash \mathsf{union}\ u}$$

$$\frac{\Gamma \vdash_b \tau_b}{\Gamma \vdash_* \tau_b} \qquad \frac{}{\Gamma \vdash_* \mathsf{void}} \qquad \frac{\Gamma \vdash \tau \quad 0 < n}{\Gamma \vdash_* \tau[n]} \qquad \frac{}{\Gamma \vdash_* \mathsf{struct}\ s} \qquad \frac{}{\Gamma \vdash_* \mathsf{union}\ u}$$

The judgment for well-formed environments Γ valid *is defined as:*

$$\frac{}{\emptyset\ \mathsf{valid}} \qquad \frac{\Gamma\ \mathsf{valid} \quad \Gamma \vdash \vec{\tau} \quad s \notin \mathsf{dom}\ \Gamma \quad 0 < |\vec{\tau}|}{(s : \vec{\tau}, \Gamma)\ \mathsf{valid}}$$

Due to the fact that C allows (mutually) recursive struct and union types, we allow pointers to struct and union types before they are declared in the $\Gamma \vdash_* \tau$ judgment. Note that $\Gamma \vdash \tau$ does not imply Γ valid.

Well-formedness of $\Gamma = \mathtt{T2} : [\mathsf{struct}\ \mathtt{T2}*]$ can be derived using the judgments $\emptyset \vdash_* \mathsf{struct}\ \mathtt{T2}$, $\emptyset \vdash_b \mathsf{struct}\ \mathtt{T2}*$, $\emptyset \vdash \mathsf{struct}\ \mathtt{T2}*$, and thus Γ valid. The environment $\mathtt{T1} : [\mathsf{struct}\ \mathtt{T1}]$ is ill-formed because we do not have $\emptyset \vdash \mathsf{struct}\ \mathtt{T1}$.

Lemma 2.3. *Given an arbitrary set* A, *and functions* $f_b : \mathsf{basetype} \to A$, $f_a : \mathsf{type} \to \mathbb{N} \to A \to A$, $f_s, f_u : \mathsf{tag} \to \mathsf{list\ type} \to \mathsf{list}\ A \to A$, *the function* type_iter $: \mathsf{env} \to \mathsf{type} \to A$ *is total for well-formed environments and types.*

$$\mathsf{type\_iter}_\Gamma\ \tau_b := f_b\ \tau_b$$
$$\mathsf{type\_iter}_\Gamma\ (\tau[n]) := f_a\ \tau\ n\ (\mathsf{type\_iter}_\Gamma\ \tau)$$
$$\mathsf{type\_iter}_\Gamma\ (\mathsf{struct}\ s) := f_s\ s\ (\Gamma\,s)\ (\mathsf{type\_iter}_\Gamma\ (\Gamma\,s))$$
$$\mathsf{type\_iter}_\Gamma\ (\mathsf{union}\ u) := f_u\ u\ (\Gamma\,u)\ (\mathsf{type\_iter}_\Gamma\ (\Gamma\,u))$$

We often lift type_iter Γ *to operate pointwise on lists of types.*

The previous lemma is used to define functions where recursion on fields of unions and structs is needed. Totality is proven by well-founded induction on the size of the type environment.

Our formalization of the C type system differs in various ways from existing work. In CompCert [10], fields of structs and unions are not stored in an environment, but are stored in the types itself. Hence, instead of having a construct struct s, they have a construct struct s $\vec{\tau}$ and a special pointer type struct_ptr s to allow recursive structs. Although this relieves one from having to carry a type environment around, the main disadvantage is that one has to roll and unroll types at certain places, and that one loses canonicity.

Affeldt and Marti [1] have also formalized a part of the C type system. Like us, they use an environment to capture the types of fields of structs, but they define non-cyclicity of type environments using a complex constraint on paths through types. Our definition Γ valid follows the structure of type environments, and seems more easy to use (for example for proving termination of the iteration function type_iter). Also, they omit union types, and do not parametrize by an abstract interface to allow multiple integer implementations.

3   Integer Arithmetic

In order to make C portable, the C standard gives compilers a lot of freedom to represent integers and to perform integer arithmetic. First of all, it does not specify the sizes of integer types. For example, signed int does not necessarily have to be 32 bits, use two's complement representation, and be able to exactly hold values between -2^{31} and $2^{31} - 1$. Only some minimum limits are described [6, 5.2.4.2.1]. Secondly, the standard puts few constraints on the way integers are represented as bits. Thirdly, overflow of signed integers is undefined behavior, whereas it wraps around modulo for the case of unsigned integers.

In order to capture different integer implementations, our memory model is parametrized by an abstract interface of *integer implementations*. This interface consists of a set K of *integer ranks* and functions:

$$
\begin{array}{ll}
\text{char} : K & \text{endianize} : K \to \text{list bool} \to \text{list bool} \\
\text{int} : K & \text{deendianize} : K \to \text{list bool} \to \text{list bool} \\
\text{ptr\_rank} : K & \text{int\_binop\_ok} : \text{inttype} \to \text{binop} \to \mathbb{Z} \to \mathbb{Z} \to \text{bool} \\
\text{char\_bits} : \mathbb{N}_{\geq 8} & \text{int\_binop} : \text{inttype} \to \text{binop} \to \mathbb{Z} \to \mathbb{Z} \to \mathbb{Z} \\
\text{rank\_size} : K \to \mathbb{N}_{>0} & \text{int\_cast\_ok} : \text{inttype} \to \mathbb{Z} \to \text{bool} \\
& \text{int\_cast} : \text{inttype} \to \mathbb{Z} \to \mathbb{Z}
\end{array}
$$

Here, binop is the inductive type of the C binary operations.

$$op \in \text{binop} ::= \texttt{+} \mid \texttt{-} \mid \texttt{*} \mid \texttt{<<} \mid \texttt{>>} \mid \texttt{/} \mid \texttt{\%} \mid \texttt{==} \mid \texttt{<=} \mid \texttt{<} \mid \texttt{\&} \mid \texttt{|} \mid \texttt{\textasciicircum}$$

Unary operations are derived from the binary operations.

The rank char is the rank of the smallest available integer type, and ptr_rank the rank of the types size_t and ptrdiff_t. At an actual machine char corresponds to a byte, and its bit size is char_bits (called CHAR_BIT in the C header files). The function rank_size k gives the byte size of an integer with rank k.

Since all modern architectures use two's complement representation, we allow representations to differ solely in endianness. The function endianize takes a list of bits in little endian order and permutes them according to the implementation's endianness. The function deendianize performs the inverse.

Since we restrict to two's complement, and do not allow integer representations to contain padding bits, an $x \in \mathbb{Z}$ is an integer of type signed k in case $-2^{\text{char\_bits}*\text{rank\_size }k-1} \leq x < 2^{\text{char\_bits}*\text{rank\_size }k-1}$. An $x \in \mathbb{Z}$ is an integer of type unsigned k in case $0 \leq x < 2^{\text{char\_bits}*\text{rank\_size }k}$.

In order to deal with underspecification of operations, our interface not just contains a function int_binop to perform binary operations, but also a predicate int_binop_ok τ op x y that describes when op is allowed to be performed on integers x and y of type τ. This is to allow both strict implementations that make integer overflow undefined, and those that let it wrap (as for example GCC with the `-fno-strict-overflow` flag and CompCert do). This predicate should be at least as strong as what is allowed by the C standard. Whenever an operation is allowed by the C standard, the result of int_binop τ op x y should correspond to its specification by the standard.

Integer promotions/demotions should be handled explicitly using casts, for which we use a similar treatment as for operations.

Finally, a *C environment* consists of an integer implementation with integer ranks K, a valid typing environment Γ, and functions sizeof : type $\to \mathbb{N}_{>0}$ and fieldsizes : list type $\to$ list $\mathbb{N}$. These functions should satisfy:

$$\text{sizeof } (si\ k) = \text{rank\_size } k \qquad \text{sizeof void} = 1 \qquad \text{sizeof } (\tau[n]) = n * \text{sizeof } \tau$$

$$\text{sizeof } (\text{struct } s) = \Sigma \text{ fieldsizes } \vec{\tau} \qquad \text{if } \Gamma s = \vec{\tau}$$

$$\text{sizeof } \tau_i \leq z_i \qquad \text{for each } i < |\vec{\tau}| \text{ and fieldsizes } \vec{\tau} = \vec{z}$$

$$\text{sizeof } \tau_i \leq \text{sizeof } (\text{union } u) \qquad \text{for each } i < |\vec{\tau}| \text{ and } \Gamma u = \vec{\tau}$$

We define bitsizeof τ as sizeof $\tau \cdot$ char_bits. We let sizeof void $= 1$ so as to capture that a void pointer can point to individual bytes.

Although the definition of a C environment does not explicitly state anything about alignment, it is implicitly there. If an implementation has constraints on alignment, it should set up the function fieldsizes in such a way. Together with the dynamic typing constraints of the memory (as defined in Section 4) it is ensured that no improperly aligned stores and reads will occur.

Nita *et al.* describe a more concrete notion of a *C platform* than our notion of a C environment [13]. Important difference are that alignment is implicit in our definition, that we allow pointers $\tau*$ whose size can depend on τ, and that we restrict to 2's complement.

4   Bits, Bytes and Memory Values

This section defines the internals of our memory model, and the representation of pointers. In the remainder of this paper we implicitly parametrize all definitions and proofs by a C environment with ranks K and typing environment Γ.

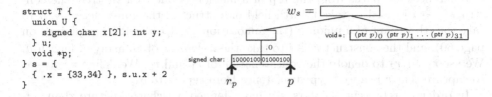

Fig. 2. A memory value w_s with pointer $p = (x_s, r_p, 2)_{\text{signed short}>\text{void}}$ on x86

Definition 4.1. Bits, memory values, objects, *and* memories *are defined as:*

$$b \in \text{bit} ::= \beta \mid (\text{ptr } p)_i \mid \text{indet}$$

$$w \in \text{mval} ::= \text{base}_{\tau_b} \vec{b} \mid \text{array } \vec{w} \mid \text{struct}_s \vec{w} \mid \text{union}_u (i, w) \mid \overline{\text{union}_u} \vec{b}$$

$$o \in \text{obj} ::= w \mid \text{freed } \tau$$

Memories ($m \in \text{mem}$) *are finite partial functions of a countable set of* memory indexes ($x \in \text{index}$) *to objects.*

A bit is either a concrete bit β (with β a Boolean), the ith fragment bit $(\text{ptr } p)_i$ of a pointer p (see Definition 4.2 for pointers), or the indeterminate bit indet. As shown in Figure 2, integers are represented using concrete sequences of bits, and pointers as sequences of fragments. This way of representing pointers is similar to Leroy *et al.* [10], but is on the level of bits instead of bytes.

Memory values are decorated with types, so that we can read off the type typeof w of each memory value w. As empty arrays are prohibited, we do not store the element type of the array $\vec{w}$ construct. We define the following partial function:

$$\text{indextype}_m \, x := \begin{cases} \text{typeof } w & \text{if } m\,x = w \\ \tau & \text{if } m\,x = \text{freed } \tau \end{cases}$$

We consider two kinds of union values. The construct $\text{union}_u (i, w)$ represents unions that are in a particular variant i, and the construct $\overline{\text{union}_u} \vec{b}$ represents unions whose variant is unknown. Unions of the latter kind can be obtained by byte-wise copying, and will appear in uninitialized memory. Note that the variant of a union is internal to the memory model, and should not be exposed through the operational semantics (as an actual machine does not store it).

Leroy *et al.* [10] represent pointers as pairs (x, i) where x identifies the object in the whole memory, and i the offset into that object. Since we use trees as the contents of objects, we use paths through these trees to represent pointers.

Definition 4.2. References, addresses *and* pointers *are defined as:*

$$r \in \text{ref} ::= \text{T} \mid r \overset{s}{\leadsto} i \mid r \overset{u}{\leadsto}_q i \mid r \overset{n}{\leadsto} i$$

$$a \in \text{addr} ::= (x, r, i)_{\tau > \sigma}$$

$$p \in \text{ptr} ::= \text{NULL } \tau \mid a$$

References are paths from the top of a memory value to a subtree: the construct $r \overset{s}{\rightsquigarrow} i$ is used to take the ith field of a struct s, the construct $r \overset{s}{\rightsquigarrow}_q i$ to take the ith variant of a union u (the annotation $q \in \{\circ, \bullet\}$ will be explained on page 60), and the construct $r \overset{n}{\rightsquigarrow} i$ to take the i element of an array of length n. We use $r_1 +\!\!\!+ r_2$ to denote the concatenation of r_1 and r_2. We define $r : \tau \rightarrowtail \sigma$ to capture that r is a well-typed reference from type τ to σ.

In order to represent pointers, we have defined a richer structure than references, namely *addresses*. An address $(x, r, i)_{\tau > \sigma}$ consists of: (a) an object identifier x, (b) a reference r to a subtree of the memory value in the object at x, (c) an offset i to refer to a particular byte in the subtree at r (note that one cannot address individual bits in C), (d) the type τ of the subtree, and (e) the type σ to which the address is cast. The type τ is stored so we do not have to recompute it when performing a pointer cast.

Typing. We define typing judgments for all of the previously defined structures. As array indexing in C is performed using pointer arithmetic, we need some auxiliary operations on references to define the typing judgment of addresses.

$$\mathsf{refoffset}\ r := \begin{cases} i & \text{if } r = r' \overset{n}{\rightsquigarrow} i \\ 0 & \text{otherwise} \end{cases} \qquad \mathsf{refsize}\ r := \begin{cases} n & \text{if } r = r' \overset{n}{\rightsquigarrow} i \\ 0 & \text{otherwise} \end{cases}$$

$$r \oplus j := \begin{cases} r' \overset{n}{\rightsquigarrow} i + j & \text{if } r = r' \overset{n}{\rightsquigarrow} i \\ r & \text{otherwise} \end{cases}$$

Definition 4.3. *The typing judgment* $m \vdash a : \sigma$ *for addresses is defined as:*

$$\frac{\mathsf{refoffset}\ r = 0,\ \ r : \mathsf{indextype}_m\ x \rightarrowtail \tau,\ \ \tau > \sigma,\ \ i \le \mathsf{sizeof}\ \tau \cdot \mathsf{refsize}\ r,\ \ \mathsf{sizeof}\ \sigma \mid i}{m \vdash (x, r, i)_{\tau > \sigma} : \sigma}$$

Here, $i \mid j$ *means that* i *is a divisor of* j*. The relation* $\tau > \sigma$*, type* τ *is pointer castable to* σ*, is the reflexive closure of:* $\tau > \mathsf{unsigned\ char}$ *and* $\tau > \mathsf{void}$*.*

The premise $\mathsf{refoffset}\ r = 0$ ensures that r always points to the first element of an array subobject, the byte index i is then used to select an individual byte (if τ is unsigned char or void), or an element of the whole array. Adding j to $(x, r, i)_{\tau > \sigma}$ thus consists of changing the offset into $i + j \cdot \mathsf{sizeof}\ \sigma$ instead of moving r. Only when a pointer is dereferenced, or used for struct or union indexing, we use the normalized reference $r \oplus i \div \mathsf{sizeof}\ \sigma$.

An address remains well-typed after the object it points to has been deallocated (indextype is defined on freed objects as well). However, as addresses of deallocated objects are indeterminate [6, 6.2.4p2], we forbid them to be used for pointer arithmetic, *etc.* We use the non-strict inequality $i \le \mathsf{sizeof}\ \tau \cdot \mathsf{refsize}\ r$ in the typing rule to allow addresses to be "one past the last element" [6, 6.5.6p8]. We call an address *strict* if is not "one past the last element" and its object has not been deallocated.

We define judgments $m \vdash p : \tau$ for pointers, $m \vdash b$ valid for bits, $m \vdash w : \tau$ for memory values, and $m \vdash o : \tau$ for objects. We display the rule for the construct $\underline{\mathsf{union}_u\ b}$ of a union whose variant is unknown for illustration.

$$\frac{\Gamma\, u = \vec{\tau} \qquad |\vec{\tau}| \neq 1 \qquad m \vdash \vec{b} \text{ valid} \qquad |\vec{b}| = \text{bitsizeof (union } u)}{m \vdash \overline{\text{union}_u\, \vec{b}} : \text{union } u}$$

We exclude unions with only one variant in this rule because their variant is always known. Validity of memories, notation m valid, is defined as:

$$\forall x\, w\,.\, m\, x = o \rightarrow \exists \tau\,.\, m \vdash o : \tau\ \wedge\ \text{sizeof } \tau < 2^{\text{char\_bits}*(\text{rank\_size ptr\_rank})-1}$$

We need the restriction on the size to ensure that the result of pointer subtraction is representable by a signed integer of rank ptr_rank.

Conversion from and to bits. We compute the bit representation mtobits w of a memory value w by flattening it and inserting padding bits (as specified by fieldsizes). The bit representation of w_s displayed in Figure 2 is thus:

mtobits w_s = 1000010001000100 indet indet . . . indet (ptr $p)_0$ (ptr $p)_1$ . . . (ptr $p)_{31}$

Likewise, given a type τ and sequence of bits $\vec{b}$, we construct a memory value mofbits $\tau\, \vec{b}$ of type τ by iteration on τ (using Lemma 2.3). In the case of a union type u, we obviously cannot guess the variant as that information is not stored in the bit representation, so we use the $\overline{\text{union}_u\, \vec{b}}$ construct.

Notice that mtobits and mofbits are neither left nor right cancellative. We do not have mofbits τ (mtobits w) = w for each $m \vdash w : \tau$ as variants of unions may have gotten lost, nor mtobits (mofbits $\tau\, \vec{b}$) = $\vec{b}$ for each $\vec{b}$ with $|\vec{b}|$ = bitsizeof τ as padding bits become indeterminate during the conversion.

Operations. In Section 6 we will define the following memory operations:

1. alloc : mem $\rightarrow$ index $\rightarrow$ type $\rightarrow$ mem allocates a new object.
2. free : mem $\rightarrow$ index $\rightarrow$ mem deallocates an object.
3. _ !! _ : mem $\rightarrow$ addr $\rightarrow$ val^{opt} yields a stored value or fails in case it does not exist or effective types are violated.
4. _[_ := _] : mem $\rightarrow$ addr $\rightarrow$ val $\rightarrow$ mem stores a value.

Here, val is the data type of *abstract values* (see Section 5). Many of the above operations are partial, but are defined using a total function that assigns a default behavior to ease formalization. For example, alloc should only be used on fresh indexes, and _[_ := _] should only be used if the address is accessible (*i.e.* _ !! _ succeeds). Notice that we model an unbounded memory as we consider a countable set of memory indexes. Formalizing a bounded memory is orthogonal to the strict-aliasing restrictions, and thus left for future work.

So as to define these operations, we first define variants on memory values, and lift those to whole memories in Section 6.

Definition 4.4. *The* empty memory value new : type → mval *is defined as:*

$$\text{new } \tau_b := \text{base}_{\tau_b} (\text{indet} \dots \text{indet}) \text{ (bitsizeof } \tau_b \text{ times)}$$

$$\text{new } (\tau[n]) := \text{array} (\text{new } \tau \dots \text{new } \tau) \text{ } (n \text{ times)}$$

$$\text{new } (\text{struct } s) := \text{struct}_s (\text{new } \tau_0 \dots \text{new } \tau_{n-1}) \quad \text{if } \Gamma s = \tau_0 \dots \tau_{n-1}$$

$$\text{new } (\text{union } u) := \begin{cases} \text{union}_u (0, \text{new } \tau) & \text{if } \Gamma u = \tau \\ \text{union}_u (\text{indet} \dots \text{indet}) \text{ (bitsizeof (union } u) \text{ times)} & \text{otherwise} \end{cases}$$

The operation new is used to create an empty memory value to implement alloc. The definition is well-defined for valid types by Lemma 2.3.

Definition 4.5. *The operation* _ !! _ : mval → ref → mval^{opt} *is defined as:*

$$w \text{ !! } \mathsf{T} := w$$

$$(\text{array } \vec{w}) \text{ !! } (\mathsf{T} \stackrel{n}{\leadsto} i \mathbin{+\!\!+} r) := w_i \text{ !! } r$$

$$(\text{struct}_s \vec{w}) \text{ !! } (\mathsf{T} \stackrel{s}{\leadsto} i \mathbin{+\!\!+} r) := w_i \text{ !! } r$$

$$(\text{union}_u (i, w)) \text{ !! } (\mathsf{T} \stackrel{u}{\leadsto}_q i \mathbin{+\!\!+} r) := w \text{ !! } r$$

$$(\text{union}_u (j, w)) \text{ !! } (\mathsf{T} \stackrel{u}{\leadsto}_\bullet i \mathbin{+\!\!+} r) := (\text{mofbits } \tau_i (\text{mtobits } w)) \text{ !! } r \text{ if } \Gamma u = \vec{\tau}, i \neq j$$

$$(\overline{\text{union}_u \vec{b}}) \text{ !! } (\mathsf{T} \stackrel{u}{\leadsto}_q i \mathbin{+\!\!+} r) := (\text{mofbits } \tau_i \vec{b}) \text{ !! } r \qquad \text{if } \Gamma u = \vec{\tau}$$

The look up operation is taking annotations q on union references $r \stackrel{s}{\leadsto}_q i$ into account: $q = \bullet$ means that we may access a union using a different variant than the current one (this is called *type-punning* [6, 6.5.2.3]), and $q = \circ$ means that this is prohibited. To enable type-punning, we convert back and forth to bits so as to interpret the memory value using a different type.

To ensure that type-punning is merely allowed when the memory is accessed "through the union type" [4], we change all annotations into $\circ$ whenever a pointer is stored in memory (see the definition of the function btobits in Section 5) or used as the argument of a function. This operation is called *freezing* a pointer, and the pointers whose annotations are all of the shape $\circ$ are called *frozen*. Frozen pointers cannot be used for type-punning by definition of _ !! _.

The strict-aliasing restrictions [6, 6.5p6-7] state that an access affects the effective type of the accessed object. Since the word "access" covers both reads and stores [6, 3.1], this means that not only a store has a side-effects, but also a read. We factor these side-effects out using a function force : ref → mval → mval that changes the effective types after a succeeded look up. To define the force and store operation, we define an auxiliary operation alter f : ref → mval → mval that applies f : mval → mval to a subtree and changes the effective types accordingly. The interesting cases for unions are as follows (where $\Gamma u = \vec{\tau}$ and $i \neq j$):

$$\text{alter } f (\mathsf{T} \stackrel{u}{\leadsto}_q i \mathbin{+\!\!+} r) (\text{union}_u (j, w)) := \text{union}_u (i, \text{alter } f r (\text{mofbits } \tau_i (\text{mtobits } w)))$$

$$\text{alter } f (\mathsf{T} \stackrel{u}{\leadsto}_q i \mathbin{+\!\!+} r) (\overline{\text{union}_u \vec{b}}) := \text{union}_u (i, \text{alter } f r (\text{mofbits } \tau_i \vec{b}))$$

Now force $r w := $ alter $(\lambda w' . w') r w$ and $w[r := w'] := $ alter $(\lambda_- . w') r w$.

5   Abstract Values

The notion of memory values, as defined in the previous section, is quite low-level and exposes implementation-specific properties as bit representations. These details should remain internal to the memory model.

Definition 5.1. Base values *and* abstract values *are defined as:*

$$v_b \in \mathsf{baseval} ::= \mathsf{indet}_{\tau_b} \mid \mathsf{int}_{\tau_i} i \mid \mathsf{ptr}\, p \mid \mathsf{byte}\, \vec{b}$$

$$v \in \mathsf{val} ::= v_b \mid \mathsf{array}\, \vec{v} \mid \mathsf{struct}_s\, \vec{v} \mid \mathsf{union}_s\, (i, v) \mid \overline{\mathsf{union}_u}\, \vec{v}$$

Abstract values contain mathematical integers and pointers instead of bit arrays as their leafs. As fragment bits of pointers need to be kept outside of the memory when performing a byte-wise copy, the $\mathsf{byte}\, \vec{b}$ construct still exposes some low-level details. The typing rule for this construct is:

$$\frac{\text{Not all } \vec{b} \text{ indet} \qquad \text{Not all } \vec{b} \text{ of the shape } \beta \qquad m \vdash \vec{b} \text{ valid} \qquad |\vec{b}| = \mathsf{char\_bits}}{m \vdash_b \mathsf{byte}\, \vec{b} : \mathsf{unsigned\ char}}$$

This rule ensures that the $\mathsf{byte}\, \vec{b}$ is only used if $\vec{b}$ cannot be interpreted as an integer $\mathsf{int}_{\mathsf{unsigned\ char}}\, i$ or $\mathsf{indet}_{\mathsf{unsigned\ char}}$. The judgment $m \vdash_b v_b : \tau_b$ moreover ensures that integers $\mathsf{int}_{\tau_i} i$ are within range, and pointers $\mathsf{ptr}\, p$ are typed.

The function $\mathsf{base\_binop} : \mathsf{binop} \to \mathsf{baseval} \to \mathsf{baseval} \to \mathsf{baseval}$ that performs a binary operation on base values is defined as:

$$\mathsf{base\_binop}\, op\, (\mathsf{int}_{\tau_b} i)\, (\mathsf{int}_{\tau_b} j) := \mathsf{int}_{\tau}\, (\mathsf{int\_binop}\, \tau_b\, op\, i\, j)$$

$$\mathsf{base\_binop}\, +\, (\mathsf{ptr}\, (x, r, i)_{\tau > \sigma})\, (\mathsf{int}_{\tau_b} j) := \mathsf{ptr}\, (x, r, i + j \cdot \mathsf{sizeof}\, \sigma)_{\tau > \sigma}$$

and so on ..., together with a predicate $\mathsf{base\_binop\_ok} : \mathsf{binop} \to \mathsf{baseval} \to \mathsf{baseval} \to \mathsf{bool}$ that describes when it is allowed to perform the operation. Binary operations are prohibited on indet_{τ_b} and $\mathsf{byte}\, \vec{b}$ constructs.

Base values are converted into bit sequences as follows:

$$\mathsf{btobits}\, (\mathsf{indet}_{\tau_b}) := \mathsf{indet} \ldots \mathsf{indet}\ (\mathsf{bitsizeof}\ \tau_b\ \mathrm{times})$$

$$\mathsf{btobits}\, (\mathsf{int}_{\tau_i} x) := \mathsf{endianize}\, (\tau_i\text{-little endian representation of } x)$$

$$\mathsf{btobits}\, (\mathsf{ptr}\, p) := (\mathsf{ptr}\, (\mathsf{freeze}\, p))_0 \ldots (\mathsf{ptr}\, (\mathsf{freeze}\, p))_{\mathsf{bitsizeof}\ (\mathsf{typeof}\ p*) - 1}$$

$$\mathsf{btobits}\, (\mathsf{byte}\, \vec{b}) := \vec{b}$$

This function will be used to store values in memory (see Definition 6.1), hence we freeze pointers so as to avoid prohibited type-punning. The inverse function $\mathsf{bofbits}$ is defined in such a way that invalid bit patterns yield an indet_{τ_b}.

Abstract values contain the construct $\overline{\mathsf{union}_u}\, \vec{v}$ for unions whose variant is unknown. The values $\vec{v}$ correspond to interpretations of all variants of u. Of course, these values should be consistent in the sense that they can be represented by the same bit sequence. The typing rule to ensure this is:

$$\frac{\Gamma u = \vec{\tau} \qquad |\vec{\tau}| \neq 1 \qquad m \vdash \vec{b} \text{ valid} \qquad |\vec{b}| = \mathsf{bitsizeof}\, (\mathsf{union}\, u) \qquad \forall i \,.\, v_i = \mathsf{vofbits}\, \tau_i\, \vec{b}}{m \vdash \overline{\mathsf{union}_u}\, \vec{v} : \mathsf{union}\, u}$$

The operation vofbits to obtain the bit representation of an abstract value, is defined similarly as its variant mtobits on memory values, but uses bofbits on the leafs. Obtaining a memory value ofval v and bit representation vtobits v from a value v is more challenging as the evidence of existence of a bit representation of a $\overline{\mathsf{union}_u\, \vec{v}}$ construct is only present in the typing judgment, and not in the value itself. We reconstruct the bits by "merging" the bit representations of all variants $\vec{v}$. For this, we define a join $\sqcup$ on bits satisfying indet $\sqcup\, b = b$, $b\, \sqcup$ indet $= b$, and $b \sqcup b = b$. The case for the unknown union construct is

$$\mathsf{ofval}\,(\overline{\mathsf{union}_u\,(v_0 \ldots v_{n-1})}) := \overline{\mathsf{union}_u}\,(\mathsf{vtobits}\, v_0 \sqcup \ldots \sqcup \mathsf{vtobits}\, v_{n-1})$$

where $\sqcup$ is applied pointwise to the bit sequences obtained from vtobits. This reconstruction is well-defined for well-typed abstract values.

6   The Memory

Now that we have all definitions in place, we can finally combine them to define the main memory operations. In order to shorten these definitions we lift the operations $_-\, !!\, _-$ and $_-[_- := _-]$ on memory values to whole memories. We define $m\, !!\, (x, r) := m\, x\, !!\, r$, and $m[(x, r) := w] := m[x := (m\, x)[r := w]]$. Notations are overloaded for conciseness of presentation.

Definition 6.1. *The* main memory operations *are defined as:*

$$m\, !!\, (x, r, i)_{\tau > \sigma} := \begin{cases} \mathbf{let}\ \hat{r} := r \oplus i \div \mathsf{sizeof}\ \sigma,\ j := i\ \mathsf{mod}\ \mathsf{sizeof}\ \sigma\ \mathbf{in} \\ \mathbf{if}\ \tau = \sigma\ \mathbf{then}\ \mathsf{toval}\,(m\, !!\, (x, \hat{r})) \\ \mathbf{else}\ \mathsf{vofbits}\,(\mathsf{unsigned\ char})\,(j\mathrm{th}\ \mathrm{byte}\ \mathrm{of}\ m\, !!\, (x, \hat{r})) \end{cases}$$

$$m[(x, r, i)_{\tau > \sigma} := v] := \begin{cases} \mathbf{let}\ \hat{r} := r \oplus i \div \mathsf{sizeof}\ \sigma,\ j := i\ \mathsf{mod}\ \mathsf{sizeof}\ \sigma\ \mathbf{in} \\ \mathbf{if}\ \tau = \sigma\ \mathbf{then}\ m[(x, \hat{r}) := \mathsf{ofval}\, v] \\ \mathbf{else}\ m[(x, \hat{r}) := \mathsf{set}\ j\mathrm{th}\ \mathrm{byte}\ \mathrm{of}\ m\, !!\, (x, \hat{r})\ \mathrm{to}\ \mathsf{vtobits}\, v] \end{cases}$$

$$\mathsf{force}\,(x, r, i)_{\tau > \sigma}\, m := \mathbf{let}\ \hat{r} := r \oplus i \div \mathsf{sizeof}\ \sigma\ \mathbf{in}\ m[x := \mathsf{force}\ \hat{r}\,(m\, x)]$$

$$\mathsf{alloc}\ x\ \tau\ m := m[x := \mathsf{new}\ \tau]$$

$$\mathsf{free}\ x\ m := m[x := \mathsf{freed}\,(\mathsf{indextype}_m\ x)]$$

The lookup operation $m\, !!\, (x, r, i)_{\tau > \sigma}$ normalizes the reference r, and then makes a case distinction on whether a whole subobject or a specific byte should be returned. In case of the former (*i.e.* $\tau = \sigma$), it converts the memory value $m\, !!\, (x, \hat{r})$ of the subobject in question into an abstract value. Otherwise, it yields abstract value representing the jth byte of $m\, !!\, (x, \hat{r})$.

In the Coq development we have proved the expected laws about the interaction between the memory operations. We list some for illustration:

- If m valid, $m \vdash a : \tau$, and $m\, !!\, a = v$, then $m \vdash v : \tau$
- If m valid, $m \vdash a : \tau$, $m \vdash v : \tau$, and $m \vdash a' : \sigma$, then $m[a := v] \vdash a' : \sigma$

– If m valid, $a_1 \perp a_2$, $m \vdash a_2 : \tau_2$, $m \vdash v_2 : \tau_2$, and $m \;!!\; a_1 = v_1$, then $m[a_2 := v_2] \;!!\; a_1 = v_1$

Here, $a_1 \perp a_2$, denotes that a_1 and a_2 are *disjoint*, which means that somewhere along their path from the top of the whole object to their subobject they take a different branch at an array of struct subobject.

Theorem 6.2 (Strict-aliasing). *Given a memory m with m valid, and frozen addresses $m \vdash a_1 : \sigma_1$ and $m \vdash a_2 : \sigma_2$ such that $\sigma_1, \sigma_2 \neq$ unsigned char *and* σ_1 not a subtype of σ_2 and vice versa. Now $a_1 \perp a_2$, or accessing a_1 after accessing a_2 and vice versa fails.*

Using this theorem, a compiler can optimize the generated code in the example below based on the assumption that p and q are not aliased.

```
float g(int *p, float *q) { float x = *q; *p = 10; return x; }
```

If these pointers are aliased, the program exhibits undefined behavior as both the read from *q, and the assignment to *p, are considered an access (captured by the operations force and $\_[\_ := \_]$ respectively).

In order to prove the correctness of program transformations one has to relate the memory states during execution of the original program to the memory states during execution of the transformed program. Leroy and Blazy [11] defined the notions of *memory extensions and injections* to facilitate this. We adapt memory extensions to our memory model, and demonstrate it by verifying an abstract version of the memcpy function that copies an object byte-wise.

A *memory extension* is a binary relation $\sqsubseteq$ on memories. The relation $m_1 \sqsubseteq m_2$ captures that m_2 makes more memory contents determinate, and that m_2 has fewer restrictions on effective types. This means that m_2 allows more behaviors. In order to define $\sqsubseteq$ we first define relations $\sqsubseteq_m$ on bits, abstract values, memory values, and objects. Some rules of these relations are:

$$\frac{}{b \sqsubseteq_m b} \qquad \frac{m \vdash b \text{ valid}}{\text{indet} \sqsubseteq_m b} \qquad \frac{w_1 \sqsubseteq_m w_2}{\text{union}_u\,(i, w_1) \sqsubseteq_m \text{union}_u\,(i, w_2)} \qquad \frac{\vec{b}_1 \sqsubseteq_m \vec{b}_2}{\overline{\text{union}_u\,\vec{b}_1} \sqsubseteq_m \overline{\text{union}_u\,\vec{b}_2}}$$

$$\frac{\Gamma u = \vec{\tau} \quad |\vec{\tau}| \neq 1 \quad w \sqsubseteq_m \text{mofbits}\,\tau_i\,\vec{b} \quad m \vdash \vec{b} \text{ valid} \quad |\vec{b}| = \text{bitsizeof}\,(\text{union}\,u)}{\text{union}_u\,(i, w) \sqsubseteq_m \overline{\text{union}_u\,\vec{b}}}$$

The relation $m_1 \sqsubseteq m_2$ is now defined as for all x and o with $m_1\,x = o_1$ there exists an $o_2 \sqsupseteq_{m_2} o_1$ s.t. $m_2\,x = o_2$. This relation is a partial order. In order to use memory extensions to reason about program transformations we have to prove that all memory operations are respected by it. For example:

If $w_1 \sqsubseteq_m w_2$ then mtobits $w_1 \sqsubseteq_m$ mtobits w_2

– If m_1 valid, $m_1 \sqsubseteq m_2$ and $m_1 \;!!\; a = v_1$, then $\exists v_2 \sqsupseteq_{m_2} v_1$ s.t. $m_2 \;!!\; a = v_2$

So as to show that a copy by assignment can be transformed into a byte-wise copy we proved that if $m \vdash w : \tau$, then ofval (toval w) $\sqsubseteq_m$ mofbits τ (mtobits w).

7   Formalization in Coq

Developing a formal version of a C11 memory model turned out to be much more difficult than we anticipated due to the complex and highly subtle nature of the aliasing restrictions introduced by the C99 and C11 standards. Hence, the use of a proof assistant has been essential for our development.

Since Coq is also a functional programming language, we can execute the memory model using it. This will be essential for the implementation of a certified interpreter in future work. We used Coq to formally prove properties such as:

- Type preservation and essential laws of the the memory operations.
- Compatibility of operations with respect to memory extensions.
- The fact that memory extensions form a partial order and respect typing.
- Correctness of an abstract `memcpy` and the Strict-aliasing Theorem 6.2.

We used Coq's notation mechanism combined with unicode symbols and type classes to let the Coq development correspond better to the definitions on paper. Type classes were also used to parametrize the whole development by an abstract interface for integer implementations and C environments (Section 3).

Although many operations on our memory model are partial, we formalized many such operations using a total function that assigns an appropriate default behavior. To account for partiality, we defined predicates that describe when these operations may be used. Alternatives include using the option monad or dependent types, but our approach turned out to be convenient as various proofs could be done easily by induction on the aforementioned predicate.

Our Coq code, available at `http://robbertkrebbers.nl/research/ch2o/`, is about 8.500 lines of code including comments and white space. Apart from that, we developed a library on general purpose theory (finite sets, finite functions, lists, the option monad, *etc.*) of about 10.000 lines.

8   Conclusion

The eventual goal of this work is to develop a formal semantics for a large part of the C11 programming language [8]. In previous work [9] we have developed a concise operational and axiomatic semantics for non-local control flow (goto and return statements). Recently, we have extended this work to include sequence points and non-deterministic expressions with side-effects [7]. The next step is to integrate our memory model into our operational semantics. Once integrated, we intend to develop a verified interpreter so we can test the memory model using actual C programs.

There are many other obvious extensions to our memory model: support for floating points, bit fields, variable length arrays, concurrency, *etc.* Bit fields are presumably easy to integrate as bits are already the smallest available unit in our memory model. Concurrency in C and C++ has received a lot of attention in formal developments (see *e.g.* Batty *et al.* [2]), but is extremely challenging on its own. Treating the weaker aliasing restrictions on base types (*e.g.* reading a `signed int` using an `unsigned int`) is left for future work too.

In order to integrate the memory model into our axiomatic semantics based on separation logic [9], we have to be able to split memory objects into disjoint subobjects. This requires a disjoint union operation on memory values. Besides, the axiomatic semantics should take types seriously as our memory model is typed. The work of Tuch *et al.* [18] may be interesting for this even though they do not consider the aliasing restrictions of C.

Acknowledgments. I thank Freek Wiedijk, Herman Geuvers, Michael Nahas, and the anonymous referees for their helpful suggestions. I thank Xavier Leroy for many discussions on the CompCert memory model. This work is financed by the Netherlands Organisation for Scientific Research (NWO).

References

1. Affeldt, R., Marti, N.: Towards formal verification of TLS network packet processing written in C. In: PLPV, pp. 35–46 (2013)
2. Batty, M., Owens, S., Sarkar, S., Sewell, P., Weber, T.: Mathematizing C++ concurrency. In: POPL, pp. 55–66 (2011)
3. Ellison, C., Rosu, G.: An executable formal semantics of C with applications. In: POPL, pp. 533–544 (2012)
4. GNU. GCC, the GNU Compiler Collection (2011), http://gcc.gnu.org/
5. International Organization for Standardization. WG14 Defect Report Summary (2008), http://www.open-std.org/jtc1/sc22/wg14/www/docs/
6. International Organization for Standardization. ISO/IEC 9899-2011: Programming languages – C. ISO Working Group 14 (2012)
7. Krebbers, R.: An operational and axiomatic semantics for non-determinism and sequence points in C. To appear in: POPL 2014 (2013)
8. Krebbers, R., Wiedijk, F.: A Formalization of the C99 Standard in HOL, Isabelle and Coq. In: Davenport, J.H., Farmer, W.M., Urban, J., Rabe, F. (eds.) Calculemus/MKM 2011. LNCS (LNAI), vol. 6824, pp. 301–303. Springer, Heidelberg (2011)
9. Krebbers, R., Wiedijk, F.: Separation Logic for Non-local Control Flow and Block Scope Variables. In: Pfenning, F. (ed.) FOSSACS 2013. LNCS, vol. 7794, pp. 257–272. Springer, Heidelberg (2013)
10. Leroy, X., Appel, A.W., Blazy, S., Stewart, G.: The CompCert Memory Model, Version 2. Research report RR-7987, INRIA (2012)
11. Leroy, X., Blazy, S.: Formal verification of a C-like memory model and its uses for verifying program transformations. JAR 41(1), 1–31 (2008)
12. Maclaren, N.: What is an Object in C Terms? Mailing list message (2001), http://www.open-std.org/jtc1/sc22/wg14/9350
13. Nita, M., Grossman, D., Chambers, C.: A theory of platform-dependent low-level software. In: POPL, pp. 209–220 (2008)
14. Norrish, M.: C formalised in HOL. PhD thesis, University of Cambridge (1998)
15. Ramananandro, T., Dos Reis, G., Leroy, X.: Formal verification of object layout for C++ multiple inheritance. In: POPL, pp. 67–80 (2011)
16. Robert, V., Leroy, X.: A Formally-Verified Alias Analysis. In: Hawblitzel, C., Miller, D. (eds.) CPP 2012. LNCS, vol. 7679, pp. 11–26. Springer, Heidelberg (2012)
17. Rossie, J.G., Friedman, D.P.: An Algebraic Semantics of Subobjects. In: OOPSLA, pp. 187–199 (1995)
18. Tuch, H., Klein, G., Norrish, M.: Types, bytes, and separation logic. In: POPL, pp. 97–108 (2007)

Proof Pearl: A Verified Bignum Implementation in x86-64 Machine Code

Magnus O. Myreen[1] and Gregorio Curello[2]

[1] Computer Laboratory, University of Cambridge, UK
[2] Autonoma University of Barcelona, Spain

Abstract. Verification of machine code can easily deteriorate into an endless clutter of low-level details. This paper presents a case study which shows that machine-code verification does not necessitate ghastly low-level proofs. The case study we describe is the construction of an x86-64 implementation of arbitrary-precision integer arithmetic. Compared with closely related work, our proofs are shorter and, more importantly, the reasoning is at a more convenient high level of abstraction, e.g. pointer reasoning is largely avoided. We achieve this improvement as a result of using an abstraction for arrays and previously developed tools, namely, a proof-producing decompiler and compiler. The work presented in this paper has been developed in the HOL4 theorem prover. The case study resulted in 800 lines of verified 64-bit x86 machine code.

1 Introduction

Hardware executes all software in the form of machine code. As a result, program verification ought to, ultimately, provide guarantees about the execution of the machine code. However, direct manual verification of machine code is to be avoided as such verification proofs easily become lengthy and unmaintainable.

Recent advances in compiler verification seem promising (e.g. [10]), making it possible to relate verification results from the source code to the compiler-generated machine code. Unfortunately, current verified compilers do not support source code with real inlined assembly (since the semantics of inlined assembly is difficult to state in terms of the semantics of the source language). Inlined assembly is a natural component of certain programs: programs that need direct access to hardware peripherals (e.g. operating systems), hand-optimised code or special-purpose machine instructions.

This paper presents a case study which shows that machine-code verification does not always require ghastly unmaintainable proofs. This paper describes how a proof-producing decompiler and compiler can together make it easy to produce verified machine code that essentially contains inlined assembly.

The case study we describe is the construction of an x86-64 implementation of arbitrary-precision arithmetic (bignum) functions. We have implemented the basic integer arithmetic operations (i.e. $+, -, \times, \mathsf{div}, \mathsf{mod}, <, =$) for arbitrary sized integers (represented as arrays in memory) and have proved that this x86-64 implementation correctly performs the desired arithmetic operations and leaves

G. Gonthier and M. Norrish (Eds.): CPP 2013, LNCS 8307, pp. 66–81, 2013.
© Springer International Publishing Switzerland 2013

memory untouched outside the result array. The implementation makes use of special-purpose instructions for multi-word arithmetic.

This paper makes the following contributions.

- The proofs presented in this paper have produced a *reusable* verified x86-64 implementation of bignum integer operations. We envisage that this implementation will be of use in construction of larger bodies of verified code, for example, verified language runtimes that provide support for bignum arithmetic. For the purpose of reuse, we keep all interfaces clean and simple.

- Compared with closely related work (Section 7), our proofs are shorter and, more importantly, the reasoning is at a more convenient high level of abstraction, e.g. pointer reasoning is largely avoided. This improvement in the length and level of detail in the proofs is due to the use of a convenient abstraction for arrays and use of previously developed tools, namely a proof-producing decompiler and compiler, that can easily be made to operate over this *domain-specific abstraction*.

- To the best of our knowledge, we are the first to have formally verified functional correctness of machine code that implements bignum integer division.

The case study resulted in 800 lines of verified 64-bit x86 machine code. The proof development[1] presented in this paper has been carried out in the HOL4 theorem prover [20].

2   Method

The method by which we construct the verified x86 implementation consists of three steps:

1. We start by defining the algorithms involved as functions in logic. The functions operate over lists of binary words. We prove that these functions correctly implement integer arithmetic, e.g. given two lists that represent the 'digits' of two integer numbers, the function for an arithmetic operation returns a list that is the representation of the 'digits' of the resulting integer. These high-level functions summarise the operations of the algorithm separately from any architecture details, e.g. the machine-word length is kept as a (type) variable throughout. (Section 3)

2. In order to generate and reason about machine code that implements the functions from above, we instantiate a proof-producing compiler and decompiler with information about how lists of 64-bit 'digits' can be represented in memory as arrays. Concretely, we define a separation logic assertion about arrays and prove theorems about x86 machine instructions for load and store instructions that access and update arrays. The decompiler and compiler can use these theorems to make it seem as if the underlying machine has a memory that consists of arrays. (Section 4)

---

[1] The HOL4 scripts are available at http://www.cl.cam.ac.uk/~mom22/cpp13/

3. Finally, we use the decompiler to prove theorems for hand-written x86 assembly, and then use the array-aware compiler to produce x86 machine code that uses hand-written assembly and the x86 instructions that we proved to have array-like behaviour. The compiler takes as input functions that are restricted in format, but otherwise operate over the same types as the algorithm specifications from Step 1 (instantiated to 64-bit word length). The compiler produces as output a proof of a theorem which states that the input function is an accurate description of the behaviour of the generated machine code. We manually prove that the input to the compiler perform the same steps as the algorithm specifications from Step 1. (Section 5)

The result of combining all of the correctness theorems together is a single theorem (Section 6) describing the behaviour of a single chunk of x86 machine code, for which we have a top-level correctness theorem: given an operation identifier (referring to one of $+, -, \times, \mathsf{div}, \mathsf{mod}, <, =$), pointers to two immutable input arrays and a pointer to a separate mutable array, where the result is to be stored, execution of the verified x86 code terminates with the result of the arithmetic operation stored in the mutable array.

3   Algorithm Specification and Verification

As mentioned above, the first step is to specify the bignum algorithms as functions in logic and verify that they correctly compute integer arithmetic. This section provides details on this first step.

3.1   Abstract Representation of Bignums

The algorithms operate over lists of machine words. In order to make sure these algorithm specifications do not get tied to any particular architecture, we use a variable as the length of the machine word. In HOL, machine words are most conveniently modelled as finite cartesian products of booleans, a neat idea by Harrison [7], which allows (the cardinality of) a type to define the size of the word. We will write bool^α for the type of words of width α and bool^{64} for the type of words with 64 bits. In this section, all words will have a variable width, i.e. have type bool^α. In subsequent sections, all words will be specialised to be 64 bits wide, i.e. have type bool^{64}.

For this representation, we have the usual word operations and mappings for turning a natural into a word (n2w) and back (w2n):

$$\mathsf{n2w} : \mathbb{N} \to \mathsf{bool}^\alpha$$

$$\mathsf{w2n} : \mathsf{bool}^\alpha \to \mathbb{N}$$

Note that n2w and w2n have polymorphic types and their definition depends on this type. The following theorems describe their relationship:

$$\forall n. \quad \mathsf{w2n} \ (\mathsf{n2w} \ n) = n \ \mathsf{MOD} \ 2^\alpha \qquad\qquad \forall w. \quad \mathsf{n2w} \ (\mathsf{w2n} \ w) = w$$

The algorithms operate over lists of such words, i.e. lists of type bool$^\alpha$ list. We have functions that map natural numbers to lists of multiple words (n2mw) and back (n2mw). Here and throughout : : is list cons.

```
n2mw n = if n = 0 then [] else n2w (n MOD 2^α) :: n2mw (n DIV 2^α)

mw2n [] = 0
mw2n (w::ws) = w2n w + 2^α × mw2n ws
```

We also define functions which translate between integers and a representation of integers as a pair consisting of a sign and a list of machine words.

```
i2mw i = (i < 0, n2mw (abs i))

mw2i (sign, ws) = if sign then 0 - mw2n ws else mw2n ws
```

Thus, the algorithm functions operate over bignum integers as represented by terms of type bool × (bool$^\alpha$ list).

3.2 Algorithm Specifications

The algorithm specification for each arithmetic function is a function of the following type. The comparison operations, of course, return bool.

$$(\texttt{bool} \times (\texttt{bool}^\alpha \texttt{ list})) \rightarrow (\texttt{bool} \times (\texttt{bool}^\alpha \texttt{ list})) \rightarrow (\texttt{bool} \times (\texttt{bool}^\alpha \texttt{ list}))$$

The following presents our specification of the long-multiplication algorithm (mwi_mul). Multiplication will be our running example, since it is neat and simple compared with the tedium of dealing with alternating signs and variable length arguments for bignum integer addition or subtraction.

Our specification of multiplication describes the operations of the standard school-book long-multiplication.

```
      6 2 3 5 1
          2 4 6
    ─────────────
      3 7 4 1 0 6
      2 4 9 4 0 4
      1 2 4 7 0 2
    ─────────────
    1 5 3 3 8 3 4 6
```

There are, of course, a number of more sophisticated and better algorithms [9], e.g. the Karatsuba and Tom-Cook algorithms are significantly faster for large inputs; and Montgomery multiplication is better suited for multiplications that are to be performed modulo a prime number.

When modelling the multiplication algorithm, we start by defining a few primitive operations that we can expect to implement in custom assembly. For example, we define a function for word addition with a carry-in and carry-out.

```
single_add (x:boolᵅ) (y:boolᵅ) (c:bool) =
  (n2w (w2n x + w2n y + if c then 1 else 0),
   2ᵅ ≤ w2n x + w2n y + if c then 1 else 0)
```

And a similar function for multiplication, which given three words, x, y, z, computes w2n x × w2n y + w2n z and returns two words describing this result. We expect either to find such a machine instruction in each architecture or implement this operation using a few instructions.

```
single_mul (x:boolᵅ) (y:boolᵅ) (z:boolᵅ) =
  (n2w (w2n x × w2n y + w2n z),
   n2w ((w2n x × w2n y + w2n z) DIV 2ᵅ))
```

Equipped with the functions from above, we can define a function for the body of the inner loop of multiplication. We follow the standard school-book long-multiplication algorithm almost exactly. The only minor optimisation is that the additions that are done on paper last are done by this algorithm in conjunction with the rest of the computation. The function describing the body of the inner loop takes word, p and q, from each input and a word k from the accumulated result. The body performs a multiplication and two additions:

```
single_mul_add p q k s =
  let (x1,x2) = single_mul p q k in
  let (y1,c1) = single_add x1 s false in
  let (y2,c2) = single_add x2 0 c1 in
    (y1,y2)
```

The function describing the inner loop traverses one of the inputs ys and the accumulated result zs for one word from the other input x.

```
mw_mul_pass x [] zs k = [k]
mw_mul_pass x (y::ys) (z::zs) k =
  let (y1,k1) = single_mul_add x y k z in
    y1 :: mw_mul_pass x ys zs k1
```

The outer loop calls the inner loop for each word in the first input.

```
mw_mul [] ys zs = zs
mw_mul (x::xs) ys zs =
  let zs2 = mw_mul_pass x ys zs 0 in
    HD zs2 :: mw_mul xs ys (TL zs2)
```

The entire multiplication algorithm comes together in mwi_mul, which computes the resulting sign and initialises the accumulated result to all zeros before starting the loop. Below, mw_rm_zero deletes the possible leading zero from the result.

```
mwi_mul (s,xs) (t,ys) =
  if (xs = []) ∨ (ys = []) then (false,[]) else
    (s ≠ t, mw_rm_zero (mw_mul xs ys (MAP (λx. 0) ys)))

mw_rm_zero [] = []
mw_rm_zero (xs ++ [x]) = if x = 0 then xs else xs ++ [x]
```

In the eventual machine code, mw_rm_zero shortens the length of the array where the result is stored so that that array has no leading zero.

3.3   Algorithm Verification

The top-level correctness theorem for each arithmetic operation is easy to state using the function i2mw for converting an integer into a signed list of words. For multiplication, the correctness statement relates mwi_mul to multiplication over the integers ($\times$).

$$\forall i \; j. \quad \text{mwi\_mul (i2mw i) (i2mw j)} = \text{i2mw (i} \times \text{j)}$$

Such statements guarantee that zero will never have the negative sign set and that mwi_mul never returns a list of words with redundant leading zeros.

Although the correctness theorem is stated in terms of i2mw, it seems easiest to arrive at the correctness theorem via proofs about mw2n. Each component in the algorithm has a neat description in terms of mw2n and w2n.

$$\forall \text{p q k1 k2 x1 x2.}$$
$$\text{single\_mul\_add p q k1 k2} = (\text{x1,x2}) \implies$$
$$\text{w2n x1} + 2^{\alpha} \times \text{w2n x2} = \text{w2n p} \times \text{w2n q} + \text{w2n k1} + \text{w2n k2}$$

$$\forall \text{ys zs x k.}$$
$$\text{LENGTH ys} = \text{LENGTH zs} \implies$$
$$\text{mw2n (mw\_mul\_pass x ys zs k)} = \text{w2n x} \times \text{mw2n ys} + \text{mw2n zs} + \text{w2n k}$$

$$\forall \text{xs ys zs.}$$
$$\text{LENGTH ys} = \text{LENGTH zs} \implies$$
$$\text{mw2n (mw\_mul xs ys zs)} = \text{mw2n xs} \times \text{mw2n ys} + \text{mw2n zs}$$

4   Instantiation of Proof Tools for Arrays

With the bignum arithmetic algorithms specified and verified in the previous section, this section describes the Hoare logic and proof tools that are used in the next section for construction of the verified machine-code implementation.

4.1   Hoare Logic for Machine Code

We will skip a detailed description of the operational semantics for x86 used in this paper, since that semantics has been described previously [15]. Instead, a few examples will be used to explain features of a machine-code Hoare logic [14] that sits on top of the bare operational semantics.

All our reasoning about x86 machine code is performed through a machine-code Hoare logic, which can be instantiated to different instruction set architectures. Here we consider only an instantiation to 64-bit x86.

The following is a Hoare triple describing an x86 instruction add r8,r9, encoded as 4D01C8, that adds the content of 64-bit register 8 with register 9 and stores the result in register 8. The following Hoare triple can be read informally as follows: *for any state where* the program counter (PC) is p, register 8 and 9 are r8 and r9, respectively, the flags have some value (S _), and 4D01C8 is at

location p in memory, *execution will reach a state where* the program counter
is set to p + 3, register 8 contains the value r8 + r9 and the flags again have
some value (S _). Here * is a form of separating conjunction [18,14]. Details of
this separating conjunction are unimportant for this paper. However, it is worth
noting that these Hoare triples are part of a separation logic and, in particular,
that all other resources not mentioned in the precondition of such a Hoare triple
must have been kept unchanged (e.g. the theorem below implicitly states that
the value of register 10 was unaffected by the add r8,r9 instruction).

```
{ PC p * R8 r8 * R9 r9 * S _ }
  p : 4D01C8
{ PC (p + 3) * R8 (r8 + r9) * R9 r9 * S _ }
```

An unusual feature of these Hoare triples is that the pre- and postconditions
include the value of the program counter. Its inclusion makes it easy to specify
branch instructions. Example: a jump-if-equal instruction, je -40 encoded as
48EBD5, is described by the following Hoare-triple theorem. The jump is conditional on the x86 z flag, which is set by most arithmetic operations.

```
{ PC p * S (a,c,o,p,z) }
  p : 48EBD5
{ PC (if z then p - 40 else p + 3) * S (a,c,o,p,z) }
```

Memory accesses are specified using a memory assertion memory m, which
states that a part of memory (the set of addresses in domain m) are described
by the partial function m. The following is a Hoare triple for a store instruction,
mov [r8],r9 encoded as 4D8908, which stores the content of register 9 at an
address given in register 8. This instruction is independent of the flags (S).

```
r8 ∈ domain m ∧ word_aligned r8 ⟹
{ PC p * R8 r8 * R9 r9 * memory m }
  p : 4D8908
{ PC (p + 3) * R8 r8 * R9 r9 * memory (m[r8 ↦ r9]) }
```

Note that the underlying model of x86 treats code as part of memory, but here
the Hoare triples separate 'data' memory from the code. This is achieved by internally separating the precondition from the code segment using the separating
conjunction *, for details see Myreen [15].

This Hoare logic supports the usual inference rules. As a result, one can perform proofs directly using these Hoare triples, as was done in previous work [13].
However, it is significantly easier if tools are used which automate much of the
routine reasoning.

4.2   Proof-Producing Decompiler and Compiler

Tool support, developed and carefully explained in previous work [14], is able to
automate much of the routine Hoare logic reasoning. An example will illustrate
what our decompiler can do. The HOL4 syntax below calls our decompiler for
assembly code that computes, in r9, Knuth's D constant ahead of his bignum
division algorithm [9].

```
val (x64_calcd_cert,x64_calcd_def) = x64_decompile "x64_calcd"
' LOOP: cmp r8,0
        js EXIT
        add r8,r8
        add r9,r9
        jmp LOOP
  EXIT:                           '
```

This call to x64_decompile first runs an assembler to turn the assembly into concrete machine code, it then derives Hoare-triple theorems for each of the instructions and finally composes the Hoare triples together. The result of this composition is a theorem that describes one pass through the loop:

```
{ PC p * R8 r8 * R9 r9 * S _ }
  p : 4983F80 48789 4D1C0 4D1C9 48EBF0
{ let (p,r8,r9) =
    (if word_sign_bit r8 then (p + 14, r8, r9) else (p, r8+r8, r9+r9))
  in
    PC p * R8 r8 * R9 r9 * S _ }
```

By applying a special loop rule [14], the decompiler turns this theorem into a theorem describing a full terminating execution of the loop. The decompiler returns two functions describing the behaviour of the machine code: the first function describes the data update performed by the code, the second function encodes a side condition on termination. Loops always produce tail-recursive functions, which can be defined HOL without termination proofs.

```
x64_calcd (r8,r9) =
  if word_sign_bit r8 then (r8,r9)
  else let r8 = r8 + r8 in let r9 = r9 + r9 in x64_calcd (r8,r9)

x64_calcd_pre (r8,r9) =
  if word_sign_bit r8 then true
  else let r8 = r8 + r8 in let r9 = r9 + r9 in x64_calcd_pre (r8,r9)
```

The result of running the decompiler is the extraction of functions that describe the behaviour of the given machine code and a separate Hoare-triple theorem, which states that the function x64_calcd is an accurate description of the effect of executing the x86-64 machine code, if the side-condition x64_calcd_pre (r8,r9) is provable (which, in this case, is true if r8 is initially non-zero). We call such Hoare-triple theorems *certificate theorems*.

```
x64_calcd_pre (r8,r9)  ⟹
{ PC p * R8 r8 * R9 r9 * S _ }
  p : 4983F80 48789 4D1C0 4D1C9 48EBF0
{ let (r8,r9) = x64_calcd (r8,r9) in
    PC (p + 14) * R8 r8 * R9 r9 * S _ }
```

The beauty of using the decompiler is that all subsequent reasoning can be done on the extracted function x64_calcd, since any result proved for this function is related back to the machine code through the certificate theorem.

Writing assembly code manually is tiresome. To help with this, a proof-producing *compiler* has been constructed using the decompiler. This compiler essentially takes as input tail-recursive functions of the form x64_calcd, it then: (1) generates (without proof) assembly code based on the input function, (2) decompiles the assembly as above, and (3) proves that the function decompilation produced is identical to the function that was to be compiled, i.e. the compiler can return the certificate theorem produced by the underlying decompiler.

4.3   Array Support in the Compiler

As explained above, the decompiler and compiler produce their proofs by simply composing machine-code Hoare triples together. By default these tools use only automatically derived Hoare triples that provide a cumbersome flat functional view of the memory of the underlying x86 machine semantics.

The technique by which we instantiate the tools to the problem domain of bignum-array programs is to supply the tools with custom Hoare-triple theorems that are stated in terms of a domain-specific bignum-memory assertion. With such an assertion the decompiler and compiler can make the machine seem as if it has a memory containing arrays (in which we will store bignums).

We define the domain-specific assertion bignums based on the default memory assertion memory as explained below. The definition of bignums uses a few basic concepts of separation logic defined next. The separating conjunction $\star$ is defined as usual, taking the disjoint union ($\uplus$) of two memory segments. The emp assertion is true only for the empty memory segment. The unusual part is our definition of the maps-to assertion: a $\mapsto$ x is true for a memory segment if the bytes of 64-bit word x are stored from 64-bit address a onwards. Here [7--0]x is notation for selecting bits 7 to 0 from x.

```
(p * q) m = ∃m1 m2. p m1 ∧ q m2 ∧ (m = m1 ⊎ m2)

emp m = (domain m = ∅)

(a ↦ x) m = (domain m = {a,a+1,a+2,a+3,...,a+7}) ∧
            (m a = [7--0]x) ∧ (m (a+1) = [15--8]x) ∧ ...
```

These basic concepts of separation logic are enough to define an array assertion for memory segments: array a xs is true for a memory segment m if the 64-bit words in list xs are stored in order from address a onwards.

```
array a [] = emp
array a (x::xs) = a ↦ x * array (a + 8) xs
```

The bignum code that we produce uses three arrays: we call the content of these arrays xs, ys and zs, and have pointers, xa, ya, za, point to the (word-aligned) base of these arrays. We allow pointers xa and ya to alias. If they do alias, then the content of xs and ys must be identical. The intention is that xs and ys hold bignum inputs and zs is the mutable result array.

```
bignum_memory m xa xs ya ys za zs =
  word_aligned xa ∧ word_aligned ya ∧ word_aligned za ∧
  if xa = ya then
    (xs = ys) ∧ (array xa xs * array za zs) m
  else
    (array xa xs * array ya ys * array za zs) m
```

The definition of the bignums assertion constrains the default memory assertion with the bignum_memory condition and states that pointers xa, ya, za are kept in registers 13, 14 and 15, respectively.

```
bignums (xa,xs,ya,ys,za,zs) =
  ∃m. memory m * R13 xa * R14 ya * R15 za *
      ⟨bignum_memory m xa xs ya ys za zs⟩
```

Using this bignums assertion, we can now manually verify a number of Hoare-triple theorems which make certain machine instructions seem as if they operate over arrays directly. For example the load instruction mov r0,[8*r10+r13], encoded as 4B8B44D500, loads the list element (EL) at index w2n r10 from list xs, if w2n r10 is within the size of xs.

```
w2n r10 < LENGTH xs ⟹
{ PC p * R0 r0 * R10 r10 * bignums (xa,xs,ya,ys,za,zs) }
  p : 4B8B44D500
{ let r0 = EL (w2n r10) xs in
    PC (p + 5) * R0 r0 * R10 r10 * bignums (xa,xs,ya,ys,za,zs) }
```

Similarly, mov [8*r10+r15],r0, encoded as 4B8904D7, updates (LUPDATE) list index w2n r10 of list zs, if w2n r10 is within the size of zs.

```
w2n r10 < LENGTH zs ⟹
{ PC p * R0 r0 * R10 r10 * bignums (xa,xs,ya,ys,za,zs) }
  p : 4B8904D7
{ let zs = LUPDATE r0 (w2n r10) zs in
    PC (p + 4) * R0 r0 * R10 r10 * bignums (xa,xs,ya,ys,za,zs) }
```

Supplied with such Hoare-triple theorems, the compiler can compile functions which contain the following lines:

```
let r0 = EL (w2n r10) xs in

let zs = LUPDATE r0 (w2n r10) zs in
```

By supplying enough such Hoare-triple theorems, we can exclusively use only statements about recognised list/array operations and thus never, in manual proofs, require pointer reasoning beyond this point. Examples of compiled array accessing functions are given in Section 5.2.

5   Construction of Verified Machine Code

With a proof-producing compiler that understands basic operations over a few arrays, we are ready to describe how one can construct verified implementations for the algorithms from Section 3. This section continues with the running example of multiplication.

5.1   Verification of Hand-Written Assembly

Certain parts of the algorithms in Section 3 are best implemented in custom hand-written assembly. The following call to x64_decompile decompiles an assembly implementation of single_mul_add from Section 3.2. The assembler, that we use, aliases r0 with rax, r1 with rcx, r2 with rdx and r3 with rbx.

```
val (_, x64_single_mul_add_def) = x64_decompile "x64_single_mul_add"
 ' mul r2
   add r0,r1
   adc r2,0
   add r0,r3
   adc r2,0  '
```

This call results in a function x64_single_mul_add (r0,r1,r2,r3), which is easily proved to be an implementation of single_mul_add:

```
∀p k q s.
  x64_single_mul_add_pre (p,k,q,s) = true ∧
  x64_single_mul_add (p,k,q,s) =
    let (x1,x2) = single_mul_add p q k s in (x1,k,x2,s)
```

5.2   Using Inlined Assembly in Compilations

Each run of the decompiler produces a certificate theorem. The certificate theorem produced for the decompilation above can be used in subsequent decompilations and compilations. Concretely, this means that the compiler can produce code for functions involving the line:

```
let (r0,r1,r2,r3) = x64_single_mul_add (r0,r1,r2,r3) in
```

Such lines result in code where the implementation of x64_single_mul_add is inlined in the generated machine code. The decompiler uses the certificate theorem for x64_single_mul_add at the point where it encounters the inlining.

This inlining feature allows writing an implementation of the inner loop, mw_mul_pass, of the multiplication algorithm. The function that we compile in order to generate machine code for mw_mul_pass is called x64_mul_pass. Its definition is shown in Figure 1. The compiler-generated machine code, shown in Figure 2, uses the custom assembly code and the list/array operations EL and LUPDATE from Section 4.3. A disassembly of the generated machine code is listed in Figure 2. The entire bignum library implementation is produced via such compilations that inline the result of previous compilations and decompilations.

```
val (_,x64_mul_pass_def,x64_mul_pass_pre_def) = x64_compile '
  x64_mul_pass (r1,r8,r9,r10,r11,ys,zs) =
  if r9 = r11 then
    let zs = LUPDATE r1 (w2n r10) zs in
    let r10 = r10 + 1w in
      (r1,r9,r10,ys,zs)
  else
    let r3 = EL (w2n r10) zs in
    let r2 = EL (w2n r11) ys in
    let r0 = r8 in
    let (r0,r1,r2,r3) = x64_single_mul_add (r0,r1,r2,r3) in
    let zs = LUPDATE r0 (w2n r10) zs in
    let r1 = r2 in
    let r10 = r10 + 1w in
    let r11 = r11 + 1w in
      x64_mul_pass (r1,r8,r9,r10,r11,ys,zs) '
```

Fig. 1. HOL4 syntax for a call to the compiler for x64_mul_pass

```
00: 4D39D9       L1:  cmp r9, r11
03: 48742C            je L2
06: 4B8B1CD7          mov r3,[8*r10+r15]      // EL (w2n r10) zs
0A: 4B8B14DE          mov r2,[8*r11+r14]      // EL (w2n r11) ys
0E: 498BC0            mov r0, r8
11: 48F7E2            mul r2                  // inlined part
14: 4801C8            add r0,r1               // inlined part
17: 4883D20           adc r2,0                // inlined part
1B: 4801D8            add r0,r3               // inlined part
1E: 4883D20           adc r2,0                // inlined part
22: 4B8904D7          mov [8*r10+r15],r0      // LUPDATE r0 (w2n r10) zs
26: 488BCA            mov r1, r2
29: 49FFC2            inc r10
2C: 49FFC3            inc r11
2F: 48EBCE            jmp L1
32: 4B890CD7     L2:  mov [8*r10+r15],r1      // LUPDATE r1 (w2n r10) zs
36: 49FFC2            inc r10
```

Fig. 2. Annotated disassembly of machine code generated for x64_mul_pass

5.3   Verification of the Generated Machine Code

Since the compiler produces a certificate theorem relating the given input function to the generated machine code, it suffices to prove properties of the input functions (and generated precondition functions) in order to prove the correctness of the machine code. For x64_mul_pass, this means that we need to prove that `x64_mul_pass` implements `mw_mul_pass`. The statement we prove, below, might seem hard to comprehend, but look closer and it becomes clear that this is a reasonably straight forward property. The length of the proof of this goal is less than twice the length of the goal statement.

```
∀ys x zs k zs1 zs2 z2.
  length zs = length ys ∧ length (zs1 ++ zs) < 2^64  ⟹
  ∃r1.
    x64_mul_pass_pre
      (k,x,n2w (length ys),n2w (length zs1),n2w 0,ys,
       zs1 ++ zs ++ z2::zs2) = true ∧
    x64_mul_pass
      (k,x,n2w (length ys),n2w (length zs1),n2w 0,ys,
       zs1 ++ zs ++ z2::zs2) =
    (r1,n2w (length ys),n2w (length (zs1 ++ zs) + 1),ys,
     zs1 ++ mw_mul_pass x ys zs k ++ zs2)
```

6   Results

The result of this verification effort is a verified library of bignum integer arithmetic functions implemented in 64-bit x86 machine code. The intention was to make this case study as reusable as possible so that future verified language implementations, e.g. future version of our verified Lisp implementation [16], can make use of arbitrary-precision integer arithmetic.

6.1   Top-Level Theorem

The verified library of integer arithmetic operations has a top-level entry point which implements a clean and simple interface: as inputs, it expects three pointers, pointers to two input arrays and one array for the result, it expects the length and sign of the input numbers to be provided in specific registers and it reads the operation identifier from a register. If the output array is long enough and disjoint from the input arrays, then the verified machine-code implementation will terminate with the result of the arithmetic operation of choice produced in the result array and the sign and length of the result return in a register. The input arrays are left unchanged.

6.2   In Numbers

In order to give some measure of the effort involved, the table below lists how many lines of proof scripts were produced for each part of this project. The three middle columns list the length of our HOL4 proof scripts and the last column lists the number of instructions in the verified machine code that was produced.

part / operation	alg.	impl.	total	x86
prelude & tool setup	398	357	755	0
comparison	138	118	256	58
addition & subtraction	307	655	962	122
multiplication	149	266	415	105
division & modulus	2149	1482	3631	447
conversion to decimal	113	95	208	57
all parts together	3254	2973	6227	779

alg. — lines for specification and verification of algorithms (Section 3)
impl. — lines for construction and verification of machine code (Sections 4, 5)
total — sum of alg. and impl. columns
x86 — number of instructions in the verified 64-bit x86 machine code

One can (correctly) read from this table that the algorithm proofs were roughly as time consuming as the construction and verification of the machine code.

The verified algorithms are the obvious single-pass algorithms for comparison, addition and subtraction; the algorithm for multiplication was described in Section 3; the algorithm for division and modulus was taken from Knuth [9]; and the conversion into decimal form performs repeated division by 10.

7   Related Work

The most closely related work on verified implementation of arithmetic functions is that of Affeldt [2], Fischer [6], Berghofer [4] and Moore [11]. We will also compare with the first author's early poster on this topic [13], and reflect on recent trends in programming logics for assembly verification.

Affeldt has constructed and verified SmartMIPS assembly code that implements the basic arithmetic functions: $+, -, \times, <, =$, notably excluding div and mod, but including Montgomery multiplication. Affeldt uses separation logic [18] and explicit reasoning about pointers in his verification proofs, which appear to be more low-level and labour intensive than the proofs reported on in this paper. Affeldt uses the GMP [1] library's bignum integer representation (which includes indirection) and, as a result, can not use the convenient array abstraction that was used in this paper. Affeldt proposes the use of a simulation relation to lift reasoning of compound operations to a more manageable level of detail.

Fischer and Berghofer both use the Isabelle/HOL theorem prover and both verify implementations written in a higher-level language. Fischer verified a C-like implementation of arbitrary-precision integer arithmetic, including division and modulus, using manual application of a separation-logic instantiation of Schirmer's Hoare logic framework [19]. Fischer reports that her proofs required significant manual effort to deal with selection of frames for the separation-logic reasoning. Her bignums were represented as linked lists. Berghofer verified a bignum library, which includes Montgomery multiplication but not division, written in SPARK/ADA using a combination of the SPARK/ADA tool suite and the Isabelle/HOL prover.

The first author's early poster, Myreen and Gordon [13], on the topic of machine-code verification showed that it is possible to use a Hoare logic directly to manually verify, in the HOL4 theorem prover, the correctness of ARM machine code implementing an optimised version of Montgomery multiplication.

Moore seems to have been the first to have formally verified the correctness of a bignum assembly routine, using the Nqthm prover. In his paper on the verified implementation of the Piton language, Moore explains that it is possible to verify an assembly routine for addition for bignums stored as arrays.

In terms of future direction, there seems to be a trend of making high-level language reasoning seamlessly available in the context of assembly verification. Significant recent work in this area include the programming logic by Jensen et al. [8], which has a powerful 'macro feature'. This macro feature makes it possible to define functions in the logic that operate over the assembly syntax and thus introduce, say, a while-loop macro and derive neat and familiar-looking proof rules for such, even though the reasoning is still about assembly code. Another noteworthy recent result in this area is Chlipala's Bedrock framework [5]. The Bedrock framework neatly fits into the Coq prover and provides proof tools which automate most routine separation-logic reasoning for assembly programs. The current paper has shown that our previously developed tools [14] are capable of providing convenient verification environment for the HOL4 theorem prover and, for this case study, explicit proofs about pointers can be avoided.

The work of this paper has focused on proof of full functional correctness. However, great strides have also been made in proofs of safety properties. Necula's work on proof-carrying code [17] spurred a lot of interest in low-level code [3,21]. An exciting recent result in this area is a new method for software-fault isolation for real machine code [12].

8   Summary

This paper has demonstrated how a proof-producing decompiler and compiler can be used in the construction of verified machine-code implementations of bignum arithmetic. By careful instantiation of the previously developed tools, the entire verification effort is kept at a manageable complexity with proofs involving pointer reasoning nearly completely avoided (only present in Section 4.3). The resulting 64-bit x86 machine code was produced from both inlined custom assembly and functions written at a higher level of abstraction.

Acknowledgements. The first author was funded by the Royal Society, UK. The second author was a summer intern supported by the University of Cambridge Computer Laboratory, UK.

References

1. GMP, the GNU multiple precision arithmetic library, http://gmplib.org/

2. Affeldt, R.: On construction of a library of formally verified low-level arithmetic functions. Innovations in Systems and Software Engineering 9(2) (2013)
3. Appel, A.W.: Foundational proof-carrying code. In: Logic in Computer Science (LICS). IEEE Computer Society (2001)
4. Berghofer, S.: Verification of dependable software using spark and isabelle. In: Brauer, J., Roveri, M., Tews, H. (eds.) Systems Software Verification (SSV). OASICS, Schloss Dagstuhl - Leibniz-Zentrum fuer Informatik (2011)
5. Chlipala, A.: Mostly-automated verification of low-level programs in computational separation logic. In: Hall, M.W., Padua, D.A. (eds.) Programming Language Design and Implementation (PLDI). ACM (2011)
6. Fischer, S.: Formal verification of a big integer library. In: DATE 2008: Workshop on Dependable Software Systems (2008),
   http://busserver.cs.uni-sb.de/publikationen/Fi08DATE.pdf
7. Harrison, J.: A HOL theory of euclidean space. In: Hurd, J., Melham, T. (eds.) TPHOLs 2005. LNCS, vol. 3603, pp. 114–129. Springer, Heidelberg (2005)
8. Jensen, J.B., Benton, N., Kennedy, A.: High-level separation logic for low-level code. In: Principles of Programming Languages (POPL). ACM (2013)
9. Knuth, D.E.: The art of computer programming, 2nd edn. Seminumerical Algorithms, vol. 2. Addison Wesley Longman Publishing (1981)
10. Leroy, X.: Formal certification of a compiler back-end, or: programming a compiler with a proof assistant. In: Morrisett, J.G., Jones, S.L.P. (eds.) Principles of Programming Languages (POPL). ACM (2006)
11. Moore, J.S.: A mechanically verified language implementation. Journal of Automated Reasoning 5 (1989)
12. Morrisett, G., Tan, G., Tassarotti, J., Tristan, J.B., Gan, E.: RockSalt: better, faster, stronger SFI for the x86. In: Vitek, J., Lin, H., Tip, F. (eds.) Programming Language Design and Implementation (PLDI). ACM (2012)
13. Myreen, M., Gordon, M.J.C.: Verification of machine code implementations of arithmetic functions for cryptography. In: Schneider, K., Brandt, J. (eds.) Theorem Proving in Higher Order Logics, Emerging Trends Proceedings (TPHOLs, Poster Session), University of Kaiserslautern, Internal Report 364/07 (2007)
14. Myreen, M.O.: Formal verification of machine-code programs. Ph.D. thesis, University of Cambridge (2009)
15. Myreen, M.O.: Verified just-in-time compiler on x86. In: Hermenegildo, M.V., Palsberg, J. (eds.) Principles of Programming Languages (POPL). ACM (2010)
16. Myreen, M.O., Davis, J.: A verified runtime for a verified theorem prover. In: van Eekelen, M., Geuvers, H., Schmaltz, J., Wiedijk, F. (eds.) ITP 2011. LNCS, vol. 6898, pp. 265–280. Springer, Heidelberg (2011)
17. Necula, G.C.: Proof-carrying code. In: Principles of Programming Languages (POPL). ACM (1997)
18. Reynolds, J.: Separation logic: A logic for shared mutable data structures. In: Logic in Computer Science (LICS). IEEE Computer Society (2002)
19. Schirmer, N.: Verification of Sequential Imperative Programs in Isabelle/HOL. Ph.D. thesis, Technical University of Munich (2006)
20. Slind, K., Norrish, M.: A brief overview of HOL4. In: Mohamed, O.A., Muñoz, C., Tahar, S. (eds.) TPHOLs 2008. LNCS, vol. 5170, pp. 28–32. Springer, Heidelberg (2008)
21. Tan, G., Appel, A.W.: A compositional logic for control flow. In: Emerson, E.A., Namjoshi, K.S. (eds.) VMCAI 2006. LNCS, vol. 3855, pp. 80–94. Springer, Heidelberg (2006)

A Constructive Theory
of Regular Languages in Coq

Christian Doczkal, Jan-Oliver Kaiser, and Gert Smolka

Saarland University, Saarbrücken, Germany
{doczkal,jokaiser,smolka}@ps.uni-saarland.de

Abstract. We present a formal constructive theory of regular languages consisting of about 1400 lines of Coq/Ssreflect. As representations we consider regular expressions, deterministic and nondeterministic automata, and Myhill and Nerode partitions. We construct computable functions translating between these representations and show that equivalence of representations is decidable. We also establish the usual closure properties, give a minimization algorithm for DFAs, and prove that minimal DFAs are unique up to state renaming. Our development profits much from Ssreflect's support for finite types and graphs.

Keywords: regular languages, regular expressions, finite automata, Myhill-Nerode, Coq, Ssreflect.

1 Introduction

The theory of regular languages is a standard topic in the computer science curriculum [10,13]. We are interested in an elegant and instructive formalization of this theory in constructive type theory. We prove Kleene's theorem [12], the Pumping Lemma, uniqueness of minimal deterministic automata, the Myhill-Nerode theorem [16,17], and various closure properties of regular languages. For our formalization [7], we use the Ssreflect [9] extension to Coq [20] which features an extensive library with support for reasoning about finite structures such as finite types and finite graphs. Building on top of the Ssreflect infrastructure, we can establish all of our results in about 1400 lines of Coq, half of which are specifications. So our development may be considered a longish proof pearl.

The largest part of our formalization deals with translations between different representations of regular languages: regular expressions [12] (RE), deterministic finite automata [15] (DFA), minimal DFAs [11] (mDFA), nondeterministic finite automata [18] (NFA), Myhill partitions [16] (MP), and Nerode partitions [17] (NP). We formalize all these representations and construct computable conversion functions between them. The conversion functions can be summarized by the following diagram:

G. Gonthier and M. Norrish (Eds.): CPP 2013, LNCS 8307, pp. 82–97, 2013.
© Springer International Publishing Switzerland 2013

The triangle corresponds to Kleene's theorem. For our representation of finite automata, we make use of the fact that in Coq, unlike HOL-based systems, types are first-class values. This allows us to formalize the sets of states of finite automata as finite types (i.e, types with finitely many inhabitants). Finite types are closed under many type constructors, and Ssreflect offers excellent support for finite types. So this representation is close to the usual mathematical definition and easy to work with.

For the conversion from regular expressions to finite automata (dashed lines), we need one construction for every constructor of regular expressions. By first establishing conversions between DFAs and NFAs, we can carry out each of these constructions on the automata model that fits best. The translation from DFAs to REs is based on Kleene's algorithm [12].

The rectangle in the diagram corresponds to the Myhill-Nerode theorem. Our constructive version of the Myhill-Nerode theorem differs from the usual formulation. We define Myhill and Nerode representations using functions from words into a finite type. We then give conversions to and from finite automata. This is similar to the formalization of the Myhill-Nerode theorem in Nuprl by Constable et al. [5], where decidability of Myhill-Nerode relations and finiteness of the corresponding quotient types are assumed.

Finite Myhill and Nerode partitions can be seen as abstract representations of DFAs where Nerode partitions correspond to minimal DFAs [13]. So the conversions from Nerode partitions to Myhill partitions and from Myhill partitions to DFAs are fairly direct. For the conversion from DFAs to Nerode partitions, we rely on automata minimization. For this, we prove correct a variant of Huffman's table filling algorithm for DFA minimization [11].

We also prove that regular languages are closed under images and preimages of homomorphisms, and we prove decidability of language equivalence for all representations. We prove our decidability results by defining, in Coq, functions to bool that decide the problem. This is sufficient since we work in the constructive logic of Coq without axioms, so every function into bool is total and computable.

1.1 Related Work

There are several recent publications concerned with the formalization of the theory of regular languages. Wu et al. [21] describe a formalization of the Myhill-Nerdode theorem in Isabelle/HOL based directly on regular expressions. The authors explain this unusual choice with limitations of Isabelle/HOL, which lacks a good graph library and does not allow quantification over types. This makes it difficult to define a type of automata that is easy to work with. Neither of these restrictions apply to Coq with Ssreflect. However, we could not find a formalization of automata theory in the literature that makes good use of the fact that types are first-class values in Coq. For example, Braibant and Pous [4] formalize automata as part of a larger development building a certified decision procedure for Kleene algebras. They use numbered states, an approach rejected as "clunky" in [21].

There are a number of other papers describing certified implementations of decision procedures. These include decision procedures for regular expression equivalence [1,6] in Coq or Matita. Berghofer et al. [3] formalize automata over bitstrings in Isabelle/HOL. Building on this formalization, they obtain a certified decision procedure for Presburger arithmetic. In these developments the focus is on efficiency. The formalizations are significantly longer than our development and formalize only those results relevant to the respective decision procedures. Kraus and Nipkow [14] develop a certified decision procedure for regular expression equivalence in Isabelle/HOL. Their formalization is very short, but they only show partial correctness. In contrast to most of the papers mentioned above, we are interested in simple proofs rather than practical executability. This gives us more freedom in our choice of representations.

1.2 Outline

After we introduce some notations and basic definitions, we explain our formalization of languages and give a decidable semantics for regular expressions. In Section 4, we prove Kleene's theorem. We also prove that equivalence of DFAs is decidable. In Section 5, we certify a minimization algorithm for DFAs and show that minimal DFAs are unique up to isomorphism. In Section 6, we prove our constructive Myhill-Nerode theorem. In Section 7, we prove that regular languages are closed under images and preimages of homomorphisms. In Section 8, we demonstrate how our Myhill-Nerode theorem can be used to show that a language is not regular.

2   Preliminaries

We review the basic mathematical definitions of formal languages and regular expressions. An *alphabet* Σ is a finite set of symbols. The letters a, b, etc. denote symbols. For simplicity, we fix some alphabet Σ throughout the paper. A *word* w is a finite sequence of symbols from Σ. We write w_n for the n-th symbol of w. The variables u, v and w always denote words and ε denotes the empty word. A *language* is a set of words. We write Σ^* for the language of all words and $\overline{L}$ for $\Sigma^* \setminus L$. We write $|w|$ to denote the *length* of the word w and $v \cdot w$ or just vw for the concatenation of v and w. Here, uv binds tighter than function application which in turn binds tighter than $\cdot$.

We consider simple *regular expressions* as defined by the following grammar

$$r, e := \emptyset \mid \varepsilon \mid a \mid r \cdot e \mid r + e \mid r^*$$

The language of a regular expression is defined as follows:

$$\mathcal{L}(\emptyset) = \emptyset \qquad \mathcal{L}(\varepsilon) = \{\varepsilon\} \qquad \mathcal{L}(a) = \{a\}$$

$$\mathcal{L}(r + e) = \mathcal{L}(r) \cup \mathcal{L}(e)$$

$$\mathcal{L}(r \cdot e) = \mathcal{L}(r) \cdot \mathcal{L}(e) = \{v \cdot w \mid v \in \mathcal{L}(r) \text{ and } w \in \mathcal{L}(e)\}$$

$$\mathcal{L}(r^*) = \mathcal{L}(r)^* = \{w_1 \cdot \ldots \cdot w_n \mid n \geq 0 \text{ and } \forall 0 < i \leq n. \ w_i \in \mathcal{L}(r)\}$$

3   Decidable Languages and Regular Expressions

Set theoretically, a language is just a set of words. Let char be a type of characters. We write word char for sequences over char and represent languages as predicates of type word char → Prop. Note that languages are not necessarily decidable, i.e., there are languages L for which we can prove neither $w \in L$ nor $w \notin L$. Regular languages, however, are always decidable. So for most of our development we use decidable languages word char → bool, which are more convenient to work with. We formalize decidable languages using Ssreflect's boolean predicates.

Definition dlang char := pred (word char).

Note that our representation of languages as predicates is intensional, i.e., equivalent languages are not necessarily equal. We state equivalences between decidable languages using Ssreflect's extensional equality operator, which satisfies

$$\text{L1} =_i \text{L2} \leftrightarrow \forall w, (w \in \text{L1}) = (w \in \text{L2})$$

Here, $\in$ is Ssreflect's generic membership operator for everything that can be seen as a boolean predicate. All our constructions respect this equivalence, so for our informal explanations we consider equivalent languages as equal.

We assign decidable languages to every representation of regular languages. To do this for regular expressions, we have to show that decidable languages are closed under all regular operations. Proving this in Coq amounts to defining $\emptyset, \varepsilon, a, \cdot, +,$ and * as operators on decidable languages. We use the definitions from Coquand and Siles [6] who also work with decidable languages. All operators but $\cdot$ and * are easily defined. The operator for $\cdot$ is

Definition conc (L1 L2: dlang char) : dlang char :=
  fun v ⇒ [exists i : 'I_(size v).+1, L1 (take i v) && L2 (drop i v)].

where 'I_(size v).+1 is the type of natural numbers smaller than (size v) + 1. This type has only finitely many inhabitants and is therefore called a finite type. Decidability is preserved by quantification over finite types and Ssreflect's boolean quantifier [exists x : T, p x] yields a boolean result provided that T is a finite type and p is a boolean predicate. Hence, conc L1 L2 is a decidable language.

For the concatenation operator, we prove the correctness lemma

Lemma concP {L1 L2 : dlang char} {w : word char} :
  reflect (∃ w1 w2, w = w1 ++ w2 ∧ w1 ∈ L1 ∧ w2 ∈ L2) (w ∈ conc L1 L2).

which reflects[1] the boolean membership statement into a Coq proposition. The definitions of the remaining operators and the associated correctness lemmas can be found in the file regexp.v of our formalization [7].

We represent regular expressions using an inductive type. Having defined all the regular operations on languages, we can associate a decidable language to every regular expression.

---

[1] For P : Prop and p : bool, the statement reflect P p asserts that the Coq proposition P and p = true are logically equivalent.

```
Fixpoint re_lang (e : regexp char) : dlang char :=
  match e with
  | Void ⇒ void char
  | Eps ⇒ eps char
  | Atom x ⇒ atom x
  | Star e1 ⇒ star (re_lang e1)
  | Plus e1 e2 ⇒ plus (re_lang e1) (re_lang e2)
  | Conc e1 e2 ⇒ conc (re_lang e1) (re_lang e2)
  end.
```

Theorem 1. *The matching problem for regular expressions is decidable.*

Proof. This is an immediate consequence of defining the semantics of regular expressions in terms of decidable languages.   □

We now call a general language regular if it is equivalent to the language of some regular expression.

Definition regular char (L : word char → Prop) := ∃ e : regexp char, ∀ w, L w ↔ w ∈ e.

Defining regularity on general languages has the advantage that one can prove regularity by giving a regular expression or a finite automaton without first proving that the language is decidable.

4 Kleene's Theorem

While regular expressions can be seen as the natural characterization of regular languages, finite automata can be seen as an operational characterization of the same class of languages. Several theorems about regular languages can be proven more easily using automata rather than regular expressions. This includes the closure of regular languages under complement and the Myhill-Nerode theorem. We formalize nondeterministic and total deterministic automata and show that both have the same expressive power as regular expressions.

Definition 1.   – A nondeterministic finite automaton *(NFA)* is a tuple (Q, s, F, δ) *where Q is a finite set of states, $s \in Q$ is the starting state, $F \subseteq Q$ is the set of accepting states, and $\delta \subseteq (Q \times \Sigma) \times Q$ is the transition relation.*
   – *A deterministic finite automaton (DFA) is a tuple (Q, s, F, δ) as above, except that $\delta : Q \times \Sigma \to Q$ is a total function.*

We formalize NFAs and DFAs as two separate record types.

```
Record nfa : Type := {
  nfa_state :> finType;
  nfa_s : nfa_state;
  nfa_fin : pred nfa_state;
  nfa_trans : nfa_state → char → nfa_state → bool }.
```

The definition of dfa is the same, except that the transition function dfa_trans is a function of type dfa_state → char → dfa_state.

For these definitions, we make use of the fact that types are first-class objects in Coq. This allows us to represent the set of states as a type. We require that the type of states has a finType structure [8], i.e., is a finite type. The collection of types with a finType structure can be thought of as a type class. It is closed under the product, sum, option, and set type constructors. That means that if we have finType structures for T and T', the type checker can infer finType structures for $T*T'$, $T+T'$, option T, and {set T}, the type of sets over T. So the finType structures of all our automata constructions are inferred automatically. The annotation :> for nfa_state registers nfa_state as a coercion. So if A:nfa, x:A means that x is a state of A.

We follow [13] and define acceptance for all states of an automaton by recursion on words.

Fixpoint nfa_accept (A : nfa) (x : A) w :=
    if w is a :: w' then [exists y, nfa_trans A x a y && nfa_accept A y w']
            else x ∈ nfa_fin A.

Fixpoint dfa_accept (A : dfa) (x : A) w :=
    if w is a :: w' then dfa_accept A (dfa_trans A x a) w' else x ∈ dfa_fin A.

The language of an automaton is the set of words accepted by the starting state. Every DFA can easily be converted into an NFA accepting the same language. For the converse direction we formalize the usual powerset construction. Thus we obtain:

Theorem 2. *For every NFA (DFA) we can construct a DFA (NFA) accepting the same language.*

4.1 Regular Expressions to Finite Automata

Our next result about finite automata is the construction of an automaton for every regular expression. For this we need constructions on finite automata corresponding to constructors of regular languages. Since dfa_accept does not use existential quantification, DFAs are generally easier to work with. However, some constructions become much easier when done on NFAs.

Since we have already established conversions between NFAs and DFAs, we can carry out each construction on the automata model that fits best. We define the constructions for A^*, $A \cdot B$, and $'a'$ on NFAs and we define the constructions for $\emptyset$, ε, and $A+B$ on DFAs. All six constructions are fairly straightforward. Our NFA constructions differ slightly from Kozen's [13] since our NFAs do not admit ε-transitions. Whenever one would usually use an ε-transition from a state x to a state y, we instead duplicate all incoming transitions from x as incoming transitions of y. See the file `automata.v` [7] for details.

Theorem 3. *For every regular expression r, we can construct a DFA accepting the same language.*

Proof. Induction on r using the respective constructions on automata.   □

For DFAs it is also easy to show closure under complement and intersection. In fact, the DFA for $A \cap B$ and the DFA for $A + B$ used above are both instances of one generic construction for binary boolean operations. Further, we can give a function that decides whether the language of a DFA is empty.

Definition dfa_lang_empty A := [forall (x | reachable A x), x ∉ dfa_fin A].

The function simply checks that none of the states of A which are reachable from the starting state are final states of A. Here, reachable is defined with respect to the reflexive transitive closure of the relation below:

Definition dfa_trans_some (x y : A) := [exists a, dfa_trans x a == y].

The reflexive transitive closure of a decidable relation over a finite type is again a decidable relation and this construction is contained in the Ssreflect libraries. So we obtain:

Theorem 4.   *1. Language emptiness for DFAs is decidable.*
2. Language equivalence for DFAs is decidable.

Proof. (1) is decided by dfa_lang_empty. Part (2) reduces to (1).   □

Note that decidability of language equivalence for DFAs implies decidability of language equivalence for all representations that can be translated to DFAs. We will show that this is the case for all representations we consider.

4.2   Finite Automata to Regular Expressions

We now show that we can construct a regular expression for every deterministic finite automaton. We use Kleene's algorithm [12], because we think it is easiest to formalize. For the rest of this section, we assume that we are given a DFA $A = (Q, s, F, \delta)$.

Definition 2. *Let w be a word and $x \in Q$. We call the state sequence $x_1 \ldots x_n$ the run from x on w, written $\mathrm{run}(x, w)$, if $x \xrightarrow{w_1} x_1 \ldots \xrightarrow{w_{|w|}} x_{|w|}$ where $y \xrightarrow{a} z$ abbreviates $\delta(y, a) = z$. We define $\hat{\delta}(x, w)$ to be the last element of $x :: \mathrm{run}(x, w)$ and write $\hat{\delta}_s$ for $\lambda w. \hat{\delta}(s, w)$.*

Based on the notion of run, we can define the languages of runs from one state to another. Restricting the set of states that may be traversed in between, we obtain the following indexed collection of languages:

Definition 3. *Let $X \subseteq Q$, $x, y \in Q$, and w a word. $w \in L_{x,y}^X$ iff (1) $\hat{\delta}(x, w) = y$ and (2) all states of $\mathrm{run}(x, w)$ except possibly the last are contained in X.*

Lemma 1. $\mathcal{L}(A) = \bigcup_{x \in F} L_{s,x}^Q$

According to Lemma 1, it suffices to construct regular expressions for $L_{s,x}^Q$ for the various states $x \in F$ in order to obtain a regular expression for $\mathcal{L}(A)$. We recursively solve the problem for all languages $L_{x,y}^Q$ by successively removing states from Q. Once we reach the empty set of states, we can directly give a regular expression.

Definition 4. *Let* $x, y \in Q$, *then*

$$R_{x,y}^{\emptyset} \stackrel{def}{\equiv} (\text{if } x = y \text{ then } \varepsilon \text{ else } \emptyset) + \sum_{\substack{a \in \Sigma \\ \delta(x,a)=y}} a$$

Lemma 2. $\mathcal{L}\left(R_{x,y}^{\emptyset}\right) = L_{x,y}^{\emptyset}$

Now consider the case of a nonempty set $X \subseteq Q$, where $z \in X$ and $w \in L_{x,y}^{X}$. Then run(x, w) may reach z, come back to z an arbitrary number of times, and then end in y. Alternatively, the run may not reach z at all. This motivates the following lemma:

Lemma 3. *Let* $x, y, z \in Q$ *and* $X \subseteq Q$, *then*

$$w \in L_{x,y}^{\{z\} \cup X} \iff w \in L_{x,z}^{X} \cdot \left(L_{z,z}^{X}\right)^{*} \cdot L_{z,y}^{X} + L_{x,y}^{X}$$

The formalization of Lemma 3 is one of the more involved parts of our development. Writing delta for $\hat{\delta}$, we formalize $L_{x,y}^{X}$ as follows:

Definition L (X : {set A}) (x y : A) :=
  [pred w | (delta x w == y) && abl (mem X) (dfa_run x w)].

Here, abl (mem X) (dfa_run x w) checks the second condition of Definition 3. Showing the direction from right to left is relatively easy. We prove the converse direction by induction on $|w|$. The essential lemma for this direction is:

Lemma L_split X x y z w : w ∈ L^(z |: X) x y →
  w ∈ L^X x y ∨ ∃ w1 w2, w = w1 ++ w2 ∧ size w2 < size w
                ∧ w1 ∈ L^X x z ∧ w2 ∈ L^(z |: X) z y.

which is itself proved by induction on w. The Notation z |: X stands for $\{z\} \cup X$. Following Lemma 2 and Lemma 3, we can give a recursive procedure R such that the language of R^X x y is L^X x y.

Function R (X : {set A}) (x y : A) {measure (fun X ⇒ #|X|) X} :=
  match [pick z ∈ X] with
  | None ⇒ R0 x y
  | Some z ⇒ let X' := X :\ z in
    Plus (Conc (R X' x z) (Conc (Star (R X' z z)) (R X' z y))) (R X' x y)
  end.

The definition employs Coq's **Function** command, which allows the definition of functions by size recursion. In our case, the measure is the size of the set X. The expression [pick z ∈ X] evaluates to None if X is empty and to Some z with $z \in X$ otherwise. Due to the match on the pick expression, Coq is, at the time of writing, not capable of generating the functional induction principle for R. However, this does not pose a problem, since the correspondence between R and L can be proved directly by induction on the size of X.

Lemma L_R n (X : {set A}) x y : #|X| = n → L^X x y =i R^X x y.

Thus we have:

Theorem 5. *For every automaton A, we can construct a regular expression accepting $\mathcal{L}(A)$.*

Corollary 1. *Let r and e be regular expressions. We can construct regular expressions accepting $\mathcal{L}(r) \cap \mathcal{L}(e)$ and $\overline{\mathcal{L}(r)}$.*

5   Minimization

We now construct a minimization function for DFAs. We follow Kozen's presentation [13] of Huffman's table filling algorithm [11]. We fix some DFA $A = (Q, s, F, \delta)$.

Definition 5. *Let $x, y \in Q$. The collapsing relation on A is defined as follows:*

$$x \approx y \stackrel{def}{\equiv} \forall w.\ \delta(x, w) \in F \iff \delta(y, w) \in F$$

Minimization merges every equivalence class of the collapsing relation into a single state. To construct this quotient automaton in Coq, we need to show that the collapsing relation is decidable. Once we have a boolean reflection collb : A $\rightarrow$ A $\rightarrow$ bool of the collapsing relation, the quotient construction follows a generic pattern.

The construction makes use of the fact that there is a constructive choice operator for finite types. Consider, for instance, a decidable equivalence relation e over a type T with choice operator. We use this choice operator to get a canonical element[2] of every equivalence class of e.

Definition repr x := choose (e x) x.

We call repr x the representative of the equivalence class of x with respect to e. The function repr is idempotent. Thus, the quotient of T modulo e can be defined as follows:

Definition quot := { x : T | x == repr x }.

The type quot is a sigma type, i.e., the type of dependent pairs of elements x and proofs of x == repr x. In particular, this type is finite if T is finite, since x == repr x has at most one proof and, thus, quot has at most as many elements as T. Using repr we can also define a function class : T $\rightarrow$ quot corresponding to the function $\lambda x.[x]_e$. The first projection of quot, written val, allows us to obtain the canonical representative of the class $[x]_e$. Taking e to be collb we can define the quotient automaton as follows:

Definition minimize : dfa := {|
    dfa_s := class (dfa_s A);
    dfa_trans x a := class (dfa_trans (val x) a);
    dfa_fin := [pred x | val x $\in$ dfa_fin A] |}.

---

[2] If p y = true, the result of choose p y satisfies p but does not depend on y.

To compute collb, we compute its complement, the distinguishable states, using a fixpoint construction. Final and non-final states are distinguishable using the empty word.

Definition dist0 : {set A*A} := [set x | (x.1 ∈ dfa_fin A) != (x.2 ∈ dfa_fin A)].

We also mark those pairs as distinguishable that transition to an already distinguishable pair of states.

Definition distS (dist : {set A*A}) :=
  [set x | [exists a, (dfa_trans x.1 a, dfa_trans x.2 a) ∈ dist]].

Now we can define a monotone function one_step_dist : {set $M * M$} → {set $M * M$} corresponding to one pass through Huffman's algorithm [11].

Definition one_step_dist dist := dist0 ∪ distS dist.

Its least fixpoint, computed by iterating the function sufficiently often on the empty set, is exactly the set of distinguishable pairs. This finishes the minimization construction. At this point, we can show:

Lemma minimize_correct A : dfa_lang (minimize A) =i dfa_lang A.

Lemma minimize_size A : #|minimize A| ≤ #|A|.

For connected DFAs the result of minimization is indeed minimal and minimal DFAs are unique up to isomorphism, i.e, up to a renaming of the states. In this context, a connected DFA is a DFA $A = (Q, s, F, \delta)$ where $\hat{\delta}_s$ is surjective. Surjectivity of $\hat{\delta}_s$ allows us to define a partial inverse $\hat{\delta}_s^{-1} : Q \to \Sigma^*$ such that $\hat{\delta}_s(\hat{\delta}_s^{-1}x) = x$. For this we exploit that there is also a choice operator for the countable type of words. In fact our inverse construction for $\hat{\delta}_s$ is just an instance of a generic inverse construction for surjective functions from types with choice operator to types with decidable equality.

Definition cr {X : choiceType} {Y : eqType} {f : X → Y} (Sf : surjective f) y : X :=
  xchoose (Sf y).

We call cr Sf x (with types as above) the *canonical representative* of x. The construction uses xchoose, a stronger variant of constructive choice, which for a decidable predicate p turns a proof of ∃ x, p x into an element satisfying p.

We now show that all connected and collapsed (i.e., the collapsing relation is the identity) DFAs are isomorphic. Consider two collapsed DFAs A and B accepting the same language and a proof A_conn : connected A for A and likewise for B. We use cr to define an isomorphism between A and B.

Definition iso (x : A) : B := delta (dfa_s B) (cr A_conn x).

To show that iso is a bijection we define its inverse iso_inv by swapping the role of A and B in the definition of iso and show that iso and iso_inv cancel each other in both directions. The proofs make use of the following fact about the interaction of delta and iso:

Lemma delta_iso w x : delta (iso x) w ∈ dfa_fin B = (delta x w ∈ dfa_fin A).

Using the fact that both automata are fully collapsed, we can show that iso is not just a bijection but also respects the structure of the automata. Thus, we have shown:

Theorem 6. *Let* $A = (Q_1, s_1, F_1, \delta_1), B = (Q_2, s_2, F_2, \delta_2)$ *be collapsed and connected DFAs. If* $\mathcal{L}(A) = \mathcal{L}(B)$, *then there exists a bijection* $i : Q_1 \to Q_2$ *satisfying*

$$\forall q \in Q_1. \, i(\delta_1(q, a)) \quad = \quad \delta_2(i(q), a)$$
$$\forall q \in Q_1. \, i(q) \in F_2 \iff q \in F_1$$
$$i(s_1) \quad = \quad s_2$$

It is easy to show that minimization preserves connectedness and yields collapsed DFAs. In particular, this entails that the result of minimizing a connected automaton is indeed minimal and that minimal DFAs are unique up to isomorphism.

6 Myhill-Nerode Theorem

Myhill [16] and Nerode [17] relations characterize regular languages in terms of simple algebraic properties. We now define two additional representations of regular languages: Myhill partitions and Nerode partitions. Our constructive version of the Myhill-Nerode theorem then consists of three conversion functions: from Nerode partitions to Myhill partitions, from Myhill partitions to DFAs, and from minimal DFAs to Nerode partitions.

Definition 6. *Let* $\equiv \, \subseteq \Sigma^* \times \Sigma^*$ *be an equivalence relation. The* partition $(\mathcal{I}, E)$ *with* $\mathcal{I}$ *a finite set and* $E : \Sigma^* \to \mathcal{I}$ *surjective* represents $\equiv$ *if*

$$\forall u \, v. \, E \, u = E \, w \iff u \equiv w$$

We call $\mathcal{I}$ *the* index set *and* E *the* representation function. *Further,* $E(w)$ *represents* $[w]_{\equiv}$, *the equivalence class of* w *with respect to* $\equiv$.

Every partition $(\mathcal{I}, E)$ represents some equivalence relation of finite index. We phrase the Myhill and Nerode conditions directly in terms of partitions.

Definition 7. *Let* $P = (\mathcal{I}, M)$ *be a partition.* P *is a* Myhill partition *for* L, *if* M

1. *is* right congruent: $\forall u \in \Sigma^* v \in \Sigma^* a \in \Sigma. \, M \, u = M \, v \Rightarrow M \, ua = M \, va$
2. *refines* L: $\forall u \in \Sigma^* v \in \Sigma^*. \, M \, u = M \, v \Rightarrow (u \in L \iff v \in L)$

Definition 8. *Let* $P = (\mathcal{I}, N)$ *be a partition.* P *is a* Nerode partition *for* L *if*

$$\forall u \in \Sigma^* v \in \Sigma^*. \, N \, u = N \, v \iff \forall w \in \Sigma^*. \, (uw \in L \iff vw \in L)$$

We refer to the representation functions of Myhill partitions as Myhill functions and similarly for Nerode partitions. We formalize finite partitions using records:

Record finPar := {
  finpar_classes : finType;
  finpar_fun :> word → finpar_classes;
  finpar_surj : surjective finpar_fun }.

In addition to finpar_fun, we manually register finpar_classes as a second coercion. So if E:finPar, we write x:E to mean that x is in the index set just as we did for the states of automata, but we can also write E w for the class of a word w.

We then formalize Myhill and Nerode partitions as more constrained versions of the above type. For example, Nerode partitions for L are defined as follows:

Definition nerode (X : eqType) (L : dlang char) (E : word → X) :=
  ∀ u v, E u = E v ↔ ∀ w, (u++w ∈ L) = (v++w ∈ L).

Record nerodePar L := {
  nerode_par :> finPar;
  nerodeP : nerode L nerode_par }.

We now define the translation functions we need for our Myhill-Nerode theorem. Myhill partitions can be seen as abstract representations of connected DFAs and Nerode partitions can be seen as abstract representations of minimal DFAs.

The translation from Nerode partitions to Myhill partitions is particularly easy. The representation function of a Nerode partition for a language L refines L and is right congruent. Thus the representation function of a Nerode partition can also serve as representation function for a Myhill partition.

For the direction from Myhill partitions to DFAs consider some Myhill partition $(\mathcal{I}, M)$ representing a Myhill relation $\equiv$ for some language L. We construct an automaton A with $Q := \mathcal{I}$ and $s := M\,\varepsilon$. For the final states and the transition function, we make use of the fact that $M : \Sigma^* \to \mathcal{I}$ is surjective. Due to surjectivity, every $x \in \mathcal{I}$ represents a class $[w]_\equiv$ for some w. Using cr we can now define a transition function on $\mathcal{I}$ which in terms of the represented classes satisfies $\delta([w]_\equiv, a) = [wa]_\equiv$.

Definition fp_trans (E : finPar) (x : E) a := E (cr E x ++ [:: a]).

The conversion from Myhill partitions to DFAs is then defined as follows:

Definition myhill_to_dfa L (M : myhillPar L) :=
  {| dfa_s := M [::]; dfa_fin x := cr M x ∈ L; dfa_trans := @fp_trans M |}.

Lemma myhill_to_dfa_correct L (M : myhillPar L) : dfa_lang (myhill_to_dfa M) =i L.

For the conversion from DFAs to Nerode partitions we can rely on our minimization algorithm. Hence, it is sufficient to convert minimal DFAs to Nerode partitions. Consider, for instance, a minimal DFA $A = (Q, s, F, \delta)$. Since A is minimal and thus connected, the function $\hat{\delta}_s : \Sigma^* \to Q$ is surjective. Further A is collapsed and therefore $(Q, \hat{\delta}_s)$ is a Nerode partition. Thus we have:

Theorem 7

1. *For every Nerode partition we can construct an equivalent Myhill partition.*
2. *For every Myhill partition we can construct an equivalent DFA.*
3. *For every DFA we can construct an equivalent Nerode partition.*

7   Closure under Homomorphisms

In Section 4.2, we have shown that regular languages are closed under intersection and complement. We now show that regular languages are closed under preimages and images of homomorphisms. For this section we assume a second alphabet Γ.

Definition 9. *Let* $h : \Sigma^* \to \Gamma^*$ *be a function.*

- h *is a* homomorphism *if* $h\,uv = h\,u \cdot h\,v$ *for all* $u \in \Sigma^*$ *and* $v \in \Sigma^*$.
- *The* preimage *of* L *under* h *is* $h^{-1}(L) := \{w \mid h\,w \in L\}$
- *The* image *of* L *under* h *is* $h(L) := \{v \mid \exists w \in L.\ h\,w = v\}$

The closure under taking the preimage of a homomorphism is easy to show. We use Kozen's [13] construction on DFAs.

The closure of regular languages under taking the image of a homomorphism is more interesting. Unlike all the closure properties of regular languages we have shown so far, it is not a closure property of decidable languages in general. This means, that we cannot define an image operator on decidable languages. We can only express the image as a predicate in Prop.

Definition image (h : word char → word char') (L : word char → Prop) v :=
    ∃ w, L w ∧ h w = v.

For a homomorphism h, Kozen [13] gives a construction of a regular expression e_h from a regular expression e satisfying $\mathcal{L}(e_h) = h(\mathcal{L}(e))$. The construction works by replacing all atoms a in e with the string $h\,a$. We can prove this construction correct, but we can only state the correctness as a reflection lemma and not as a quantified boolean equation as we did for all the other constructions so far.

Lemma re_imageP e v : reflect (image h (re_lang e) v) (v ∈ re_image e).

Once we abstract away the concrete constructions, this difference disappears and we obtain:

Lemma preim_regular (char char' : finType) (h : word char → word char') L :
    homomorphism h → regular L → regular (preimage h L).

Lemma im_regular (char char' : finType) (h : word char → word char') L :
    homomorphism h → regular L → regular (image h L).

Theorem 8. *Regular languages are closed under taking preimages and images of homomorphisms.*

8 Proving Languages Non-regular

So far, we have formalized a number of constructions that can be used to prove that a language is regular. One option to prove that a language is not regular is using the Pumping Lemma, which is included in our formalization. More interesting in our constructive setting is the use of Nerode partitions. If we want to prove that a language L is not regular, we can assume that L is regular. This provides us with a corresponding decidable language L' and, using our translation functions, with a Nerode partition for L'.

Lemma regularE (L : word char → Prop) : regular L →
   ∃ L' : dlang char, (∀ w, L w ↔ w ∈ L') ∧ inhabited (nerodePar L').

Hence, it is sufficient to have a reasoning principle to show non-regularity of decidable languages. One such criterion is the existence of an infinite collection of words that are not related by the Nerode relation.

Lemma nerodeIN (f : nat → word char) (L : dlang char) :
   (∀ n1 n2, (∀ w, (f n1 ++ w ∈ L) = (f n2 ++ w ∈ L)) → n1 = n2) →
   ~ regular L.

We use this principle to prove that $\{w \mid \exists n.\ w = a^n b^n\}$ is not regular.

So for proofs of non-regularity, the restriction of Nerode partitions to decidable languages is irrelevant. For proofs of regularity, it does pose a restriction, and this restriction is unavoidable in a constructive setting. Consider some independent proposition P, i.e., some P for which we can prove neither P nor $\neg P$. Then regularity of the language $L := \{w \mid P\}$ is equivalent to the unprovabe proposition $P \vee \neg P$. However, we can still prove that L has exactly one Nerode class. So except for the decidability requirement on the language itself, we have a Nerode partition and thus a proof of regularity. This also shows that there are some languages on which the constructive interpretation of regularity differs from the set theoretic interpretation.

9 Conclusion

We have formalized a number of fundamental results about regular languages. Our selection of results corresponds roughly to the first part of Kozen's "Automata and Computability" [13]. We added the uniqueness result for minimal DFAs and skipped NFAs with ε-transitions, stopping right before the definition of 2DFAs.

Concerning ε-NFAs, we came to the conclusion that the added flexibility is not worth the effort. In fact, we had a formalization of ε-NFAs. However, in our formalization the correctness proof for the conversion function from ε-NFAs to NFAs was just reflexivity and thus not interesting. So the only advantage of working with ε-NFAs would be more compact definitions, but at the cost of having to consider, in proofs, possible sequences of ε-transitions at every transition step.

A more interesting addition to our development would be 2DFAs. Since 2DFAs may move back and forth on the input word and even have infinite runs, their language cannot be defined by a simple recursion on the input word. Instead, the boolean acceptance predicate would be computed using a finite fixpoint construction similar to the one we employed in the minimization algorithm.

To make our formalization more instructive, we obtain our results with many small lemmas. This way at least the overall structure of the proofs can be understood without stepping through the proofs. In total, we prove about 170 lemmas and almost 50% of the 1400 lines of our development are specifications.

All of our proofs are carried out in the constructive type theory of Coq, and there are a number of places where we have to deviate from the textbook presentation [13] of the material we formalize. We have to show at several places that the predicates and relations we define are decidable, which is supported very well by Ssreflect.

For Kleene's Theorem, staying constructive did not cause any difficulties, since the textbook proof [13] of this theorem is also constructive. For automata minimization, the lack of general quotient types in Coq forces us to compute the collapsing relation before we can define the quotient automaton. However, the collapsing algorithm is interesting in its own right. For the Myhill-Nerode Theorem, we represent equivalence relations of finite index by making the finite set of equivalence classes explicit in the form of a finite type. This turns Myhill and Nerode partitions into objects we can compute with. Consequently, we have to rely on minimization to obtain Nerode partitions.

Many of the constructions we use in our development are applicable in a wide range of situations. Notably, this includes finite types, constructive quotients, and finite fixpoints. While we are certainly not the first to use these techniques, we believe they deserve a wider recognition. Since the material we formalize is fairly standard, we hope that our proofs can serve as examples for teaching theory development in Coq.

Acknowledgments. We thank the anonymous reviewers of the preliminary versions of this paper for their helpful comments.

References

1. Asperti, A.: A compact proof of decidability for regular expression equivalence. In: Beringer, L., Felty, A. (eds.) ITP 2012. LNCS, vol. 7406, pp. 283–298. Springer, Heidelberg (2012)
2. Berghofer, S., Nipkow, T., Urban, C., Wenzel, M. (eds.): TPHOLs 2009. LNCS, vol. 5674. Springer, Heidelberg (2009)
3. Berghofer, S., Reiter, M.: Formalizing the logic-automaton connection. In: Berghofer, et al. (eds.) [2], pp. 147–163
4. Braibant, T., Pous, D.: Deciding kleene algebras in coq. Logical Methods in Computer Science 8(1) (2012)
5. Constable, R.L., Jackson, P.B., Naumov, P., Uribe, J.C.: Constructively formalizing automata theory. In: Plotkin, G.D., Stirling, C., Tofte, M. (eds.) Proof, Language, and Interaction, pp. 213–238. The MIT Press (2000)

6. Coquand, T., Siles, V.: A decision procedure for regular expression equivalence in type theory. In: Jouannaud, J.-P., Shao, Z. (eds.) CPP 2011. LNCS, vol. 7086, pp. 119–134. Springer, Heidelberg (2011)
7. Doczkal, C., Kaiser, J.O., Smolka, G.: Formalization accompanying this paper, http://www.ps.uni-saarland.de/extras/cpp13/
8. Garillot, F., Gonthier, G., Mahboubi, A., Rideau, L.: Packaging mathematical structures. In: Berghofer, et al. (eds.) [2], pp. 327–342
9. Gonthier, G., Mahboubi, A., Tassi, E.: A Small Scale Reflection Extension for the Coq system. Rapport de recherche RR-6455, INRIA (2008), http://hal.inria.fr/inria-00258384
10. Hopcroft, J.E., Motwani, R., Ullman, J.D.: Introduction to automata theory, languages, and computation - international edition, 2nd edn. Addison-Wesley (2001)
11. Huffman, D.: The synthesis of sequential switching circuits. Journal of the Franklin Institute 257(3), 161–190 (1954)
12. Kleene, S.C.: Representation of events in nerve nets and finite automata. In: Shannon, McCarthy (eds.) [19], pp. 3–42
13. Kozen, D.: Automata and computability. Undergraduate texts in computer science. Springer (1997)
14. Krauss, A., Nipkow, T.: Proof pearl: Regular expression equivalence and relation algebra. J. Autom. Reasoning 49(1), 95–106 (2012)
15. Moore, E.F.: Gedanken-experiments on sequential machines. In: Shannon, McCarthy (eds.) [19], pp. 129–153
16. Myhill, J.R.: Finite Automata and the Representation of Events. Tech. Rep. WADC TR-57-624, Wright-Paterson Air Force Base (1957)
17. Nerode, A.: Linear automaton transformations. Proceedings of the American Mathematical Society 9(4), 541–544 (1958)
18. Rabin, M.O., Scott, D.: Finite automata and their decision problems. IBM J. Res. Dev. 3(2), 114–125 (1959)
19. Shannon, C., McCarthy, J. (eds.): Automata Studies. Princeton University Press (1956)
20. The Coq Development Team: http://coq.inria.fr
21. Wu, C., Zhang, X., Urban, C.: A formalisation of the Myhill-Nerode theorem based on regular expressions (Proof pearl). In: van Eekelen, M., Geuvers, H., Schmaltz, J., Wiedijk, F. (eds.) ITP 2011. LNCS, vol. 6898, pp. 341–356. Springer, Heidelberg (2011)

Certified Parsing of Regular Languages

Denis Firsov and Tarmo Uustalu

Institute of Cybernetics, Tallinn University of Technology
Akadeemia tee 21, 12618 Tallinn, Estonia
{denis,tarmo}@cs.ioc.ee

Abstract. We report on a certified parser generator for regular languages using the Agda programming language. Specifically, we programmed a transformation of regular expressions into a Boolean-matrix based representation of nondeterministic finite automata (NFAs). And we proved (in Agda) that a string matches a regular expression if and only if the NFA accepts it. The proof of the if-part is effectively a function turning acceptance of a string into a parse tree while the only-if part gives a function turning rejection into a proof of impossibility of a parse tree.

1   Introduction

Parsing is the process of structuring a linear representation (sentence, a computer program, etc.) in accordance with a given grammar. Parsers are procedures which perform parsing. They are being used extensively in a number of disciplines: in computer science (for compiler construction, database interfaces, artificial intelligence), in linguistics (for text analysis, corpora analysis, machine translation, textual analysis of biblical texts), in typesetting chemical formulae, in chromosome recognition, and so on [4]. It is therefore clear that having correct parsers is important for all these disciplines. Surprisingly, relatively little research had been done in field of certified parsing.

However, with the recent development of programming languages with dependent types it has become possible to encode useful invariants in the types and prove properties of a program while implementing it. In this paper we use the system of dependent types of the Agda language [8], which is based on Martin-Löf's type theory. One of the basic ideas behind Martin-Löf's type theory is the Curry-Howard interpretation of propositions as types. A proposition is proved by writing a program of the corresponding type.

Dependent types allow types to talk about values. A classical example of a dependent type is the type of lists of a given length: Vec A n. Here A is the type of the elements and n is the length of the list. Having such a definition of vector, we can define "safe" functions. Let us look at definition of a safe head function:

```
head : ∀ {A n} → Vec A (suc n) → A
head (x :: xs) = x
```

This definition says that the function head accepts only vectors with at least one element. So, it is now the responsibility of the programmer to provide non-empty vectors, otherwise compilation will fail.

G. Gonthier and M. Norrish (Eds.): CPP 2013, LNCS 8307, pp. 98–113, 2013.
© Springer International Publishing Switzerland 2013

In this paper, we have adopted the same general technique of expressing properties of data in their types to implement a library of matrix operations, program a transformation of regular expressions into a Boolean matrix based representation of nondeterministic finite automata (NFAs) and prove it correct. The correctness proof turns NFAs effectively into parsers: the proof of acceptance or rejection of a string gives us a parse tree (a witness of matching) or a proof of impossibility of one.

Our Agda development can be found online at http://cs.ioc.ee/~denis/cert-reg .

2   Regular Expressions

Before we start, let us take care of the alphabet (Σ) of our interest. Since many functions require Σ as a parameter and an NFA must share its alphabet with the regular expressions, we define Σ as a global parameter for each module of the parser-generator library.

module ModuleName (Σ : Set) ($\_\overset{?}{=}\_$: Decidable ($\_\equiv\_$ {A = Σ}))

It states that the module is parametrized by an alphabet and also a decidable equality on the alphabet.

The datatype for regular expressions is called RegExp and is defined as

data RegExp : Set **where**
   ε   : RegExp
   '_   : $\Sigma \rightarrow$ RegExp
   _∪_ : RegExp $\rightarrow$ RegExp $\rightarrow$ RegExp
   _·_ : RegExp $\rightarrow$ RegExp $\rightarrow$ RegExp
   _$^+$: RegExp $\rightarrow$ RegExp

The base cases are regular expressions for the empty string (ε) and for single-character strings ('_). The step cases are given by the regular operations: _∪_ for union, _·_ for concatenation and _$^+$ for iteration at least once. Note that instead of the Kleene star (_*) we use plus (_$^+$). This is more convenient for us and does not restrict generality, as star is expressible as choice between the empty string and plus.

Now, we need to specify when a string (an element of type List Σ) is in the language of the regular expression, which is also called matching. This is done by introducing a *parsing (or matching) relation* (denoted by _▶_) between strings and regular expressions.

String : Set
String = List Σ
data _▶_ : String $\rightarrow$ RegExp $\rightarrow$ Set **where**
  empt : [] ▶ ε
  symb : {x : Σ} $\rightarrow$ [x] ▶ ' x
  unionl : {u : String} {e_1 e_2 : RegExp} $\rightarrow$ u ▶ e_1 $\rightarrow$ u ▶ $e_1 \cup e_2$
  unionr : {u : String} {e_1 e_2 : RegExp} $\rightarrow$ u ▶ e_2 $\rightarrow$ u ▶ $e_1 \cup e_2$

$$\begin{aligned}
\mathsf{con} \quad &: \{u\ v\ :\ \mathsf{String}\}\ \{e_1\ e_2\ :\ \mathsf{RegExp}\} \to u \blacktriangleright e_1 \to v \blacktriangleright e_2 \\
&\to u +\!\!+ v \blacktriangleright e_1 \cdot e_2 \\
\mathsf{plus1} \quad &: \{u\ :\ \mathsf{String}\}\ \{e\ :\ \mathsf{RegExp}\} \to u \blacktriangleright e \to u \blacktriangleright e^+ \\
\mathsf{plus2} \quad &: \{u\ v\ :\ \mathsf{String}\}\ \{e\ :\ \mathsf{RegExp}\} \\
&\to u \blacktriangleright e \to v \blacktriangleright e^+ \to u +\!\!+ v \blacktriangleright e^+
\end{aligned}$$

(Arguments enclosed in curly braces are implicit. The type checker will try to figure out the argument value for you. If the type checker cannot infer an implicit argument, then it must be provided explicitly, e.g., symb $\{x\}$.)

Let us now examine the constructors of the relation $\_ \blacktriangleright \_$:

1. The constructor empt states that the empty string is in the language of the regular expression ε.
2. The constructor symb states that the string consisting of a single character x is in the language of the regular expression $' x$.
3. The constructor unionl (unionr) states that if a string u is in the language defined by e_1 (e_2), then u is also in the language of $e_1 \cup e_2$ for any e_2 (e_1).
4. The constructor con states that if a string u is in the language e_1 and a string v is in the language of e_2 then the concatenation of both strings is in the language of $e_1 \cdot e_2$.
5. The constructor plus1 states that if a string u is in the language of e, then it is also in the language of e^+.
6. The constructor plus2 states that if a string u is in the language of e and a string v is in the language of e^+, then concatenation of both strings is in the language of e^+.

Note that a proof that a string is in the parsing relation with a regular expression is a parse tree. Note that we do not introduce any notion of "raw" parse trees, a parse tree is always a parse tree of a specific string.

3   A Matrix Library

The transition relation of a nondeterministic finite automaton (NFA) can be viewed as a labeled directed graph. So it can be expressed as a family of incidence matrices (one matrix per label). In addition to the nice algebraic properties that this approach highlights, it allows us to compose automata (expressed with matrices) in various ways by using block operations.

We therefore formalize matrices and some important matrix operations and their properties.

3.1   Matrices and Matrix Operations

What sort of elements can a matrix contain? Our approach abstracts from the type of elements. But in order for matrix addition and multiplication to be well-defined and satisfy the standard properties, it must form a commutative semiring. In Agda we introduce a parametrized module

module Data.Matrix (sr : CommSemiRing)

This declaration says that the module Data.Matrix is parametrized by a commutative semiring sr. A semiring will be a record containing the carrier type R, the operations, and the proofs of the laws of the semiring.

Next we define a representation for matrices. A matrix is a vector of vectors, therefore the matrix type can be defined as

$$\_\times\_ \ : \ \mathbb{N} \to \mathbb{N} \to \mathsf{Set}$$
$$k \times l \ = \ \mathsf{Vec}\ (\mathsf{Vec}\ R\ l)\ k$$

where k denotes the number of rows and l the number of columns in a matrix. Let us implement some of the most important operations on matrices:

Null Matrix. The zero or null matrix is a matrix with all entries zero.

$$\mathsf{null} \ : \ \{k\ l \ : \ \mathbb{N}\} \to k \times l$$
$$\mathsf{null} \ = \ \mathsf{replicate}\ (\mathsf{replicate}\ \mathsf{zero})$$

Here zero is the additive identity element of the semiring. Note that the arguments k and l are implicit (enclosed in curly braces), so in the most cases we can omit them and the type checker will try to infer their values automatically.

Identity Matrix. The identity or unit matrix of size k is the k × k square matrix with ones on the main diagonal and zeros elsewhere.

$$\mathsf{id} \ : \ \{k \ : \ \mathbb{N}\} \to k \times k$$
$$\mathsf{id}\ \{0\} \ = \ []$$
$$\mathsf{id}\ \{\mathsf{suc}\ k\} \ = \ (\mathsf{one} :: \mathsf{replicate}\ \mathsf{zero}) :: \mathsf{zipWith}\ \_+\!\!+\_\ \mathsf{null}\ (\mathsf{id}\ \{k\})$$

Here one is the multiplicative identity element.

Addition. Matrix addition is the operation of adding two matrices by adding corresponding entries together.

$$\_\oplus\_ \ : \ \{k\ l \ : \ \mathbb{N}\} \to k \times l \to k \times l \to k \times l$$
$$\_\oplus\_ \ []\ [] \ = \ []$$
$$\_\oplus\_ \ (\mathsf{rowA} :: A')\ (\mathsf{rowB} :: B') \ = \ \mathsf{zipWith}\ \_+\_\ \mathsf{rowA}\ \mathsf{rowB} :: A' \oplus B'$$

Note, that signature of addition requires the dimensions of the two input matrices to be equal.

Transposition. The transpose of a matrix A is another matrix $\langle\, A\, \rangle$ (or A^T in mathematical notation) created by writing the rows of A as the columns of $\langle\, A\, \rangle$.

$$\langle\_\rangle \ : \ \{k\ l \ : \ \mathbb{N}\} \to k \times l \to l \times k$$
$$\langle\, []\, \rangle \ = \ \mathsf{replicate}\ []$$
$$\langle\, \mathsf{rowA} :: A'\, \rangle \ = \ \mathsf{zipWith}\ \_::\_\ \mathsf{rowA}\ \langle\, A'\, \rangle$$

Multiplication. Matrix multiplication is a binary operation that takes a pair of matrices and produces another matrix. If A is an k × l matrix and B is an l × m matrix, the result A ⊗ B of their multiplication is an k × m matrix defined only if the number l of columns of the first matrix A is equal to the number of rows of the second matrix B.

```
_⊗_ : {k l m : ℕ} → k × l → l × m → k × m
_⊗_ [] _ = []
_⊗_ (rowA :: A') B = multRow :: A' ⋆ B
  where
    multRow = map
      (λ colB → (foldr (_+_) zero (zipWith (_*_) rowA colB))) ⟨ B ⟩
```

The result of matrix multiplication is a matrix whose elements are found by multiplying the elements within a row from the first matrix by the associated elements within a column from the second matrix and summing the products.

In our library we have proved a number of basic properties of matrix transposition, addition, multiplication.

3.2 Block Operations

How to create matrices from smaller matrices systematically? This section describes an approach to block operations on matrices that has been advocated by Macedo and Oliveira [6]. It is centered around four operations corresponding to the injections and projections of the biproducts on the category of natural numbers and matrices.

1. ι_1 : $\{k\,l : \mathbb{N}\} \to (k+l) \times k$
   ι_1 = id ⧺ null
2. ι_2 : $\{k\,l : \mathbb{N}\} \to (k+l) \times l$
   ι_2 = null ⧺ id
3. π_1 : $\{k\,l : \mathbb{N}\} \to k \times (k+l)$
   π_1 = zipWith _⧺_ id null
4. π_2 : $\{k\,l : \mathbb{N}\} \to l \times (k+l)$
   π_2 = zipWith _⧺_ null id

Example 1. Let us show some instances of these operations:

$$\iota_1\{3\}\{1\} = \begin{bmatrix} 1 & 0 & 0 \\ 0 & 1 & 0 \\ 0 & 0 & 1 \\ \hline 0 & 0 & 0 \end{bmatrix} \qquad \iota_2\{2\}\{2\} = \begin{bmatrix} 0 & 0 \\ 0 & 0 \\ \hline 1 & 0 \\ 0 & 1 \end{bmatrix}$$

$$\pi_1\{3\}\{1\} = \begin{bmatrix} 1 & 0 & 0 & 0 \\ 0 & 1 & 0 & 0 \\ 0 & 0 & 1 & 0 \end{bmatrix} \qquad \pi_2\{2\}\{2\} = \begin{bmatrix} 1 & 0 & 0 & 0 \\ 0 & 1 & 0 & 0 \end{bmatrix}$$

The main block operations are now defined as follows:

Concatenation (copairing) is the operation of placing two matrices next to each other.

$$[\_|\_] : \{k\,l\,m : \mathbb{N}\} \to k \times l \to k \times m \to k \times (l+m)$$
$$[\,A\,|\,B\,] = A \otimes \pi_1 \oplus B \otimes \pi_2$$

Stacking (pairing) is the operation of placing two matrices on top of each other.

$$[\_/\_] : \{k\,l\,m : \mathbb{N}\} \to k \times m \to l \times m \to (k+l) \times m$$
$$[\,A \;/\; B\,] = \iota_1 \otimes A \oplus \iota_2 \otimes B$$

Example 2. Let $A = \begin{bmatrix} 1 & 2 \\ 3 & 4 \end{bmatrix}$ and $B = \begin{bmatrix} 5 & 6 \\ 7 & 8 \end{bmatrix}$. Then

$$[\,A \;|\; B\,] = \left[\begin{array}{cc|cc} 1 & 2 & 5 & 6 \\ 3 & 4 & 7 & 8 \end{array}\right] \text{ and } [\,A \;/\; B\,] = \left[\begin{array}{cc} 1 & 2 \\ 3 & 4 \\ \hline 5 & 6 \\ 7 & 8 \end{array}\right]$$

3.3 Properties of Block Operations

The main advantage of working with injections and projections is that they structure proofs about block operations. Next, we list some properties of block operations.

– Multiplying the concatenation of A and B by ι_1 (ι_2) yields A (B).

$$m_1 m_2\text{-con-}\iota_1 : \{k\,l\,m : \mathbb{N}\}\,(A : k \times l)\,(B : k \times m)$$
$$\to [\,A\,|\,B\,] \otimes (\iota_1\,\{l\}\,\{m\}) \equiv A$$
$$m_1 m_2\text{-con-}\iota_2 : \{k\,l\,m : \mathbb{N}\}\,(A : k \times l)\,(B : k \times m)$$
$$\to [\,A\,|\,B\,] \otimes (\iota_2\,\{l\}\,\{m\}) \equiv B$$

– Multiplying π_1 (π_2) by the stacking of A and B gives A (B).

$$\pi_1\text{-}m_1 m_2\text{-stack} : \{k\,l\,m : \mathbb{N}\}\,(A : k \times l)\,(B : m \times l)$$
$$\to \pi_1\,\{k\}\,\{m\} \otimes [\,A\;/\;B\,] \equiv A$$
$$\pi_2\text{-}m_1 m_2\text{-stack} : \{k\,l\,m : \mathbb{N}\}\,(A : k \times l)\,(B : m \times l)$$
$$\to \pi_2\,\{k\}\,\{m\} \otimes [\,A\;/\;B\,] \equiv B$$

– The product of C and the concatenation of A and B is equal to the concatenation of $C \otimes A$ and $C \otimes B$.

$$\text{distrib-lft} : \{k\,l\,m\,n : \mathbb{N}\}\,(A : k \times l)\,(B : k \times m)\,(C : n \times k)$$
$$\to C \otimes [\,A\,|\,B\,] \equiv [\,C \otimes A\,|\,C \otimes B\,]$$

– The product of a stacking reduces to a stacking of products.

$$\text{distrib-rgt} : \{k\,l\,m\,n : \mathbb{N}\}\,(A : k \times l)\,(B : m \times l)\,(C : l \times n)$$
$$\to [\,A\;/\;B\,] \otimes C \equiv [\,A \otimes C\;/\;B \otimes C\,]$$

– Multiplying the concatenation of A and B by the stacking of C and D yields the sum of $A \otimes C$ and $B \otimes D$.

$$\text{con-}\otimes\text{-stack} : \{k\,l\,m\,n : \mathbb{N}\}\,(A : k \times l)\,(B : k \times m)$$
$$\to (C : l \times n) \to (D : m \times n)$$
$$\to [\,A\,|\,B\,] \otimes [\,C\;/\;D\,] \equiv A \otimes C \oplus B \otimes D$$

4   NFAs and Parsing with NFAs

We are now in the position to implement a parser generator for regular languages. We parse strings with nondeterministic finite automata and represent them in terms of Boolean matrices.

From now on we therefore use matrices over the commutative semiring of Booleans, with false as zero, disjunction as addition, true as one, conjunction as multiplication.

4.1   Nondeterministic Finite Automata

A Σ-NFA can be defined as a record with four fields.

record NFA : Set **where**
   field ∇ : $\mathbb{N}$
      δ : $\Sigma \to \nabla \times \nabla$
      I : $1 \times \nabla$
      F : $\nabla \times 1$

- ∇ is the *size of the state space*. We do not name states. Instead we identify them with positions in rows and columns of matrices.
- δ specifies the *transition function*. In our implementation δ is a total function from letters of the alphabet to incidence matrices such that for any x : Σ the function call δ x will return an incidence matrix D of size $\nabla \times \nabla$ where $D_{ij} = 1$ iff q_j is a successor of q_i for character x.
- I specifies the *set of initial states*. The initial states can be represented by a $1 \times \nabla$ matrix (row vector) where the element I_{1i} is 1 iff q_i is an initial state.
- F specifies the *set of final states*. The final states can be represented by a $\nabla \times 1$ matrix (column vector) where the element F_{i1} is 1 iff q_i is a final state.

4.2   Running an NFA

Running an NFA on the string $x_0 \ldots x_n$ from a set of states X represented by a row vector can be implemented as the series of multiplications $X \otimes \delta$ nfa $x_0 \otimes \ldots \otimes \delta$ nfa x_n. This computes the row vector of all states reachable from the states X by following the transitions corresponding to the individual letters of the string $x_0 \ldots x_n$.

   run : (nfa : NFA) $\to$ String $\to$ $1 \times \nabla$ nfa $\to$ $1 \times \nabla$ nfa
   run nfa u X = foldl (λ A x $\to$ A $\otimes$ δ nfa x) X u

If we take X to be I and multiply the matrix further with the column vector F of the final states, we get the 1×1 matrix id $\{1\}$ (i.e. [[true]]), if there is an overlap between the states reachable from some initial state and the final states, i.e., if the string is accepted, and null (i.e. [[false]]) otherwise.

   runNFA : NFA $\to$ String $\to$ 1×1
   runNFA nfa u = (run nfa u (I nfa)) $\otimes$ F nfa

4.3   Converting Regular Expressions to NFAs (Parsers)

We have introduced and defined the types RegExp and NFA. We now implement a conversion from RegExp to NFA, which we will use as a parser generator.

$$
\begin{aligned}
&\text{reg2nfa} \ : \ \text{RegExp} \rightarrow \text{NFA} \\
&\text{reg2nfa } \varepsilon &&= \ \varepsilon' \\
&\text{reg2nfa } ('\,a) &&= \ '' \, a \\
&\text{reg2nfa } (e_1 \cup e_2) &&= \ (\text{reg2nfa } e_1) \cup' (\text{reg2nfa } e_2) \\
&\text{reg2nfa } (e^+) &&= \ (\text{reg2nfa } e) \ ^{+\prime} \\
&\text{reg2nfa } (e_1 \cdot e_2) &&= \ (\text{reg2nfa } e_1) \cdot' (\text{reg2nfa } e_2)
\end{aligned}
$$

This function recurses over the regular expression and replaces every constructor with a corresponding operation on NFAs. We describe each case:

$-\ e = \varepsilon$:

$$
\begin{aligned}
\varepsilon' = \textbf{record } \{ \\
\nabla &= 1 \ ; \\
\delta &= \lambda\,x \rightarrow \text{null}; \\
I &= \text{id}; \\
F &= \text{id} \}
\end{aligned}
$$

Clearly, an NFA that accepts only the empty string can be given by one state 0 that is both initial and final.

$-\ e = '\,a$:

In this case, the regular expression describes the single-character string a. So, the corresponding NFA should accept only this string.

$$
\begin{aligned}
'' \, a = \textbf{record } \{ \\
\nabla &= 2; \\
\delta &= \lambda\,x \rightarrow \text{if } a \stackrel{?}{=} x \text{ then } [\,[\,\text{null} \mid \text{id} \,\{1\}\,] \ / \\
&\qquad\qquad\qquad\qquad\qquad [\,\text{null} \mid \text{null} \ ]\,] \\
&\qquad\quad\ \text{else null}; \\
I &= [\,\text{id} \,\{1\} \mid \text{null}\,]; \\
F &= [\,\text{null} \,/\, \text{id} \,\{1\}\,] \}
\end{aligned}
$$

This NFA has two states 0 and 1 such that 0 is an initial and 1 a final state. The transition function compares the character x with the expected character a. If they coincide, then it returns the incidence matrix for the graph with a single edge from 0 to 1, otherwise, the null matrix for the empty graph.

$-\ e = e_1 \cup e_2$:

Recall that strings of both languages must be accepted. So we must run both NFAs.

$$
\begin{aligned}
\text{nfa}_1 \cup' \text{nfa}_2 = \textbf{record } \{ \\
\nabla &= \nabla\,\text{nfa}_1 + \nabla\,\text{nfa}_2; \\
\delta &= \lambda\,x \rightarrow [\,[\,\delta\,\text{nfa}_1\,x \mid \text{null}\ ] \ / \\
&\qquad\qquad\quad [\ \text{null} \mid \delta\,\text{nfa}_2\,x\,]\,]; \\
I &= [\,I\,\text{nfa}_1 \mid I\,\text{nfa}_2\,]; \\
F &= [\,F\,\text{nfa}_1 \,/\, F\,\text{nfa}_2\,] \}
\end{aligned}
$$

The resulting NFA is built of nfa_1 and nfa_2 as follows:

(∇) The state space of the resulting NFA must contain states from nfa_1 and nfa_2. So $\nabla = \nabla\, nfa_1 + \nabla\, nfa_2$.

(δ) The transition function of the resulting NFA is composed of four blocks.
   1. The top left block is the incidence matrix of the first NFA.
   2. The top right block is null. So, no transitions from nfa_1 to nfa_2.
   3. The bottom left block is null. In terms of incidence matrices this signals that there are no transitions from nfa_2 to nfa_1.
   4. The bottom right block is the incidence matrix of the second NFA.

(I) The set of initial states in the resulting NFA is the union of the sets of initial states of nfa_1 and nfa_2.

(F) The set of final states in the resulting NFA is the union of the sets of final states of nfa_1 and nfa_2.

$- e = e'^{+}$:

$$nfa^{+\prime} = \textbf{record}\ \{$$
$$\nabla = \nabla\, nfa;$$
$$\delta = \lambda\, x \to (\text{id} \oplus F\, nfa \otimes I\, nfa) \otimes (\delta\, nfa\, x)$$
$$I = I\, nfa;$$
$$F = F\, nfa\,\}$$

The difference between $nfa^{+\prime}$ and nfa is in δ only. Specifically, we add a new edge from each final state to each successor of an initial state. This is achieved by $F\, nfa \otimes I\, nfa \otimes \delta\, nfa\, x$, where $I\, nfa \otimes \delta\, nfa\, x$ stands for edges reachable from initial state by reading the token x. And $F\, nfa \otimes I\, nfa \otimes \delta\, nfa\, x$ puts an edge from each final state to each successor of an initial state.

$- e = e_1 \cdot e_2$:

$$nfa_1 \cdot{'} nfa_2 = \textbf{record}\ \{$$
$$\nabla = \nabla\, nfa_1 + \nabla\, nfa_2;$$
$$\delta = \lambda\, x \to [\ [\ \delta\, nfa_1\, x\ |\ F\, nfa_1 \otimes I\, nfa_2 \otimes \delta\, nfa_2\, x\ ]\ /$$
$$[\quad \text{null}\quad |\qquad \delta\, nfa_2\, x \qquad\quad]\ ];$$
$$I = [\, I\, nfa_1\ |\ \text{null}\,];$$
$$F = [\, F\, nfa_1 \otimes I\, nfa_2 \otimes F\, nfa_2\ /\ F\, nfa_2\,]\}$$

The fields of nfa_1 and nfa_2 are combined in the following way:

(∇) The state space of the resulting NFA consists of the disjoint union of state spaces of nfa_1 and nfa_2, i.e. $\nabla = \nabla\, nfa_1 + \nabla\, nfa_2$.

(δ) The transition function constructs incidence matrices from four blocks.
   1. The top left block contains the incidence matrix of nfa_1. Hence, the transition relation between the states of nfa_1 is not changed.
   2. The top right block is $F\ nfa_1 \otimes I\, nfa_2 \otimes \delta\, nfa_2\, x$. This expression constructs an incidence matrix with transitions from all final states of nfa_1 to all successors of initial states in nfa_2. In other words, it says that upon reaching a final state of nfa_1, it is time to transition to nfa_2.
   3. The bottom left block is null. Hence, there are no transitions from nfa_2 back to nfa_1.

4. The bottom right block consists of the transition function of nfa_2. So, the transitions between the states of nfa_2 are unchanged.

(**I**) The resulting NFA's initial states must contain only initial states of nfa_1.

(**F**) Clearly, the final states of the resulting NFA must contain all final states of the second NFA. But what if the second NFA accepts the empty string? Then the resulting NFA must also accept the language of the first NFA, hence, contain all its final states. The desired behaviour is achieved by

$$F = [\, F\ nfa_1 \otimes I\ nfa_2 \otimes F\ nfa_2\ /\ F\ nfa_2\,]$$

The top block of this column vector is equal to either the empty vector or $F\ nfa_1$ depending on the result of $I\ nfa_2 \otimes F\ nfa_2$. The latter multiplication is equal to id $\{1\}$, if nfa_2 accepts the empty string, and null, if it does not. The bottom block of F is always equal to $F\ nfa_2$. So, the desired behaviour is achieved.

4.4   Correctness

The correctness of an NFA with respect to a regular expression consists of completeness and soundness. Completeness guarantees that every string matching a regular expression will be accepted by the NFA. However, completeness alone is not enough, since an NFA accepting all strings is also complete. Soundness in turn guarantees that, if the NFA accepts, the string matches the regular expression. Similarly to the case of completeness, soundness is not sufficient alone, because an NFA rejecting every string is sound.

Completeness. Completeness states that, if a string is matched by the given regular expression, then the constructed NFA reg2nfa e accepts it.

complete : (e : RegExp) $\rightarrow$ (u : String) $\rightarrow$ u ▶ e
$\rightarrow$ runNFA (reg2nfa e) u $\equiv$ id $\{1\}$

We prove this theorem by induction on the proof of u ▶ e. Hence, all shapes of parse trees must be considered. We describe only the cases for union and concatenation, the others can be found in the Agda code.

Union In this case e $=$ $e_1 \cup e_2$.

complete ($e_1 \cup e_2$) u parseTree $= \ldots$

We start with pattern matching on parseTree.

complete ($e_1 \cup e_2$) u (unionl parseTree$'$) $= \ldots$
complete ($e_1 \cup e_2$) u (unionr parseTree$'$) $= \ldots$

This yields two cases for the last rule used in the parse tree: unionl or unionr. Since both cases are proved in the same way, we describe only the first one.

The main idea is to show that a run of $nfa_1 \cup' nfa_2$ can be split into runs of nfa_1 and nfa_2. It is proved by the lemma union-split:

union-split : $(nfa_1 \ nfa_2 : NFA) (u : String) (X_1 : 1 \times \_) (X_2 : 1 \times \_)$
$\quad \to \mathsf{run} \ (nfa_1 \cup' nfa_2) \ s \ [\ X_1 \ | \ X_2 \ ] \equiv [\ \mathsf{run} \ nfa_1 \ s \ X_1 \ | \ \mathsf{run} \ nfa_2 \ s \ X_2 \ ]$

Next, by using the previously described property

con-$\otimes$-stack : $\{k \ l \ m \ n : \mathbb{N}\} (A : k \times l) (B : k \times m) (C : l \times n) (D : m \times n)$
$\quad \to [\ A \ | \ B \ ] \otimes [\ C \ / \ D \ ] \equiv A \otimes C \oplus B \otimes D$

we complete the proof:

$$\cfrac{\cfrac{\cfrac{\overline{\mathsf{id} \ \{1\} \oplus \mathsf{id} \ \{1\} \equiv \mathsf{id} \ \{1\}} \ \text{Boolean arithm.}}{(\mathsf{run} \ nfa_1 \ s \ X_1 \otimes (F \ nfa_1)) \oplus (\mathsf{run} \ nfa_2 \ s \ X_2 \otimes (F \ nfa_2)) \equiv \mathsf{id} \ \{1\}} \ \text{IHs}}{[\ \mathsf{run} \ nfa_1 \ s \ X_1 \ | \ \mathsf{run} \ nfa_2 \ s \ X_2 \ ] \otimes [\ F \ nfa_1 \ / \ F \ nfa_2 \ ] \equiv \mathsf{id} \ \{1\}} \ \text{con-}\otimes\text{-stack}}{\mathsf{run} \ (nfa_1 \cup' nfa_2) \ s \ [\ X_1 \ | \ X_2 \ ] \otimes [\ F \ nfa_1 \ / \ F \ nfa_2 \ ] \equiv \mathsf{id} \ \{1\}} \ \text{union-split}$$

Plus In this case $e = e'^+$.

complete (e'^+) u parseTree $= \ldots$

Pattern matching on parseTree yields two cases. We examine them in turn:

1. In the first case the last rule of the parse tree is plus1.

   complete (e'^+) u (plus1 parseTree$')$ $= \ldots$

   Recall that plus1 is a constructor which states that, if u is in language of e', then it is also in language of (e'^+). Hence, the main lemma for this case can be stated as

   plus-weak : $(nfa : NFA) (u : String) (X : 1 \times (\nabla \ nfa))$
   $\quad \to \mathsf{run} \ nfa \ u \ X \otimes F \ nfa \equiv \mathsf{id} \ \{1\}$
   $\quad \to \mathsf{run} \ (nfa^{+'}) \ u \ X \otimes F \ nfa \equiv \mathsf{id} \ \{1\}$

   It is proved by induction on the length of the string u.

2. In the second case the last rule of the parse tree is plus2.

   complete $(e'^+) \ . \ (u_1 +\!\!+ u_2)$ (plus2 $\{u_1\} \ \{u_2\}$ tree$_1$ tree$_2$) $= \ldots$

Note that string u is now split into u_1 and u_2 such that u_1 is in the language of e' and u_2 is in the language of e'^+. We must prove that $u_1 +\!\!+ u_2$ are in the language of e'^+. To do so, we first introduce some useful lemmas.

– We show that run on $u_1 +\!\!+ u_2$ can be split into a run on u_1 and a run on u_2.

   plus-split : $(nfa : NFA) (u_1 \ u_2 : String)$
   $\quad \to \mathsf{run} \ (nfa^{+'}) \ (u_1 +\!\!+ u_2) \ (I \ nfa) \otimes F \ nfa$
   $\quad\quad \equiv \mathsf{run} \ (nfa^{+'}) \ u_2 \ (\mathsf{run} \ (nfa^{+'}) \ u_1 \ (I \ nfa)) \otimes F \ nfa$

– We also show that, if the automaton $nfa^{+'}$ accepts a string from the initial states, then it will also accept that string from any final state.

   plus-fin : $(nfa : NFA) (u : String) (X : 1 \times \nabla \ nfa)$
   $\quad \to \mathsf{run} \ (nfa^{+'}) \ u \ (I \ nfa) \otimes F \ nfa \equiv \mathsf{id} \ \{1\}$
   $\quad \to X \otimes F \ nfa \equiv \mathsf{id} \ \{1\}$
   $\quad \to \mathsf{run} \ (nfa^{+'}) \ u \ X \otimes F \ nfa \equiv \mathsf{id} \ \{1\}$

Finally, the big picture of the proof looks like this:

$$
\cfrac{
  \cfrac{
    \text{run (nfa}^{+\prime}\text{) } u_2 \text{ (I nfa)} \otimes \text{F nfa} \equiv \text{id } \{1\}
    \qquad
    \cfrac{
      \cfrac{\text{run nfa } u_1 \text{ (I nfa)} \otimes \text{F nfa} \equiv \text{id } \{1\}}{\text{run (nfa}^{+\prime}\text{) } u_1 \text{ (I nfa)} \otimes \text{F nfa} \equiv \text{id } \{1\}}\ \text{IH}
    }{}\ \text{plus-weak}
  }{\text{run (nfa}^{+\prime}\text{) } u_2 \text{ (run (nfa}^{+\prime}\text{) } u_1 \text{ (I nfa))} \otimes \text{F nfa} \equiv \text{id } \{1\}}
}{\text{run (nfa}^{+\prime}\text{) } (u_1 +\!\!+ u_2) \text{ (I nfa)} \otimes \text{F nfa} \equiv \text{id } \{1\}}\ \text{plus-split}
$$

(IH, plus-fin)

Soundness. Showing that our NFA generation is sound is more complicated, but also more interesting. Let us look at the signature of the soundness theorem:

```
sound : (e : RegExp) → (u : String)
  → runNFA (reg2nfa e) u ≡ id {1} → u ▶ e
```

It states that, if the NFA accepts a string, then it matches the regular expression. sound is a proposition, but it is also a type! Its proof is a function that delivers parse trees. We prove this theorem by induction on the argument e : RegExp.

As in case of completeness, our aim is to explain the high-level ideas of the proof. We skip most of the details and describe only two cases.

Single character. This case is interesting because it demonstrates the essence of all soundness cases. We are given an accepting run of the automaton. Using this fact we must construct a parse tree. However, most of the cases generated by pattern matching are discharged by showing that they contradict with the accepting run we have at our disposal.

```
sound (′ a) u run = ...
```

We pattern match on the string u and examine three different cases in turns:

1. u *is the empty string.*

   ```
   sound (′ a) u run = ...
   ```

   We must show that in this case it is impossible to give run. We do so by pattern matching on run and the rest is taken care of by Agda's type checker.
2. u *is a string of one symbol.*

   ```
   sound (′ a) (x :: []) run = ...
   ```

   This is the only situation when the automaton can accept u. Still, we must check if x is equal to a. We case analyse on the decidable equality of a and x:

   ```
   sound (′ a) (x :: []) run with a ≟ x
   sound (′ a) (x :: []) | eq = ...
   sound (′ a) (x :: []) | neq = ...
   ```

   Then two cases must be discharged.
   (a) x *is equal to* a.
       This is exactly the case when the automaton finishes in the accepting state. To close this case, we rewrite the context using a ≡ x and provide the constructor symb {x} as the required proof.

(b) x *is not equal to* a.
Then Agda computes that runNFA (reg2nfa (' a)) [x] is equal to null, but this contradicts the assumptions. Hence, the case is discharged.

3. u *is a string of two or more characters.*

$$\text{sound } ('\, a)\ (x_1 :: x_2 :: xs)\ run\ =\ ...$$

Similarly to the first case, our goal is to show that runNFA (reg2nfa (' a)) u will never accept a string consisting of two or more characters. This is done by observing the fact that, even if the automaton reaches the second state by reading the first character, then by reading the second character the automaton will lose all active states, since there are no transitions going out of the second state. Hence, runNFA (reg2nfa (' a)) $(x_1 :: x_2 :: xs)$ cannot return id $\{1\}$. Therefore, the case is discharged.

Concatenation. We are in the case

$$\text{sound } (e_1 \cdot e_2)\ u\ run\ =\ ...$$

Fortunately, there is only one possible constructor for this case in the parsing relation, namely con. It states that to prove $u \blacktriangleright e_1 \cdot e_2$ we must show that $u_1 \blacktriangleright e_1$ and $u_2 \blacktriangleright e_2$ for some splitting of u into u_1 and u_2. Hence, we must be able to extract from run two shorter runs and use them to get $u_1 \blacktriangleright e_1$ and $u_2 \blacktriangleright e_2$ by induction hypothesis. To express this in Agda we use sigma-types, corresponding to existentials.

$$
\begin{aligned}
&\text{cons-split} : (nfa_1\ nfa_2 : NFA)\ (u : String) \\
&\quad \rightarrow run\ (nfa_1 \cdot' nfa_2)\ u\ [\ I\ nfa_1\ |\ null\ ] \\
&\qquad \otimes [\ F\ nfa_1 \otimes I\ nfa_2 \otimes F\ nfa_2\ /\ F\ nfa_2\ ] \equiv id\ \{1\} \\
&\quad \rightarrow \exists\,[u_1 : String]\ \exists\,[u_2 : String]\ u \equiv u_1 +\!\!+ u_2 \\
&\qquad \wedge\ run\ nfa_1\ u_1\ (I\ nfa_1) \otimes F\ nfa_1 \equiv id\ \{1\} \\
&\qquad \wedge\ run\ nfa_2\ u_2\ (I\ nfa_2) \otimes F\ nfa_2 \equiv id\ \{1\}
\end{aligned}
$$

Note that we split the string u into u_1 and u_2, but must also provide a proof that $u \equiv u_1 +\!\!+ u_2$. To prove it, we will need a variant of cons-split.

$$
\begin{aligned}
&\text{cons-split-state} : (nfa_1\ nfa_2 : NFA)\ (x : \Sigma)\ (u : String)\ (X : 1 \times \_) \\
&\quad \rightarrow run\ (nfa_1 \cdot' nfa_2)\ (x :: u)\ [\ X\ |\ null\ ] \otimes (null +\!\!+ F\ nfa_2) \equiv id\ \{1\} \\
&\quad \rightarrow \exists\,[u_1 : String]\ \exists\,[u_2 : String]\ (x :: u) \equiv u_1 +\!\!+ u_2 \\
&\qquad \wedge\ run\ nfa_1\ u_1\ X \otimes F\ nfa_1 \equiv id\ \{1\} \\
&\qquad \wedge\ run\ nfa_2\ u_2\ (I\ nfa_2) \otimes F\ nfa_2 \equiv id\ \{1\}
\end{aligned}
$$

The important differences between cons-split and cons-split-state are the following:

- In cons-split, the given run starts from $[\ I\ nfa_1\ |\ null\ ]$, where I nfa_1 is the set of initial states of nfa_1, but in cons-split-state, we use a more general variant $[\ X\ |\ null\ ]$, where X is a given parameter.
- In cons-split, the run of the automaton can terminate either in the final states of nfa_1 or in the final states of nfa_2, but in cons-split-state, the set of final states is limited to those of nfa_2.

– In contrast with cons-split-state, the string xs can be empty in cons-split.

The restrictions present in cons-split-state force it to address the specific and most complicated case when the given run starts in some states of nfa_1, but terminates in nfa_2. The lemma states that in this case we can break an accepting run of $nfa_1 \cdot' nfa_2$ down into two smaller accepting runs.

We will not describe here how cons-split-state is proved. Instead we show how to reduce cons-split to cons-split-state. To do so, we perform multiple case analyses. First, we distinguish the empty string case from the cons-case.

1. If $u \equiv [\,]$, then $u_1 \equiv [\,]$ and $u_2 \equiv [\,]$ and we must show that $I\ nfa_1 \otimes F\ nfa_1 \equiv$ id $\{1\}$ and $I\ nfa_2 \otimes F\ nfa_2 \equiv$ id $\{1\}$. Both proofs are easily derived from the given premise:

$$[\,I\ nfa_1\ |\ \mathsf{null}\,] \otimes [\,F\ nfa_1 \otimes (I\ nfa_2 \otimes F\ nfa_2)\ /\ F\ nfa_2\,] \equiv \mathsf{id}\ \{1\}$$

2. If $u \equiv x :: xs$, then we perform an additional case analysis: whether the second automaton has final states among its initial states:
   (a) $I\ nfa_2 \otimes F\ nfa_2 \equiv$ null. In this case, the problem is clearly an instance of cons-split-state.
   (b) $I\ nfa_2 \otimes F\ nfa_2 \equiv$ id $\{1\}$. In this case, we perform a third level of case analysis: whether the first automaton nfa_1 accepts the whole string.
      i. run $nfa_1\ u\ (I\ nfa_1) \otimes F\ nfa_1 \equiv$ id $\{1\}$.
         If we take $u_1 \equiv u$ and $u_2 \equiv [\,]$, then case is immediately discharged.
      ii. run $nfa_1\ u\ (I\ nfa_1) \otimes F\ nfa_1 \equiv$ null.
         In this case, we need an additional lemma

$$\mathsf{consnd2}\ :\ (nfa_1\ nfa_2\ :\ \mathsf{NFA})\ (u\ :\ \mathsf{String})$$
$$\to (\mathsf{run}\ nfa_1\ u\ (I\ nfa_1)) \otimes F\ nfa_1 \equiv \mathsf{null}$$
$$\to (\mathsf{run}\ (nfa_1 \cdot' nfa_2)\ u\ [\,I\ nfa_1\ |\ \mathsf{null}\,]) \otimes [\,F\ nfa_1\ /\ F\ nfa_2\,]$$
$$\equiv (\mathsf{run}\ (nfa_1 \cdot' nfa_2)\ u\ [\,I\ nfa_1\ |\ \mathsf{null}\,]) \otimes [\,\mathsf{null}\ /\ F\ nfa_2\,]$$

   It states that, if the first automaton does not accept the whole string, then running the automaton $nfa_1 \cdot' nfa_2$ with the final states of nfa_1 is equivalent to running the automaton $nfa_1 \cdot' nfa_2$ without the final states of nfa_1. So, this branch of cons-split is also reduced to cons-split-state.

sound returns one parse tree. If there are multiple parse trees for a single string, it prefers unionl over unionr and also plus1 over plus2. It also never invokes plus2 with the first string empty, as in this case no progress is made. In fact makes sense to restrict the first string argument of plus2 to be a cons-string—this removes the possibility for a string to have an infinite number of parse trees.

4.5 Parsing

Correctness of reg2nfa turns the NFA for a regular expression effectively into a parser. Since we can decide whether a 1×1 matrix contains true or false, using sound and complete, for any string we can have a parse tree or a proof of that there cannot be one.

$$\text{parse} : (e : \text{RegExp}) \to (u : \text{String}) \to u \blacktriangleright e \uplus (u \blacktriangleright e \to \bot)$$

5   Related Work

Braibant and Pous [1] implement a Coq tactic for deciding equational theory of Kleene algebras. The work is based on checking if two regular expressions represent the same language. This is done in four steps. First, regular expressions are converted into ε-NFAs. Then ε-transitions are removed to get NFAs. Next, determinisation procedure converts NFAs into DFAs. Finally, they check whether the DFAs are equivalent. This results in a general decision procedure for Kleene algebras. In principle, it can be used to solve the recognition problem: to check whether a word w is in the language defined by a regular expression r, we can check if w ∪ r and r define the same language. But as this requires go through all four steps for each query, it is impractical.

In contrast, we focus only on the recognition problem. The main difference of our work is that we convert regular expressions directly into NFAs without ε-transitions. This makes the overall process simpler, since then we do not have to find ε-closures and remove ε-transitions afterward and pepper all that, as Braibant and Pous confirm, with quite tricky proofs of correctness.

Many works formalizing recognition of regular languages are based on the concept of the derivative of a language [5,2,3,7]. This is not accidental, since derivatives have nice algebraic properties which make them attractive for a formal development.

It seems that the alternative approach of converting regular expressions to finite automata is believed to be a messy procedure with too much low-level detail involved. For instance, Krauss and Nipkow [5] discuss the link between regular expressions and finite automata in the context of lexing, but point out that encoding finite automata as graphs involves a painful amount of detail and a higher-level approach is desirable.

Wu et al. [9] show how to formalize the Myhill-Nerode theorem by only using regular expressions and the motivation behind this approach is again to avoid the trouble of representing automata as graphs.

We have shown that conversion from regular expressions to finite state automata encoded as Boolean matrices can be done in a concise and high-level way by using block operations on matrices. Proofs in this setting benefit significantly from lemmas about block operations.

6   Conclusion

We presented an implementation of a certified parser generator for regular languages. In particular we showed how to reduce operations and proofs about NFAs into linear-algebra operations and proofs. The practical part of this work was divided into two parts. In the first part, we implemented a generic library for matrices, focusing on block operations. Besides an implementation of basic matrix operations, we also proved many well-known properties of these functions.

In the second part of the practical work, we implemented a transformation of regular expressions to NFAs and proved its correctness. A string is parsed by checking whether the NFA accepts it, as soundness turns the positive answer into a parse tree while completeness can be used to conclude impossibility of a parse tree in the negative case.

This work could be continued in several directions. The implemented framework (the matrix library, RegExp and NFA libraries) can be used to formalize different aspects of regular language theory: minimizing NFAs, showing equivalence of regular expressions, conversion of NFAs to regular expressions, etc.

One variation on the theme of this work would be to consider matrices over natural numbers instead of Booleans. This would allow counting of accepting runs of an NFA (paths from an initial to a final state). Soundness and completeness of the transformation of regular expressions to NFAs would establish a bijection between the parse trees of a given string and the accepting runs of the NFA.

Acknowledgements. This work was supported by the ERDF funded CoE project EXCS, the Estonian Ministry of Education and Research target-financed theme no. 0140007s12 and the Estonian Science Foundation grant no. 9475.

References

1. Braibant, T., Pous, D.: An efficient Coq tactic for deciding Kleene algebras. In: Kaufmann, M., Paulson, L.C. (eds.) ITP 2010. LNCS, vol. 6172, pp. 163–178. Springer, Heidelberg (2010)
2. Coquand, T., Siles, V.: A decision procedure for regular expression equivalence in type theory. In: Jouannaud, J.-P., Shao, Z. (eds.) CPP 2011. LNCS, vol. 7086, pp. 119–134. Springer, Heidelberg (2011)
3. Danielsson, N.A.: Total parser combinators. In: Proc. of 15th ACM SIGPLAN Int. Conf. on Functional Programming, ICFP 2010, pp. 285–296. ACM (2010)
4. Grune, D.: Parsing Techniques: A Practical Guide, 2nd edn. Springer (2010)
5. Krauss, A., Nipkow, T.: Proof pearl: Regular expression equivalence and relation algebra. J. of Autom. Reasoning 49(1), 95–106 (2012)
6. Macedo, H.D., Oliveira, J.N.: Typing linear algebra: A biproduct-oriented approach. Sci. of Comput. Program. 78(11), 2160–2191 (2013)
7. Morrisett, G., Tan, G., Tassarotti, J., Tristan, J.B., Gan, E.: Rocksalt: better, faster, stronger SFI for the x86. In: Proc. of 33rd ACM SIGPLAN Conf. on Programming Language Design and Implementation, PLDI 2012, pp. 395–404. ACM (2012)
8. Norell, U.: Dependently typed programming in Agda. In: Koopman, P., Plasmeijer, R., Swierstra, D. (eds.) AFP 2008. LNCS, vol. 5832, pp. 230–266. Springer, Heidelberg (2009)
9. Wu, C., Zhang, X., Urban, C.: A formalisation of the Myhill-Nerode theorem based on regular expressions (Proof pearl). In: van Eekelen, M., Geuvers, H., Schmaltz, J., Wiedijk, F. (eds.) ITP 2011. LNCS, vol. 6898, pp. 341–356. Springer, Heidelberg (2011)

Nonfree Datatypes in Isabelle/HOL
Animating a Many-Sorted Metatheory

Andreas Schropp[1,3] and Andrei Popescu[1,2]

[1] Technische Universität München, Germany
[2] Institute of Mathematics Simion Stoilow of the Romanian Academy
[3] COMSA GmbH

Abstract. Datatypes freely generated by their constructors are well supported in mainstream proof assistants. Algebraic specification languages offer more expressive datatypes on axiomatic means: nonfree datatypes generated from constructors modulo equations. We have implemented an Isabelle/HOL package for nonfree datatypes, without compromising foundations. The use of the package, and its nonfree iterator in particular, is illustrated with examples: bags, polynomials and λ-terms modulo α-equivalence. The many-sorted metatheory of nonfree datatypes is formalized as an ordinary Isabelle theory and is animated by the package into user-specified instances. HOL lacks a type of types, so we employ an ad hoc construction of a universe embedding the relevant parameter types.

1 Introduction

Free datatypes are at the heart of logic and computer science and are well supported in most proof assistants. Equational theories over them are often less convenient. Finite multisets or "bags" are a popular construction and can be regarded as finite lists modulo the permutation of elements. This results in the following nonfree datatype of bags over the type α, with "empty bag" and "bag-insert" constructors:

$$\text{datatype } \alpha \text{ bag} = \text{BEmp} \mid \text{BIns } \alpha \, (\alpha \text{ bag})$$
$$\text{where } \text{BIns } a_1 \, (\text{BIns } a_2 \, B) = \text{BIns } a_2 \, (\text{BIns } a_1 \, B)$$

where the equation, left-commutativity, is (implicitly) universally quantified over a_1, a_2 and B. Bags are thus specified by list-like constructors and an identification of differently constructed terms based on (all consequences of) the indicated equation.

This style of definition is standard in the world of algebraic specifications [6, 7]. Nonfree datatypes and suitable recursors for them allow one to express many concepts at the appropriate level of abstraction, as opposed to encoding them in more concrete free types. For instance, bags are encodable as lists, but the price is a loss of abstraction, hence more error-prone processing methods. This is equally true for programming [23] and theorem proving. However, mainstream proof assistants based on type theory [1,5] or higher-order logic (HOL) [10, 17] currently do not provide mechanisms for specifying nonfree datatypes directly.

In HOL-based provers, such as our favorite one Isabelle/HOL [17], datatypes are not integrated into the logic, but are provided as a definitional layer on top of the logical primitives. Given a user specification, a *definitional package* produces the appropriate

G. Gonthier and M. Norrish (Eds.): CPP 2013, LNCS 8307, pp. 114–130, 2013.
© Springer International Publishing Switzerland 2013

types, terms and theorems, including induction and recursion schemes. In this paper, we present a definitional package in Isabelle/HOL for nonfree datatypes. Its expressiveness goes a little beyond standard algebraic specifications (typically, equational theories), allowing Horn clauses over equations *and predicates*.

Our package also contributes a new methodology for addressing an old problem: the incomplete, dynamic nature of typical package certification. Indeed, the mathematics behind a datatype package requires reasoning about arbitrary numbers of types and operators on them. This is not possible generically inside HOL, because it lacks a type of types. The constructions performed by HOL packages are usually certified dynamically, for each particular instance that the user requests. Our package essentially limits the amount of dynamic certification to a minimum of uniform facts concerning the transfer across isomorphisms.[1] The nontrivial part of the constructions is statically certified in a metatheory formalized in Isabelle. It is parameterized on a collection of sets over a fixed "universe" type, instead of a collection of types. This "universe" type is instantiated by ad hoc sums over the relevant types when animating the metatheory.

The paper is structured in two main parts. The first part, consisting of §2, illustrates the package by examples—bags, polynomials and λ-terms modulo α-equivalence—carefully chosen to illustrate different aspects and features of the package: nonfree recursion, interaction with Isabelle's type classes, and predicate-based Horn specifications. We also hope that these examples help popularize nonfree recursion, a standard technique that is not so standard in proof assistants. The second part describes the package design and architecture: §3 illustrates with an example the actual steps that are automated by the package, §4 presents the formalization of the metatheory up to the construction of the initial model, and §5 shows how the metatheory is automatically instantiated to user-specified datatypes. The package is compatible with Isabelle2013 and is publicly available [21].

Preliminaries. In this paper, by HOL we mean classical higher-order logic with Hilbert choice, schematic polymorphism and the typedef principle. The Isabelle/HOL proof assistant [17] is an implementation of HOL enhanced with Haskell-style type classes [9] and locales [15]. Types in HOL are either atomic types such as unit, nat and bool, or type variables α, β, or built from these using type constructors. We use postfix notation for type constructors, e.g., α list and α set denote the list and powerset types over α. Polymorphic types are not syntactically distinguished—e.g., α list also denotes the polymorphic type $\forall \alpha.\ \alpha$ list. We write $\alpha \to \beta$, $\alpha + \beta$, and $\alpha \times \beta$ for the function-space, sum and product types, respectively. All types are nonempty.[2] New types are introduced with the typedef principle by carving out nonempty subsets of existing types. A term t of type τ is indicated as $t : \tau$. (We prefer the more mathematical notations $\alpha \to \beta$ and $t : \tau$ to the Isabelle notations $\alpha \Rightarrow \beta$ and $t :: \tau$.)

Type classes are an overloading mechanism wired into Isabelle's type system. A type class C specifies for its member types, $\tau : C$, constants of composite types containing τ and axioms for these constants. Typical cases are the algebraic classes, e.g., τ : semigroup means that there exists an operation $+ : \tau \to \tau \to \tau$ assumed associative.

---

[1] Additionally we employ rewriting steps, forward chaining of facts, well-sortedness checking rules, and finite datatypes and functions over them to construct the signature instantiation.

[2] HOL is not following the propositions-as-types paradigm, so this is not troublesome.

Isabelle locales are essentially proof contexts, fixing type and term variables with assumptions. A locale can be instantiated by providing concrete types and terms for its type and term variables and then discharging its assumptions. This makes the instantiated content of the locale available in the outer context.

2 The Package in Action

Here we present the package and its different features by examples. We start with the datatype of bags, whose single-equation specification makes it easy to present in detail the package's contract: what is expected from the user and what is produced in response.

2.1 Bags

The declaration of the datatype of bags from §1 produces the type α bag and the following polymorphic constants:

– the constructors BEmp : α bag and BIns : $\alpha \to \alpha$ bag $\to \alpha$ bag,
– the iterator iter_bag : $\beta \to (\alpha \to \beta \to \beta) \to \alpha$ bag $\to \beta$.

In addition, several characteristic theorems are derived. They include facts also available for standard free datatypes:
– Case distinction: $(B = \text{BEmp} \longrightarrow \varphi) \wedge (\forall a\, C.\, B = \text{BIns}\, a\, C \longrightarrow \varphi) \longrightarrow \varphi$
– Induction: $\varphi\, \text{BEmp} \wedge (\forall a\, B.\, \varphi\, B \to \varphi\, (\text{BIns}\, a\, B)) \longrightarrow (\forall B.\, \varphi\, B)$

Note that the injectivity of the constructors, here,

$$\text{BIns}\, a_1\, B_1 = \text{BIns}\, a_2\, B_2 \longrightarrow a_1 = a_2 \wedge B_1 = B_2,$$

is not among these facts, since it does not hold for nonfree datatypes.

The interesting derived theorems are those specific to nonfree datatypes:
– The characteristic equation(s) specified by the user:

   BIns a_1 (BIns $a_2\, B$) $=$ BIns a_2 (BIns $a_1\, B$)

– The recursion principle, consisting of conditional equations for iteration:

   bag_alg $E\, I \to$ iter_bag $E\, I\, \text{BEmp} = E$
   bag_alg $E\, I \to (\forall a\, B.\, \text{iter\_bag}\, E\, I\, (\text{BIns}\, a\, B) = I\, a\, (\text{iter\_bag}\, E\, I\, B))$

where bag_alg $E\, I$ is the predicate $\forall a_1\, a_2\, b.\, I\, a_1\, (I\, a_2\, b) = I\, a_2\, (I\, a_1\, b)$.

Thus, the package produces a type α bag that satisfies the specified equation. In addition, α bag is *initial* among the algebras $(\beta,\, E : \beta,\, I : \alpha \to \beta \to \beta)$ satisfying the equation (with E and I replacing BEmp and BIns) as expressed by the predicate bag_alg $E\, I$. This means that from α bag to any such algebra there exists precisely one morphism, i.e., function commuting with the algebra operations. The existence of a morphism is expressed by the iteration equations: given such an algebra, the morphism is iter_bag $E\, I$. Its uniqueness is given by the induction principle.

As with other definitional packages for recursion, the user does not needs to employ the iterator directly—the package allows the user to inline I and E in the desired recursive equations. For example, the following specifies the map function for bags:

   nonfreerec bag_map : $(\alpha \to \beta) \to \alpha$ bag $\to \beta$ bag where

   bag_map f BEmp $=$ BEmp

   bag_map f (BIns $a\, B$) $=$ BIns $(f\, a)$ (bag_map $f\, B$)

In response to this command, the package does the following (for a fixed $f : \alpha \to \beta$):
(1) identifies E and I as being BEmp : β bag and $(\lambda a.\ \text{BIns}\ (f\ a)) : \alpha \to \beta$ bag $\to \beta$ bag;
(2) defines bag_map $f = $ iter_bag $E\ I$;
(3) prompts the user to discharge the goal bag_alg $E\ I$;
(4) infers the desired *unconditional* equations stated in the nonfreerec declaration from the conditional equations for iter_bag and the fact proved at step (3).

Thus, the user obtains the desired simplification rules for the newly introduced bag_map after discharging the bag_alg goal, here,

$$\text{BIns}\ (f\ a_1)\ (\text{BIns}\ (f\ a_2)\ B) = \text{BIns}\ (f\ a_2)\ (\text{BIns}\ (f\ a_1)\ B)$$

which is immediate from the characteristic equation for β bag.

This complication, of having to discharge goals that imply well-definedness of a function definition, is inherent in the nature of quotiented types and is shared with the quotient and nominal packages [11, 13, 14]. For this paper's examples, the conditions are easy to discharge by simplification (but this cannot be guaranteed in general). This is also the case for the sum of a numeric function over the elements of a bag:

$$\text{nonfreerec sum} : (\alpha \to \text{nat}) \to \alpha \text{ bag} \to \text{nat where}$$

$$\text{sum}\ f\ \text{BEmp} = 0$$

$$\text{sum}\ f\ (\text{BIns}\ a\ B) = \text{sum}\ f\ B + f\ a$$

which yields the goal $(m + f\ a_1) + f\ a_2 = (m + f\ a_2) + f\ a_1$. It is discharged using associativity and commutativity of $+$ on nat, which means that the definition generalizes: we can replace nat with the type class member β : comm_monoid_add, covering all types equipped with a commutative monoid structure $(\beta, 0, +)$. The multiplicity of an element in a bag, mult : $\alpha \to \alpha$ bag $\to$ nat is obtained as mult $a\ B = $ sum $(\lambda a'.\ \text{if}\ a = a'\ \text{then 1 else 0})\ B$.

2.2 Algebra

The package can be used to streamline algebraic constructions. The following example builds the ring of polynomials over a commutative ring α with variables in β, where Sc is the embedding of scalars yielding Sc 0 as the zero polynomial and Var gives the polynomial variables.

> datatype $(\alpha : \text{comm\_ring}, \beta)$ poly $= $ Sc α | Var β | Uminus $((\alpha, \beta)$ poly$)$ |
>   Plus $((\alpha, \beta)$ poly$)\ ((\alpha, \beta)$ poly$)$ | Times $((\alpha, \beta)$ poly$)\ ((\alpha, \beta)$ poly$)$
> where $(-a_1) = a \longrightarrow$ Uminus $(\text{Sc}\ a_1) = \text{Sc}\ a$
>   and $a_1 + a_2 = a \longrightarrow$ Plus $(\text{Sc}\ a_1)\ (\text{Sc}\ a_2) = \text{Sc}\ a$
>   and $a_1 * a_2 = a \longrightarrow$ Times $(\text{Sc}\ a_1)\ (\text{Sc}\ a_2) = \text{Sc}\ a$
>   and Plus $(\text{Sc}\ 0)\ P = P$
>   and Plus $(\text{Plus}\ P_1\ P_2)\ P_3 = $ Plus P_1 $(\text{Plus}\ P_2\ P_3)$
> *** Etc.: All the commutative-ring axioms for Plus, Times, Sc 0 ***

This example illustrates the nontrivial use of type class annotations in the datatype declaration: since α is a ring, it provides operations $*, +, 0$, which are used in the definition of the new type. Type class constraints in polymorphic datatype specifications are already present in Isabelle's standard datatype package, but only serve as a syntactic constraint there. The feature is essential here for performing universal extensions over an unspecified algebraic structure: we need to form a type depending on its operations.

The first three clauses ensure that the restrictions of polynomial inverse, addition and multiplication to scalars, collapse to the scalar operations $-$, $+$ and $*$. They illustrate the use of parameter variables a_1, a_2, a from type α. Strictly speaking, each of the clauses forms an infinite family of Horn clauses, indexed by $a, a_1, a_2 : \alpha$. One may employ any condition on the parameters, not just equality as here.

This direct definition of polynomials can replace the tedious standard construction based on lists. By its characteristic equations, (α, β) poly forms a commutative ring if α does and we can register this with the type-class system. Universality is established by an operator that extends morphisms $f : (\alpha : \text{comm\_ring}) \to (\gamma : \text{comm\_ring})$ (assumed to commute with $+, *, 0$) and variable interpretations $g : \beta \to \gamma$, to morphisms ext $f\, g$: (α, β) poly $\to \gamma$. In the context of such f and g, we define ext by simply writing down its desired interaction with the polynomial operators:

$$\text{nonfreerec ext} : (\alpha, \beta) \text{ poly} \to \gamma \text{ where}$$

$$\text{ext (Sc } a) = f\, a \qquad \text{ext (Var } b) = g\, b \qquad \text{ext (Uminus } P) = -\text{ext } P$$
$$\text{ext (Plus } P\, Q) = \text{ext } P + \text{ext } Q \qquad \text{ext (Times } P\, Q) = \text{ext } P * \text{ext } Q$$

where simplification with the ring axioms of γ and the morphism axioms of f immediately discharges the goals (resulting from the nonfreerec command). Polynomial evaluation is obtained from ext taking $f = \text{id}$.

2.3 λ-Terms Modulo α-Equivalence

Next we discuss a less standard example—λ-terms modulo α-equivalence—which employs the full expressive power of the package, combining parameter conditions with Horn predicates. This type can be specified as the initial model of a Horn theory if we factor in the freshness predicate and at least one of the substitution and swapping operators [18, 19]. In particular, the following provides (a type isomorphic to) the λ-calculus terms (modulo α-equivalence) over variables in α and constants in β, including the syntactic constructors, freshness and substitution:

```
datatype (α,β) lterm = Var α | Ct β | App ((α,β) lterm) ((α,β) lterm) |
                 Lam α ((α,β) lterm) | Subst ((α,β) lterm) ((α,β) lterm) α
with fresh : α → (α,β) lterm → bool
where   (Var x) [t/x] = t
  and   x ≠ y ⟶ (Var y) [t/x] = Var y
  and   (Ct c) [t/x] = Ct c
  and   (App s₁ s₂) [t/x] = App (s₁[t/x]) (s₂[t/x])
  and   x ≠ y ∧ fresh y t ⟶ (Lam y s) [t/x] = Lam y (s [t/x])
  and   x ≠ y ⟶ fresh x (Var y)
  and   fresh x (Ct c)
  and   fresh x s₁ ∧ fresh x s₂ ⟶ fresh x (App s₁ s₂)
  and   fresh x (Lam x s)
  and   fresh x s ⟶ fresh x (Lam y s)
  and   x ≠ y ∧ fresh x s ⟶ Lam y s = Lam x (s [Var x / y])
```

where we wrote $s[t/x]$ instead of Subst $s\, t\, x$. Besides operations, this type also comes with a predicate fresh, which plays a crucial role in the behavior of the capture-free substitution operators, as regulated by the above Horn clauses. Specifically, substitution can "enter" λ-abstractions only under certain freshness conditions. Nevertheless, substitution can always be reduced away from terms by using the last clause to perform a renaming to a fresh variable.

This Horn-based definition of λ-terms is easily extendable to any syntax with static bindings, but does require some tuning to become a useful framework for reasoning about bindings. In particular it lacks a substitution-free induction schema. One type of task where the Horn view of λ-terms excels are recursive definitions: besides going through modulo α-equivalence, they also yield compositionality with freshness and substitution as a bonus. This is argued in [19] with many examples, ranging from higher-order abstract syntax and semantic-domain interpretation to CPS transformations. These examples are instances of the nonfree recursion provided by our package.

For instance, occs $t\, x$ yields the number of free occurrences of a variable x in a λ-term t. It is defined stating the "naive" recursive equations (as if terms were not quotiented w.r.t. α) together with indicating the correct behavior w.r.t. freshness and substitution:

$$\text{nonfreerec occs} : (\alpha, \beta)\ \text{lterm} \to (\alpha \to \text{nat})\ \text{where}$$
$$\text{occs}\,(\text{Ct}\,c) = (\lambda x.\,0) \qquad \text{occs}\,(\text{Lam}\,y\,s) = (\lambda x.\,\text{if}\ x = y\ \text{then}\ 0\ \text{else}\ \text{occs}\,s\,x)$$
$$\text{occs}\,(\text{Var}\,y) = (\lambda x.\,\text{if}\ x = y\ \text{then}\ 1\ \text{else}\ 0) \qquad \text{occs}\,(\text{App}\,s\,t) = (\lambda x.\,\text{occs}\,s\,x + \text{occs}\,s\,y)$$
$$\text{occs}\,(s\,[t/y]) = (\lambda x.\,\text{occs}\,s\,y * \text{occs}\,t\,x + (\text{if}\ x = y\ \text{then}\ 0\ \text{else}\ \text{occs}\,s\,x))$$
$$\text{fresh}\,y\,s \longrightarrow \text{occs}\,s\,y = 0$$

Note that, while the operators require (recursive) *equations*, predicates such as fresh require *implications*. Indeed, the implication for fresh indicates that, on the target domain $\alpha \to \text{nat}$, freshness is interpreted as $\lambda y\,s.\,\text{occs}\,s\,y = 0$. The goals emerging from this definition amount to arithmetic properties known by the Isabelle simplifier.

3 Automated Constructions

Here we sketch the development required to obtain the functionality provided by the package, using our λ-term example. (1) One starts with the free datatype of "pre-terms":

datatype (α, β) lterm$'$ = Var$'\,\alpha$ | Ct$'\,\beta$ | App$'\,((\alpha, \beta)\ \text{lterm}')\,((\alpha, \beta)\ \text{lterm}')$ |
$\qquad\qquad$ Lam$'\,\alpha\,((\alpha, \beta)\ \text{lterm}')$ | Subst$'\,((\alpha, \beta)\ \text{lterm}')\,((\alpha, \beta)\ \text{lterm}')\,\alpha$

(2) Next, one defines mutually inductively the desired "equality" $\equiv$ and the "pre-fresh" predicate. (In general, mutually recursive datatypes involve n equalities, one for each type, and m predicates, one for each predicate specified by the user.)

inductive $\equiv : \alpha\ \text{lterm}' \to \alpha\ \text{lterm}' \to \text{bool}$ and fresh$' : \alpha\ \text{lterm}' \to \text{bool}$
where
*** One clause for each user-specified Horn clause: ***
$\qquad (\text{Var}'\,x)[t/x] \equiv t$
$\quad$ and $\quad x \neq y \longrightarrow (\text{Var}'\,y)\,[t/x] \equiv \text{Var}'\,y$
*** etc. ***
*** The equivalence rules: ***
$\quad$ and $\quad s \equiv s \quad$ and $\quad s_1 \equiv s_2 \longrightarrow s_2 \equiv s_1 \quad$ and $\quad s_1 \equiv s_2 \wedge s_2 \equiv s_3 \longrightarrow s_1 \equiv s_3$
*** A congruence rule for each user-specified constructor: ***
$\quad$ and $\quad s_1 \equiv t_1 \wedge s_2 \equiv t_2 \longrightarrow \text{App}'\,s_1\,s_2 = \text{App}'\,t_1\,t_2$
*** etc. ***
*** A preservation rule for each constructor-predicate combination: ***
$\quad$ and $\quad s_1 \equiv t_1 \wedge s_2 \equiv t_2 \wedge \text{fresh}'\,x\,(\text{App}'\,s_1\,s_2) \longrightarrow \text{fresh}'\,x\,(\text{App}'\,t_1\,t_2)$
*** etc. ***

(3) The type α lterm is defined by quotienting α lterm$'$ by the equivalence $\equiv$, establishing a surjection $\pi : \alpha$ lterm$' \to \alpha$ lterm, with the choice function $\varepsilon : \alpha$ lterm $\to$ α lterm$'$ as its right inverse. The operations Var, App and Lam and the predicate fresh are defined on α lterm from the corresponding ones from α lterm$'$ using π and ε. (4) The induction principle from α lterm$'$ is transported to α lterm. (5) α lterm is shown to satisfy all the desired Horn clauses. To obtain the recursion principle, we fix an arbitrary type β with operations and relations on it and assume it satisfies the Horn clauses. (6) A function $f : \alpha$ lterm$' \to \beta$ is then defined by standard recursion. (7) By induction on the derivation of $\equiv$, we get that f is invariant under equivalent arguments. (8) This allows one to define a function $g : \alpha$ lterm $\to \beta$ such that $g \circ \pi = f$. (9) Using the surjectivity of π, this function is shown to commute with the operations and preserve the relations.

All the involved constructions and proofs are fairly easy to perform by hand, but quite tedious and time-consuming. Parts (3–5) and (8,9) of this process can be eased by existing Isabelle quotient/lifting/transfer packages [12, 14].

Our package automates the whole construction. Moreover, it does not perform this construction over and over, for each newly specified nonfree datatype. We have experimented with a different methodology:
– Formalize the metatheory for an arbitrary many-sorted signature and Horn theory.
– Upon a user specification, instantiate the locale, then copy isomorphically the relevant types, operations, and theorems about them.

The next two sections describe these steps.

4   Formalized Metatheory

We have formalized the theory of Horn clauses up to the construction of the initial model. The development is parameterized by an arbitrary signature (giving sorts and sorted operations and relation symbols) and a set of Horn clauses over the signature. An example instantiation is given in §5.1. Both terms and clauses are deeply embedded. Sorts represent relevant Isabelle types. A specific feature is the consideration of parameters and parameter conditions in clauses, motivated by the desire to capture parameterized instances such as polymorphic datatypes and clausal side conditions.

We will use the following constants. lnl : $\alpha \to \alpha + \beta$ and lnr : $\beta \to \alpha + \beta$ are the left and right injections into the sum type, and islnl, islnr : $\alpha + \beta \to$ bool are their corresponding discriminators; namely, islnl c holds iff c has the form lnl a for some a, and islnr c holds iff c has the form lnr b for some b. [] is the empty list, $[a_1, \ldots, a_n]$ is the list of the n indicated elements. map : $(\alpha \to \beta) \to \alpha$ list $\to \beta$ list is the standard list-map operator, and map2 : $(\alpha \to \beta \to \gamma) \to \alpha$ list $\to \beta$ list $\to \gamma$ list is its binary counterpart, with map2 $f [a_1, \ldots, a_n] [b_1, \ldots, b_n] = [f\, a_1\, b_1, \ldots, f\, a_n\, b_n]$. Similarly, list_all : $(\alpha \to$ bool$) \to \alpha$ list $\to$ bool is the universal quantifier over lists, with list_all $\varphi [a_1, \ldots, a_n]$ meaning that $\varphi\, a_i$ holds for all i, and list_all2 : $(\alpha \to \beta \to$ bool$) \to \alpha$ list $\to \beta$ list $\to$ bool is its binary counterpart, with list_all2 φ $al\, bl$ meaning that al has the form $[a_1, \ldots, a_n]$, bl has the from $[b_1, \ldots, b_n]$, and $\varphi\, a_i\, b_i$ holds for all i. In particular list_all2 φ $al\, bl$ requires that al and bl have equal lengths. As a notational convention, we use the suffix "l" to indicate lists. E.g., if ps ranges over the type psort, then psl ranges over psort list.

4.1   Horn Clause Syntax

We define the types var, of variables, and pvar, of parameter variables (p-variables), as copies of nat. Our constructions are parameterized by the following type variables: sort, of sorts, giving the syntactic categories of terms (representing the mutually recursive datatypes); opsym, of operation symbols (representing the datatype constructors); rlsym, of relation symbols (representing relations); param, the parameter universe; psort, of parameter sorts (p-sorts; representing the parameter types in the datatype).

The type of terms is defined as follows:

  datatype (sort, opsym) trm = Var sort var |
                         Op opsym (pvar list) (((sort, opsym) trm) list)

Thus a term T is either a sorted variable Var $s\,x$ or has the form Op $\sigma\,pxl\,Tl$, applying an operation symbol σ to a list pxl of parameter variables and a list Tl of terms.

The type of atoms (or atomic statements) is defined as follows:

        datatype (sort, opsym, rlsym, psort, param) atm =
          Pcond (param list → bool) (psort list) (pvar list) |
          Eq ((sort, opsym) trm) ((sort, opsym) trm) |
          Rl rlsym (pvar list) (((sort, opsym) trm) list)

We provide an intuition of the semantics of these atoms here. §4.3 provides the details. The semantics of these atoms is relative to interpretations of sorts as subsets of a model, of variables as elements in a model, of operation symbols as functions on a model, of relation symbols as relations on a model and of p-variables as parameters:

(1) Parameter-condition atoms have the form Pcond $R\,psl\,pxl$. Semantically they will be interpreted as the predicate R on the interpretation of the p-variables pxl (where this interpretation is assumed to be consistent with the p-sorts psl).

(2) Equational atoms have the form Eq $s\,T_1\,T_2$. They will be interpreted as a specialized "equality" relation between T_1 and T_2, assumed to be of sort s.

(3) Relational atoms have the form Rl $\pi\,pxl\,Tl$. They will be interpreted as the model relation corresponding to π on the interpretations of the p-variables pxl and the interpretations of the terms Tl. The sorts of Tl are assumed to agree with the sorting of π.

Horn clauses are essentially lists of atoms: the premises are paired with one atom, the conclusion. In §4.3 we will interpret a Horn clause as the implication between the interpretations of the premises and the conclusion, schematically quantified over variable interpretations:

        datatype (sort, opsym, rlsym, psort, param) hcl =
          Horn (((sort, opsym, rlsym, psort, param) atm) list)
          ((sort, opsym, rlsym, psort, param) atm)

In what follows, we fix the type parameters and omit them when writing the various types that depend on them, e.g., writing trm instead of (sort, opsym) trm.

4.2 Signatures

We define signatures as a locale that fixes the data required to classify terms and parameters according to sorts:

> locale Signature =
>   fixes stOf : opsym → sort
>   and   arOf : opsym → sort list   and   arOfP : opsym → psort list
>   and   rarOf : rlsym → sort list   and   rarOfP : rlsym → psort list
>   and   params : psort → param → bool
>   and   prels : ((param list → bool) × psort list) set

Recall from the definition of terms that operation symbols are applied not only to terms, but also to parameters. Then arOf (read "arity of"), arOfP (read "parameter-arity of") and stOf (read "sort of"), regulate the sorts of terms (or, in general, elements of models) and parameters that an operation symbol takes and the sort of terms it returns. Similarly, rarOf and rarOfP indicate the arities and parameter-arities of relation symbols. Moreover, params classifies parameters according to sorts. Finally, prels specifies the set of relations over parameters that can be used as parameter conditions in Horn clauses, together with their intended arities. Given $(R, psl) \in$ prels, we only care about the behavior of R on lists pl of parameters having sorts psl according to params, i.e., such that list_all2 params pl psl holds. We have to represent R as a relation on the larger type param list because dependent types are not available. Similar phenomena are observable in our definitions of models below.

4.3 Models

We work in the Signature context. The (well-formed) terms of a given sort are defined as the predicate trms : sort → trm → bool, by requiring that operation symbols are applied according to their arities.

A model is a tuple $(\alpha, intSt, intOp, intRl)$, where:
- α is the carrier type,
- $intSt$: sort → α → bool classifies the elements of α according to sorts;
- $intOp$: opsym → param list → α list → α interprets the operation symbols as parameterized operations on α;
- $intRl$: rlsym → param list → α list → bool interprets the relation symbols as parameterized relations on α.

In (well-formed) models the interpretation of operation symbols has to be compatible with sorting, i.e., the following predicate compat $intSt$ $intOp$ holds:

$$\forall \sigma \, pl \, al. \text{ list\_all2 params } (\text{arOfP } \sigma) \, pl \wedge \text{ list\_all2 } intSt \, (\text{arOf } \sigma) \, al \rightarrow$$
$$intSt \, (\text{stOf } \sigma) \, (intOp \, \sigma \, pl \, al).$$

Given a model $(\alpha, intSt, intOp, intRl)$, the notions of term interpretation and atom satisfaction are defined relative to interpretations of parameter variables $intPvar$: psort → pvar → param and variables $intVar$: sort → var → α. For equational atoms, we do not require equality, but further parameterize on a relation $intEq$: α → α → bool.

intTrm *intOp intPvar intVar* (Var *s x*) = *intVar s x*
intTrm *intOp intPvar intVar* (Op σ *pxl Tl*) =
   intOp σ (map2 *intPvar* (arOfP σ) *pxl*) (map (intTrm *intOp intPvar intVar*) *Tl*)
satAtm *intOp intEq intRl intPvar intVar* (Pcond *R psl pxl*) $\longleftrightarrow$ *R* (map2 *intPvar psl pxl*)
satAtm *intOp intEq intRl intPvar intVar* (Eq *s* T_1 T_2) $\longleftrightarrow$
   intEq (intTrm *intOp intPvar intVar* T_1) (intTrm *intOp intPvar intVar* T_2)
satAtm *intOp intEq intRl intPvar intVar* (Rl π *pxl Tl*) $\longleftrightarrow$
   intRl π (map2 *intPvar* (rarOfP π) *pxl*) (map (intTrm *intOp intPvar intVar*) *Tl*)

Thus, the term interpretation is defined recursively over terms, employing interpretations of p-variables and variables. For atom satisfaction, we distinguish the three kinds of atom, employing the parameter-conditions, the equality interpretation and the relation-symbol interpretation, respectively. Note that the interpretations do not depend on the model-carrier sorting *intSt* : $\alpha \to$ sort. However, for well-formed models we prove that well-sorted interpretations of (p-)variables yield term interpretations compatible with sorting, in that they send terms of sort *s* to model elements of sort *s*:

lemma: compat *intSt intOp* $\wedge$ ($\forall$ *ps px*. params *ps* (*intPvar ps px*)) $\wedge$
($\forall$ *s x*. intSt *s* (*intVar s x*)) $\to$ (trms *s T* $\to$ *intSt s* (intTrm *intOp intPvar intVar T*)).

The above approach is pervasive in our formalization: We do not index everything by sorts, but use global (unsorted) functions and relations as much as possible, and then show that they are compatible with sorting. This optimization is particularly helpful when we quotient terms w.r.t. the Horn-induced equivalence relation building a single quotient instead of a sorted family of quotients (as customary in universal algebra).

Finally, satisfaction of a Horn clause by a model is defined as the implication between satisfaction of the premises and satisfaction of the conclusion for all well-sorted interpretations *intPvar* of the p-variables and *intVar* of the variables:

satHcl *intSt intOp intEq intRl* (Horn *atml atm*) $\longleftrightarrow$
   $\forall$ *intPvar intVar*. ($\forall$ *ps px*. params *ps* (*intPvar ps px*)) $\wedge$ ($\forall s$ *x*. intSt *s* (*intVar s x*)) $\wedge$
      list_all (satAtm *intOp intEq intRl intPvar intVar*) *atml* $\to$
         satAtm *intOp intEq intRl intPvar intVar atm*

4.4   The Initial Model of a Horn Theory

Traditionally, ground terms are simply terms with no free variables. However, in our parameterized setting, terms contain p-variables, while the ground terms will need to contain actual parameters. We define a separate type of ground terms, gtrm, built recursively from operation symbols applied to lists of parameters and list of ground terms:

datatype (opsym, param) gtrm = Gop opsym (param list) (((opsym, param) gtrm)list)

The initial model of a Horn theory will be constructed by quotienting ground terms w.r.t. an equivalence relation. Hence its carrier will be the following type of "Horn terms" defined to be sets of ground terms:

type_synonym (opsym, param) htrm = ((opsym, param) gtrm) set

In what follows, we fix a signature with assumptions guaranteeing non-emptiness of sorts and p-sorts and a well-formed Horn theory HCL. Technically, we work in the context of the following locale extending the Signature locale:

locale HornTheory = Signature + fixes HCL : hcl set
assumes ∀hcl ∈ HCL. wf hcl   and   ∀s. reach s   and   ∀ps. ∃p. params ps p

Above, wf hcl states that the Horn clause is well-formed in that all its atoms are well-formed in the expected way, e.g., in equational atoms Eq s T_1 T_2, s is the sort of T_1 and T_2. The inductively defined predicate reach s states that the sort s is reachable by operation symbols. This ensures the existence of ground terms of sort s, where sorting of ground terms gtrms : sort → gtrm → bool is defined as expected.

On gtrm we define mutually inductive relations Geq : gtrm → gtrm → bool and Grel : rlsym → param list → gtrm list → bool in a similar fashion to the example of §3, but working symbolically with the clauses in HCL instead of concrete clauses. We show that Geq is an equivalence and that both relations are compatible with sorting and with the operations. This allows us to quotient gtrm by Geq, giving the type htrm. We lift the sorting gtrms of ground terms and the interpretations Gop, Grel of the operation and relation symbols on ground terms to equivalence classes. This yields the functions htrms : sort → htrm → bool, Hop : opsym → param list → htrm list → htrm and Hrel : rlsym → param list → htrm list → bool.

The ground-term model (gtrm, gtrms, Gop, Grel) satisfies all the clauses in HCL if we interpret equality as Geq:

lemma: hcl ∈ HCL → satHcl gtrms Gop Geq Grel hcl

From this, we obtain that the Horn-term model (htrm, htrms, Hop, Hrel) satisfies the clauses with the standard interpretation of equality:

theorem satisfaction: hcl ∈ HCL → satHcl htrms Hop (=) Hrel hcl

Structural induction is easily inherited by Horn terms from ground terms:

theorem induction: (∀σ pl Hl. list_all2 params (arOfP σ) pl ∧ list_all2 htrms (arOf σ) Hl ∧ list_all2 φ (arOf σ) Hl → φ (stOf σ) (Hop σ pl Hl)) → (htrms s H → φ s H).

Moreover, the cases theorem is obtained as a degenerate induction. We are left to show that (htrm, htrms, Hop, Hrel) is initial among the models of HCL. First we define giter : (opsym → param list → $α$ list → $α$) → gtrm → $α$ that interprets ground terms with an operation symbol interpretation on a type $α$, as giter *intOp* (Gop σ pl Tl) = *intOp* σ pl (map (giter *intOp*) Tl). Then we lift giter to htrm equivalence classes, giving iter : (opsym → param list → $α$ list → $α$) → htrm → $α$. If ($α$, *intSt*, *intOp*, *intRl*) is a model that satisfies HCL, then iter *intOp* is well-sorted and behaves like an iterator, i.e., commutes with the operations, and preserves the relations:

theorem it_sort: compat *intSt intOp* ∧ (∀ hcl ∈ HCL. satHcl *intSt intOp* (=) *intRl* hcl) → htrms s H → *intSt* s (iter *intOp* H)

theorem iteration: compat *intSt intOp* ∧ (∀ hcl ∈ HCL. satHcl *intSt intOp* (=) *intRl* hcl) → iter *intOp* (Hop σ pl Hl) = *intOp* σ pl (map (iter *intOp*) Hl)

theorem it_pres: compat *intSt intOp* ∧ (∀ hcl ∈ HCL. satHcl *intSt intOp* (=) *intRl* hcl) → Hrel π pl Hl → *intRl* π pl (map (iter *intOp*) Hl).

After some lemmas concerning the interaction between the choice function and the operations on Gop, the above theorems are proved by induction on the definition of

Geq and Grel. Note that our iterator only depends on the operation part of the model, although its properties rely on the whole model and its satisfaction of HCL.

5 Animation of the Metatheory

From a purely mathematical viewpoint, having formalized the general case for arbitrary signatures and Horn theories, we did capture all the instances. But we still have to bridge the gap between the abstract characterization of the instances in the metatheory and the instance descriptions offered by users of the package. Moreover, the metatheory introduces operations over a quotient term universe, while users want to use curried datatype constructors between distinguished types for each of the mutually recursive datatypes.

5.1 Instantiation of the Metatheory

We focus on an example instantiation of the metatheory here and refer to [20] for a description of the instantiation in general.

To obtain the λ-terms modulo α from §2.3, we simply instantiate the HornTheory locale. The types are instantiated as follows:
– sort becomes a type with 1 element, lt, for the unique syntactic category of λ-terms;
– opsym becomes a type with 5 elements, var, ct, app, lam, subst, corresponding to the operations Var, Ct, App, Lam, Subst;
– rlsym becomes a type with 1 element, fr, corresponding to the predicate fresh;
– param becomes the sum type $\alpha + \beta$, embedding the type α of variables and β of constants used in λ-terms and thus forming the parameter universe;
– psort becomes a type with 2 elements, a and b, matching the 2 kinds of parameters.

The signature variables are instantiated as follows:
– stOf _ = lt; arOf var = []; arOfP var = [a]; arOf ct = []; arOfP ct = [b];
  arOf app = [lt, lt]; arOfP app = []; arOf lam = [lt]; arOfP lam = [a];
  arOf subst = [lt, lt]; arOfP subst = [a]; rarOf fr = [lt]; rarOfP fr = [a];
– params $ps\ p \leftrightarrow (ps = a \land$ islnl $p) \lor (ps = b \land$ islnr $p)$;
– prels = $\{(\text{dif}_2, [a, a])\}$, where dif_2 is the function sending any list of two parameters of the form [Inl a_1, Inl a_2] to (the truth-value of) $a_1 \neq a_2$ (and with immaterial definition elsewhere).

Finally, HCL is instantiated to the set containing the reflections of the λ-term clauses. For example, $x \neq y \land$ fresh $x\ s \longrightarrow$ Lam $y\ s =$ Lam $x\ (s\ [\text{Var}\ x\ /\ y])$ becomes Horn [atm$_1$, atm$_2$] atm$_3$, where:
– we take x and y to be distinct elements of pvar and s to be some element of var;
– atm$_1$ = Pcond dif$_2$ [a, a] [x, y],
– atm$_2$ = Rl fr [x] [Var lt s] and atm$_3$ = Eq lt $T_1\ T_2$, with $T_1 =$ Op lam [y] [Var lt s] and $T_2 =$ Op lam [x] [Op subst [y] [Var lt s, Op var [x] []]].

Then, after checking the HornTheory assumptions for this particular instances, we indeed obtain valid formulations of the satisfaction, induction and iteration theorems for λ-terms as instances of the general theorems. However, these formulations are inconvenient to use in a theorem prover. One would certainly prefer to write App $s_1\ s_2$ instead of Hop app $[]$ $[s_1, s_2]$ for λ-term application, and $\forall x\, y\, s.\ x \neq y \wedge$ fresh $x\, s \longrightarrow$ Lam $y\, s =$ Lam $x\, (s\, [\text{Var } x\, /\, y])$ instead of satHcl $intSt\ intOp\ (=)\ intRl$ (Horn $[\text{atm}_1, \text{atm}_2]\ \text{atm}_3$).

Superficially, fixing this seems to be a matter of syntactic sugar. But the situation is a little more complex, since we also want to use a more appropriate type for λ-terms. Indeed, htrm may contain junk—the general theorems only speak about sorted terms. Therefore, the type we care about needs to be carved out from htrm by restricting to those T such that htrms lt T (where lt is here the only sort). Then App needs to be defined as a copy of Hop app on the new type, also using two arguments instead of lists with two elements. These transformations are realized with the isomorphic transfer of types and terms, which we describe in the next section.

5.2 Isomorphic Transfer

Isomorphic transfer is based on establishing appropriate bijections between primitive types, lifting these bijections to composite types and mapping term constructions under the corresponding bijections away from the input types.

We shall employ relators, which are operators on predicates matching the type constructors. E.g., given $\varphi : A \to$ bool and $\psi : B \to$ bool, $\varphi \otimes \psi : A \times B \to$ bool is defined by $(\varphi \otimes \psi)\,(a, b) \leftrightarrow (\varphi\, a \wedge \psi\, b)$ and $\varphi \Rightarrow \psi : (A \to B) \to$ bool is defined by $(\varphi \Rightarrow \psi)\, f \leftrightarrow (\forall a.\ \varphi\, a \to \psi\,(f\, a))$.

htrms lt : htrm → bool	lterm
islnl : param → bool	α
islnr : param → bool	β
(list_all2 params (arOfP app)) ⊗ (list_all2 htrms (arOf app)) : param list × htrm list → bool	lterm × lterm
(list_all2 params (arOfP lam)) ⊗ (list_all2 htrms (arOf lam)) : param list × htrm list → bool	α × lterm

Fig. 1. Instance types and predicates (left) versus target types (right)

Figure 1 shows two categories of types side by side:
– on the left, the *instance types*, i.e., those obtained from the locale instantiation, where necessary together with predicates describing the relevant subset based on the sorting;
– on the right, the corresponding *target types* exported to the user.

We assume α and β have been fixed and omit spelling them out, e.g., we write lterm instead of (α, β)lterm. Also, param, htrm, etc. refer to the concrete types obtained by the locale instantiation from §5.1.

The first 3 rows show the primitive types. For the Horn terms, we have defined lterm by carving out from trmHCL the terms of sort lt (were there multiple sorts, we would have multiple target types of terms). For parameters, the instance type was defined from the target types, as their sum. In either case, we have bijections between sets of elements in the instance types satisfying corresponding predicates and the target types.

These bijections are extended to bijections between the domains of the instance operations and the intended domains of the target operations[3]—rows 4 and 5 show the

---

[3] To ease the presentation, we ignore currying and pretend that the domains are products.

extensions for two operation symbols, app and lam. To see how the extension operates, note that the instance predicates regulate the length of the lists and the sorts of their contents. E.g., since arOf app = [lt, lt], we see that list_all (arOf app) Hl requires that Hl have the form $[H_1, H_2]$ such that htrms lt H_1 and htrms lt H_2 hold—thus, the lists boil down to pairs of Horn terms of sort lt, hence correspond bijectively to lterm × lterm.

With the bijection construction in place, we proceed to copy the operations on the instance types into operations on the target types, by defining constants equal to their image under the corresponding bijection. E.g., App : lterm × lterm → lterm is defined as the image of Hop app : param list × htrm list → trmHCL restricted according to the suitable predicates. Thus, App corresponds to Hop app under the lifted bijection to lterm × lterm → lterm from the set of elements of param list × htrm list → trmHCL for which the predicate (list_all2 params (arOfP app)) ⊗ (list_all2 htrms (arOf app)) ⇒ htrms (stOf app) holds. This set contains Hop app because of the sorting of app.

Finally, the theorems about instance types, that is, the satisfaction, induction, cases and recursion theorems, are transported from the instance types to the target types. Technically this works because we choose the bijection on propositions to be the identity. For instance, let us consider the induction theorem, where we write l_2 instead of list_all2:

$$\forall \varphi\, s\, H.\ (\forall \sigma\, pl\, Hl.\ l_2\ \text{params}\ (\text{arOfP}\ \sigma)\ pl\ \wedge\ l_2\ \text{htrms}\ (\text{arOf}\ \sigma)\ Hl\ \wedge\ l_2\ \varphi\ (\text{arOf}\ \sigma)\ Hl\ \rightarrow$$
$$\varphi\ (\text{stOf}\ \sigma)\ (\text{Hop}\ \sigma\, pl\, Hl))\ \rightarrow\ \text{htrms}\ s\ H\ \rightarrow\ \varphi\, s\, H$$

To ease the presentation let us pretend that the signature only has app and lam as operation symbols. The theorem is processed as follows, into equivalent theorems. First, the quantification over σ is replaced by conjunction over all operation symbols:

$$\forall \varphi\, s\, H.\ (\forall \varphi\, pl\, Hl.\ l_2\ \text{params}\ (\text{arOfP}\ \text{app})\, pl\ \wedge\ l_2\ \text{htrms}\ (\text{arOf}\ \text{app})\ Hl\ \wedge\ l_2\ \varphi\ (\text{arOf}\ \text{app})\ Hl \rightarrow \varphi\ (\text{stOf}\ \text{app})\ (\text{Hop}\ \text{app}\, pl\, Hl))$$
$$\wedge\ (\forall \varphi\, pl\, Hl.\ l_2\ \text{params}\ (\text{arOfP}\ \text{lam})\, pl\ \wedge\ l_2\ \text{htrms}\ (\text{arOf}\ \text{lam})\ Hl\ \wedge\ l_2\ \varphi\ (\text{arOf}\ \text{lam})\ Hl \rightarrow \varphi\ (\text{stOf}\ \text{lam})\ (\text{Hop}\ \text{lam}\, pl\, Hl))$$
$$\rightarrow\ \text{htrms}\ s\ H\ \rightarrow \varphi\ s\ H$$

Computing the values of the sort and arity functions, this becomes:

$$\forall \varphi\, s\, H.\ (\forall \varphi\, pl\, Hl.\ l_2\ \text{params}\ []\ pl\ \wedge\ l_2\ \text{htrms}\ [\text{lt, lt}]\ Hl\ \wedge\ l_2\ \varphi\ [\text{lt, lt}]\ Hl \rightarrow \varphi\ \text{lt}\ (\text{Hop}\ \text{app}\, pl\, Hl))$$
$$\wedge\ (\forall \varphi\, pl\, Hl.\ l_2\ \text{params}\ [a]\ pl\ \wedge\ l_2\ \text{htrms}\ [\text{lt}]\ Hl\ \wedge\ l_2\ \varphi\ [\text{lt}]\ Hl \rightarrow \varphi\ \text{lt}\ (\text{Hop}\ \text{lam}\, pl\, Hl))$$
$$\rightarrow\ \text{htrms}\ s\ H\ \rightarrow \varphi\ s\ H$$

By isomorphic transfer over the aforementioned extended bijections, we obtain:

$$(\forall H_1\, H_2.\ \varphi\, H_1\ \wedge\ \varphi\, H_2 \rightarrow \varphi\ (\text{App}\ H_1\ H_2))\ \wedge\ (\forall x\, H.\ \varphi\, H\ \rightarrow\ \varphi\ (\text{Lam}\ x\, H))\ \rightarrow\ \varphi\, H.$$

In this step the l_2 params, l_2 htrms constraints have disappeared, since the extended bijections map the constrained variables pl, Hl to empty tuples, pairs or single elements.

5.3 General Animation Infrastructure

All these constructions, namely, defining the types and terms necessary for the instantiation, establishing bijections between primitive types, extending them to the relevant composite types, and transferring the term constructions and theorems to the target types, are automated by employing a general infrastructure for algorithmic rule systems and forward propagation of facts.

Algorithmic rule systems are collections of proven rules about moded judgments, which are defined predicates in Isabelle/HOL. The definition of such a judgment constitutes its propositional meaning, while the rules are theorems that constitute the sound algorithm we use to synthesize the outputs and establish the judgment. For details we refer to the first author's M.Sc. thesis [20]. We just note here that the animation of algorithmic rule systems can be regarded as a deterministic variant of Lambda-Prolog [16].

We employ the new concept of "forward rules" to drive the instantiation of the metatheory and invoke the term transformations. A forward rule is an implicational theorem that, algorithmically speaking, waits for input facts matching its conjunctive head premise, processes them with algorithmic rule systems indicated by judgmental premises, issues term and type definitions indicated by further premises and makes output facts available that correspond to the conclusion.

Isomorphic transfer is implemented [20] in the form of an algorithmic rule system. We want to note that currying of functions over finite products is an ad hoc higher-order transformation overriding the uniform transfer of applications. In our case the products are realized as lists over a universe. Currying an operator application f (Cons t ts) proceeds by recursion on the list argument, regarding the uncurry-image of the partially-curried operator $\psi_1\ f$ applied to the transformed first component $\psi_2\ t$, as the new operator $\psi_3^{-1}\ ((\psi_1\ f)\ (\psi_2\ t))$ in the recursive transfer of $\psi_3^{-1}\ ((\psi_1\ f)\ (\psi_2\ t))\ ts$. The general approach using algorithmic rule systems is beneficial for term transformations with nonuniform behaviour.

6   Conclusions and Related Work

We implemented the first package for nonfree datatypes in a HOL-based prover, pioneering a metatheory approach. We provide parameter conditions, relations, induction, case distinction, satisfaction of the specification and iterative recursion (i.e. initiality). The presented ideas are relevant to all HOL-based provers, but type class constraints are Isabelle-specific and essential for some nonfree datatypes (see §2.2).

The metatheories of packages in HOL usually are of an informal nature and rely on the dynamic checking of inferences for soundness. Formalizing their metatheories will make theorem provers more reliable by offering completeness guarantees. Metatheorems of a common shape can be processed uniformly, which leads to better extensibility of packages. Metatheory-based constructions are a relatively recent idea even in dependent type theories that can engage in generic programming over type universes [3]. Application of these metatheories is usually not facilitated with automated isomorphic transfer and is thus left to idealistic users.

The Isabelle package for (co)datatypes [24] based on bounded natural functors (BNFs) lacks support for equational theories. But nonfree datatypes defined with our package can be registered as a BNF and nested in later (co)datatype definitions.

Nonfree datatypes are natively supported by algebraic-specification provers such as the Maude ITP [2]. One simply declares signatures and arbitrary sets of equations in Maude, on top of which a basic mechanism for inductive reasoning is available. New function symbols can be declared together with equations defining them, but there is no compatibility check w.r.t. the other equations. This means a check of well-definedness as for our nonfree recursor is lacking.

Moca [4] translates nonfree datatype specifications with an equational theory specified in a extension of OCaml, down to implementation datatypes with private datatype constructors. These can only be used for pattern matching and inhabitants are instead constructed with construction functions that normalize w.r.t. the equational theory. Efficient construction functions are a core concern of Moca. A translation to Coq is planned.

The quotient/lifting/transfer packages of Isabelle [12, 14] overlap in functionality with our tool for isomorphic transfer. The novelty here is its realization inside a general infrastructure and the possibility of ad hoc higher-order transformations such as currying of functions on finite products. We support the transfer under setoid isomorphisms, so quotient lifting is available with the canonical surjection into the quotient type as the setoid isomorphism. Packages for quotient lifting/transfer can ease some parts of the manual construction of nonfree datatypes (see §3).

In the homotopy interpretation of type theory there is a recent trend [22] to investigate "higher inductive datatypes" that feature constructors introducing equalities. The main motivation here is to represent constructions of homotopy theory by describing their path space, but quotients similar to our package can also be introduced. The univalence axiom implies [8] that isomorphic mathematical structures are identified, so isomorphic transfer is available by substitution.

Acknowledgements. We thank Tobias Nipkow for making this collaboration possible, Jasmin Blanchette, Armin Heller, and the reviewers for providing many comments that helped improve the presentation, Ondrej Kuncar for answering questions about Isabelle's new lifting/transfer package, and the people on the Coq-Club mailing list for pointing us to related work. The work reported here is supported by the DFG project Ni 491/13–2 (part of the DFG priority program Reliably Secure Software Systems–RS3).

References

1. The Coq Proof Assistant (2013), http://coq.inria.fr
2. Maude ITP (2013), http://maude.cs.uiuc.edu/tools/itp
3. Altenkirch, T., McBride, C., Morris, P.: Generic programming with dependent types. In: Backhouse, R., Gibbons, J., Hinze, R., Jeuring, J. (eds.) SSDGP 2006. LNCS, vol. 4719, pp. 209–257. Springer, Heidelberg (2007)
4. Blanqui, F., Hardin, T., Weis, P.: On the implementation of construction functions for nonfree concrete data types. In: De Nicola, R. (ed.) ESOP 2007. LNCS, vol. 4421, pp. 95–109. Springer, Heidelberg (2007)
5. Bove, A., Dybjer, P.: Dependent types at work. In: Bove, A., Barbosa, L.S., Pardo, A., Pinto, J.S. (eds.) LerNet ALFA Summer School 2008. LNCS, vol. 5520, pp. 57–99. Springer, Heidelberg (2009)
6. Clavel, M., Durán, F., Eker, S., Lincoln, P., Martí-Oliet, N., Meseguer, J., Quesada, J.F.: The Maude system. In: Narendran, P., Rusinowitch, M. (eds.) RTA 1999. LNCS, vol. 1631, pp. 240–243. Springer, Heidelberg (1999)
7. CoFI task group on semantics, CASL — The Common Algebraic Specification Language, Semantics (1999),
   http://www.informatik.uni-bremen.de/cofi/wiki/index.php/CASL
8. Coquand, T., Danielsson, N.A.: Isomorphism is equality. Draft (2013)
9. Haftmann, F., Wenzel, M.: Constructive type classes in Isabelle. In: Altenkirch, T., McBride, C. (eds.) TYPES 2006. LNCS, vol. 4502, pp. 160–174. Springer, Heidelberg (2007)

10. Harrison, J.: HOL Light: A tutorial introduction. In: Srivas, M., Camilleri, A. (eds.) FMCAD 1996. LNCS, vol. 1166, pp. 265–269. Springer, Heidelberg (1996)

11. Homeier, P.V.: A design structure for higher order quotients. In: Hurd, J., Melham, T. (eds.) TPHOLs 2005. LNCS, vol. 3603, pp. 130–146. Springer, Heidelberg (2005)

12. Huffman, B., Kuncar, O.: Lifting and transfer: A modular design for quotients in Isabelle/HOL. In: Isabelle Users Workshop (2012)

13. Huffman, B., Urban, C.: A new foundation for Nominal Isabelle. In: Kaufmann, M., Paulson, L.C. (eds.) ITP 2010. LNCS, vol. 6172, pp. 35–50. Springer, Heidelberg (2010)

14. Kaliszyk, C., Urban, C.: Quotients revisited for Isabelle/HOL. In: SAC, pp. 1639–1644 (2011)

15. Kammüller, F., Wenzel, M., Paulson, L.C.: Locales - A sectioning concept for Isabelle. In: Bertot, Y., Dowek, G., Hirschowitz, A., Paulin, C., Théry, L. (eds.) TPHOLs 1999. LNCS, vol. 1690, pp. 149–166. Springer, Heidelberg (1999)

16. Nadathur, G., Miller, D.: An overview of Lambda-Prolog. In: ICLP/SLP, pp. 810–827 (1988)

17. Nipkow, T., Paulson, L.C., Wenzel, M.: Isabelle/HOL: A Proof Assistant for Higher-Order Logic. LNCS, vol. 2283. Springer, Heidelberg (2002)

18. Norrish, M.: Recursive function definition for types with binders. In: Slind, K., Bunker, A., Gopalakrishnan, G.C. (eds.) TPHOLs 2004. LNCS, vol. 3223, pp. 241–256. Springer, Heidelberg (2004)

19. Popescu, A., Gunter, E.L.: Recursion principles for syntax with bindings and substitution. In: ICFP, pp. 346–358 (2011)

20. Schropp, A.: Instantiating deeply embedded many-sorted theories into HOL types in Isabelle. Master's thesis, Technische Universität München (2012),
http://home.in.tum.de/~schropp/master-thesis.pdf

21. Schropp, A., Popescu, A.: Nonfree datatypes: metatheory, implementation and examples,
http://bitbucket.org/isaspecops/nonfree-data/downloads/cpp2013_bundle.zip

22. Shulman, M., Licata, D., Lumsdaine, P.L., et al.: Higher inductive types on the homotopy type theory blog,
http://homotopytypetheory.org/category/higher-inductive-types/

23. Breazu-Tannen, V., Subrahmanyam, R.: Logical and computational aspects of programming with sets/bags/lists. In: Leach Albert, J., Monien, B., Rodríguez-Artalejo, M. (eds.) ICALP 1991. LNCS, vol. 510, pp. 60–75. Springer, Heidelberg (1991)

24. Traytel, D., Popescu, A., Blanchette, J.C.: Foundational, compositional (co)datatypes for higher-order logic—Category theory applied to theorem proving. In: LICS 2012, pp. 596–605 (2012)

Lifting and Transfer: A Modular Design for Quotients in Isabelle/HOL

Brian Huffman[1] and Ondřej Kunčar[2]

[1] Galois, Inc.
[2] Technische Universität München

Abstract. Quotients, subtypes, and other forms of type abstraction are ubiquitous in formal reasoning with higher-order logic. Typically, users want to build a library of operations and theorems about an abstract type, but they want to write definitions and proofs in terms of a more concrete representation type, or "raw" type. Earlier work on the Isabelle Quotient package has yielded great progress in automation, but it still has many technical limitations.

We present an improved, modular design centered around two new packages: the *Transfer* package for proving theorems, and the *Lifting* package for defining constants. Our new design is simpler, applicable in more situations, and has more user-friendly automation.

1   Introduction

Quotients and subtypes are everywhere in Isabelle/HOL. For example, basic numeric types like integers, rationals, reals, and finite words are all quotients. Many other types in Isabelle are implemented as subtypes, including multisets, finite maps, polynomials, fixed-length vectors, matrices, and formal power series, to name a few.

Quotients and subtypes are useful as type abstractions: Instead of explicitly asserting that a function respects an equivalence relation or preserves an invariant, this information can be encoded in the function's type. Quotients are also particularly useful in Isabelle, because reasoning about equality on an abstract type is supported much better than reasoning modulo an equivalence relation.

Building a theory library that implements a new abstract type can take a lot of work. The challenges are similar for both quotients and subtypes: Isabelle requires explicit coercion functions (often "*Rep*" and "*Abs*") to convert between old "raw" types and new abstract types. Definitions of functions on abstract types require complex combinations of these coercions. Users must prove numerous lemmas about how the coercions interact with the abstract functions. Finally, it takes much effort to transfer the properties of raw functions to the abstract level. Clearly, this process needs good proof automation.

1.1   Related Work

Much previous work has been done on formalizing quotients in theorem provers. Slotosch [12] and Paulson [10] each developed techniques for defining quotient types and defining first-order functions on them. They provided limited automation for transferring properties from raw to abstract types in the form of lemmas that facilitate manual

G. Gonthier and M. Norrish (Eds.): CPP 2013, LNCS 8307, pp. 131–146, 2013.
© Springer International Publishing Switzerland 2013

proofs. Harrison [3] implemented tools for lifting constants and transferring theorems automatically, although this work was still limited to first-order constants and theorems. In 2005, Homeier [4] published a design for a new HOL package, which was the first system capable of lifting higher-order functions and transferring higher-order theorems.

Isabelle's Quotient package was implemented by Kaliszyk and Urban [5], based upon Homeier's design. It was first released with Isabelle 2009-2. The Quotient package is designed around the notion of a *quotient*, which involves two types and three constants: a raw type 'a with a partial equivalence relation $R :: 'a \Rightarrow 'a \Rightarrow bool$ and the abstract type 'b, whose elements are in one-to-one correspondence with the equivalence classes of R. The *abstraction* function $Abs :: 'a \Rightarrow 'b$ maps each equivalence class of R onto a single abstract value, and the *representation* function $Rep :: 'b \Rightarrow 'a$ takes each abstract value to an arbitrary element of its corresponding equivalence class.

The Quotient package implements a collection of commands, proof methods, and theorem attributes. Given a raw type and a (total or partial) equivalence relation R, the **quotient_type** command defines a new type with Abs and Rep that form a quotient. Given a function g on the raw type and an abstract type, the **quotient_definition** command defines a new abstract function g' in terms of g, Abs, and Rep. The user must provide a *respectfulness theorem* showing that g respects R. Finally the descending and lifting methods can transfer propositions between g and g'. Internally, this uses respectfulness theorems, the definition of g', and the quotient properties of R, Abs and Rep.

Lammich's automatic procedure for data refinement [7] was directly inspired by our packages, especially by the idea to represent types as relations.

In Coq, implementations of generalized rewriting by Coen [1] and Sozeau [13] are similar to our Transfer method—in particular, Sozeau's "signatures" for higher-order functions are like our transfer rules. Sozeau's work has better support for subrelations, but our Transfer package is more general in allowing relations over two different types.

Magaud [8] transfers Coq theorems between different types, but unlike our work, his approach is based on transforming proof terms.

1.2   Limitations of the Quotient Package

We decided to redesign the Quotient package after identifying several limitations of its implementation. A few such limitations were described by Krauss [6]: 1.) The quotient relation R and raw function f must be dedicated constants, not arbitrary terms. Thus the tool cannot be used on locale parameters and some definitions in a local theory. 2.) One cannot turn a pre-existing type into a quotient afterwards; nor can one declare a user-defined constant on the quotient type as the lifted version of another constant.

To solve problem 1 does not require major organizational changes. However, problem 2 has deeper roots and suggested splitting the Quotient package into various layers: By having separate components with well-defined interfaces, we could make it easier for users to connect with the package in non-standard ways.

Besides the problems noted by Krauss, we have identified some additional problems with the descending/lifting methods. Consider 'a fset, a type of finite sets which is a quotient of 'a list. The Quotient package can generate fset versions of the list functions map :: $('a \Rightarrow 'b) \Rightarrow 'a\ list \Rightarrow 'b\ list$ and concat :: $'a\ list\ list \Rightarrow 'a\ list$, but it has difficulty transferring the following theorems to fset:

$\text{concat } (\text{map } (\lambda x.\ [x])\ xs) = xs$
$\text{map } f\ (\text{concat } xss) = \text{concat } (\text{map } (\text{map } f)\ xss)$
$\text{concat } (\text{map concat } xsss) = \text{concat } (\text{concat } xsss)$

The problem is with the user-supplied respectfulness theorems. Note that map occurs at several different type instances here: It is used with functions of types 'a ⇒ 'b, 'a ⇒ 'a list, and 'a list ⇒ 'b list. Unfortunately a single respectfulness theorem for map will not work in all these cases—each type instance requires a different respectfulness theorem. On top of that, the user must also prove additional *preservation lemmas*, essentially alternative definitions of map_fset at different types. These rules can be tricky to state correctly and tedious to prove.

The Quotient package's complex, three-phase transfer procedure was another motivation to look for a new design. We wanted to have a simpler implementation, involving fewer separate phases. We also wanted to ease the burden of user-supplied rules, by requiring only one rule per constant. Finally, we wanted a more general, more widely applicable transfer procedure without so many hard-wired assumptions about quotients.

1.3 Overview

Our new system uses a layered design, with multiple components and interfaces that are related as shown in Fig. 1. Each component depends only on the components underneath it. At the bottom is the Transfer package, which transfers propositions between raw and abstract types (§2). Note that the Transfer package has no dependencies; it does not know anything about *Rep* and *Abs* functions or quotient predicates.

Above Transfer is the Lifting package, which lifts constant definitions from raw to abstract types (§3). It configures each new constant to work with Transfer. At the top are commands that configure new types to work with Lifting, such as **setup_lifting** and **quotient_type**. We expect that additional type definition commands might be implemented later. We conclude with the contribution and results of our packages (§4).

Our work was released in Isabelle 2013-1.

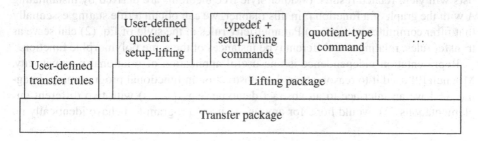

Fig. 1. Modular design of packages for formalizing quotients

2   Transfer Package

The primary function of the Transfer package is to transfer theorems from one type to another, by proving equivalences between pairs of related propositions. This process is guided by an extensible collection of *transfer rules*, which establish connections between pairs of related types or constants.

The Transfer package provides multiple user interfaces: The transfer proof method replaces the current subgoal by a logically equivalent subgoal—typically, it replaces a goal about an abstract type by a goal about the raw type. The package also provides the transferred theorem attribute, which yields a theorem about an abstract type when given a theorem involving a raw type.

2.1   Types as Relations

The design of the Transfer package is based on the idea of types as binary relations. The notions of *relational parametricity* by Reynolds [11], *free theorems* by Wadler [14], and *representation independence* by Mitchell [9] were primary sources of inspiration.

Relational parametricity tells us that different type instances of a parametrically polymorphic function must behave uniformly—that is, they must be related by a binary relation derived from the function's type. For example, the standard filter function on lists satisfies the parametricity property shown below in Eq. (2). The relation is derived from filter's type by replacing each type constructor with an appropriate relator. Relators lift relations over type constructors: Related data structures have the same shape, with pointwise-related elements, and related functions map related input to related output (see Fig. 2). For base types like bool or int we use identity relations ($\longleftrightarrow$ or $=$).

$$\text{filter} :: (\text{'a} \Rightarrow \text{bool}) \Rightarrow \text{'a list} \Rightarrow \text{'a list} \tag{1}$$

$$\forall A.\ ((A \mapsto op \longleftrightarrow) \mapsto \text{list\_all2 } A \mapsto \text{list\_all2 } A)\ \text{filter filter} \tag{2}$$

This parametricity property means that if predicates p_1 and p_2 agree on related inputs (i.e., $A\ x_1\ x_2$ implies $p_1\ x_1 \longleftrightarrow p_2\ x_2$) then filter p_1 and filter p_2 applied to related lists will yield related results. (Wadler-style free theorems are derived by instantiating A with the graph of a function f; in this manner, we can obtain a rule stating essentially that filter commutes with map.) Parametricity rules in the style of Eq. (2) can serve as transfer rules, relating two different type instances of the same polymorphic function.

Representation independence is one useful application of relational parametricity. Mitchell [9] used it to reason about data abstraction in functional programming. Imagine we have an interface to an abstract datatype (e.g. queues) with two different implementations. We would hope for any queue-using program to behave identically no

$$(\text{prod\_rel } A\ B)\ x\ y \equiv A\ (\text{fst } x)\ (\text{fst } y) \wedge B\ (\text{snd } x)\ (\text{snd } y)$$
$$(A \mapsto B)\ f\ g \equiv (\forall x\ y.\ A\ x\ y \longrightarrow B\ (f\ x)\ (g\ y))$$
$$(\text{set\_rel } A)\ X\ Y \equiv (\forall x \in X.\ \exists y \in Y.\ A\ x\ y) \wedge (\forall y \in Y.\ \exists x \in X.\ A\ x\ y)$$
$$(\text{list\_all2 } A)\ xs\ ys \equiv \text{length } xs = \text{length } ys \wedge (\forall (x, y) \in \text{set } (\text{zip } xs\ ys).\ A\ x\ y)$$

Fig. 2. Relators for various type constructors

matter which queue implementation is used—i.e., that the two queue implementations are *contextually equivalent*. Representation independence implies that this is so, as long as we can find a relation between the two implementation types that is preserved by all the corresponding operations. In our work, we refer to such a relation as a *transfer relation*.

The Transfer package is essentially a working implementation of the idea of representation independence, but in a slightly different setting: Instead of a typical functional programming language, we use higher-order logic; and instead of showing contextual equivalence of programs, we show logical equivalence of propositions.

Example: Int/nat transfer. We consider a simple use case, transferring propositions between the integers and natural numbers. We can think of type int as a concrete representation of the more abstract type nat; each type has its own implementation of numerals, arithmetic operations, comparisons, and so on. To specify the connection between the two types, we define a transfer relation ZN :: int $\Rightarrow$ nat $\Rightarrow$ bool.

$$\text{ZN } x\, n \equiv (x = \text{int } n) \tag{3}$$

We can then use ZN to express relationships between constants in the form of transfer rules. Obviously, the integer 1 corresponds to the natural number 1. The respective addition operators map related arguments to related results. Similarly, less-than on integers corresponds to less-than on naturals. Finally, bounded quantification over the non-negative integers corresponds to universal quantification over type nat.

$$(\text{ZN}) \ (1{::}\text{int}) \ (1{::}\text{nat}) \tag{4}$$
$$(\text{ZN} \Mapsto \text{ZN} \Mapsto \text{ZN}) \ (op\ +) \ (op\ +) \tag{5}$$
$$(\text{ZN} \Mapsto \text{ZN} \Mapsto op \longleftrightarrow) \ (op\ <) \ (op\ <) \tag{6}$$
$$((\text{ZN} \Mapsto op \longleftrightarrow) \Mapsto op \longleftrightarrow) \ (\text{Ball } \{0..\}) \ \text{All} \tag{7}$$

The Transfer package can use the rules above to derive equivalences like the following.

$$(\forall x{::}\text{int} \in \{0..\}.\ x < x + 1) \longleftrightarrow (\forall n{::}\text{nat}.\ n < n + 1) \tag{8}$$

If we apply the transfer method to a subgoal of the form $\forall n{::}\text{nat}.\ n < n + 1$, the Transfer package will prove the equivalence above, and then use it to replace the subgoal with $\forall x{::}\text{int} \in \{0..\}.\ x < x + 1$. The transferred attribute works in the opposite direction: Given the theorem $\forall x{::}\text{int} \in \{0..\}.\ x < x + 1$, it would prove the same equivalence, and return the theorem $\forall n{::}\text{nat}.\ n < n + 1$. In general, the Transfer package can handle any lambda term constructed from constants for which it has transfer rules.

2.2 Transfer Algorithm

The core functionality of the Transfer package is to prove equivalence theorems in the style of Eq. (8). To derive an equivalence theorem, the Transfer package uses transfer rules for constants, along with elimination and introduction rules for $\Mapsto$.

$$\frac{(A \Mapsto B)\, f\, g \qquad A\, x\, y}{B\, (f\, x)\, (g\, y)} \ (\Mapsto\text{-ELIM}) \qquad\qquad \frac{\forall x\, y.\, A\, x\, y \longrightarrow B\, (f\, x)\, (g\, y)}{(A \Mapsto B)\, f\, g} \ (\Mapsto\text{-INTRO})$$

Alternatively, these rules can be restated in the form of structural typing rules, similar to those for the simply typed lambda calculus. A typing judgment here involves two terms instead of one, and a binary relation takes the place of a type. The environment Γ collects the local assumptions for bound variables.

$$\text{APP} \quad \frac{\Gamma \vdash (A \Rrightarrow B)\, f\, g \quad \Gamma \vdash A\, x\, y}{\Gamma \vdash B\, (f\, x)\, (g\, y)}$$

$$\text{ABS} \quad \frac{\Gamma, A\, x\, y \vdash B\, (f\, x)\, (g\, y)}{\Gamma \vdash (A \Rrightarrow B)\, (\lambda x.\, f\, x)\, (\lambda y.\, g\, y)}$$

$$\text{VAR} \quad \frac{A\, x\, y \in \Gamma}{\Gamma \vdash A\, x\, y}$$

To transfer a theorem requires us to build a derivation tree using these rules, with transfer rules for constants at the leaves of the tree. For the transfer method, we are given only the abstract right-hand side; for the transferred attribute, only the left-hand side. The job of the Transfer package is to fill in the remainder of the tree—essentially a type inference problem.

Our implementation splits the process into two steps. Step one is to determine the overall shape of the derivation tree: the arrangement of APP, ABS, and VAR nodes, and the pattern of unknown term and relation variables. Step two is then to fill in the leaves of the tree using the collection of transfer rules, at the same time instantiating the unknown variables.

Step one starts by building a "skeleton" s of the known term t—a lambda term with the same structure, but with constants replaced by fresh variables. Using Isabelle's standard type inference algorithm, we annotate s with types; the inferred types determine the pattern of relation variables in the derivation tree. For step two, we set up a schematic proof state with one goal for each leaf of the tree, and then match transfer rules with subgoals. We use backtracking search in case multiple transfer rules match a given left- or right-hand side.

As an example, we will transfer the proposition $\forall n::\text{nat}.\ n \leq n$. This is actually syntax for All $(\lambda n::\text{nat}.\ \text{le}\ n\ n)$, so its skeleton has the form $t\ (\lambda x.\ u\ x\ x)$. Type inference yields a most general typing with $t :: (\text{'a} \Rightarrow \text{'b}) \Rightarrow \text{'c}$ and $u :: \text{'a} \Rightarrow \text{'a} \Rightarrow \text{'b}$, where 'a, 'b, and 'c are fresh type variables. We generate fresh relation variables $?a$, $?b$, and $?c$ corresponding to these, and use them to build an initial derivation tree following the skeleton's structure and inferred types:

$$\frac{\dfrac{\dfrac{?a\, x\, n \vdash (?a \Rrightarrow ?a \Rrightarrow ?b)\, ?u\ \text{le} \quad ?a\, x\, n \vdash ?a\, x\, n}{?a\, x\, n \vdash (?a \Rrightarrow ?b)\, (?u\, x)\, (\text{le}\, n) \quad ?a\, x\, n \vdash ?a\, x\, n}}{?a\, x\, n \vdash ?b\, (?u\, x\, x)\, (\text{le}\, n\, n)}}{\dfrac{\vdash ((?a \Rrightarrow ?b) \Rrightarrow ?c)\, ?t\ \text{All} \quad \vdash (?a \Rrightarrow ?b)\, (\lambda x.\, ?u\, x\, x)\, (\lambda n.\, \text{le}\, n\, n)}{\vdash ?c\, (?t\, (\lambda x.\, ?u\, x\, x))\, (\text{All}\, (\lambda n.\, \text{le}\, n\, n))}}$$

Note that the leaves with $?a\, x\, n$ are solved with rule VAR, but the leaves with constants All and le are as yet unsolved. Therefore this derivation tree yields a theorem with two hypotheses, $((?a \Rrightarrow ?b) \Rrightarrow ?c)\, ?t\ \text{All}$ and $(?a \Rrightarrow ?a \Rrightarrow ?b)\, ?u\ \text{le}$, and a conclusion $?c\, (?t\, (\lambda x.\, ?u\, x\, x))\, (\text{All}\, (\lambda n.\, \text{le}\, n\, n))$. In step two, we set up a proof state with the hypotheses as subgoals. The first goal is matched by Eq. (7), and the second goal by $(\text{ZN} \Rrightarrow \text{ZN} \Rrightarrow op \longleftrightarrow)\, (op \leq)\, (op \leq)$. Similarly instantiating the schematic variables in the conclusion yields the final equivalence theorem:

$$(\forall x::\text{int} \in \{0..\}.\ x \leq x) \longleftrightarrow (\forall n::\text{nat}.\ n \leq n) \tag{9}$$

2.3 Parameterized Transfer Relations

The design of the Transfer package generalizes easily to transfer relations with parameters. As an example, we define a transfer relation between lists and a finite set type; it is parameterized by a relation on the element types. We assume a function Fset :: 'a list ⇒ 'a fset that converts the given list to a finite set.

$$\mathsf{LF} :: (\text{'}a_1 \Rightarrow \text{'}a_2 \Rightarrow \mathsf{bool}) \Rightarrow \text{'}a_1 \ \mathsf{list} \Rightarrow \text{'}a_2 \ \mathsf{fset} \Rightarrow \mathsf{bool} \tag{10}$$

$$(\mathsf{LF} \ A) \ xs \ Y \equiv \exists ys. \ \mathsf{list\_all2} \ A \ xs \ ys \wedge \mathsf{Fset} \ ys = Y \tag{11}$$

If we define versions of the functions map and concat that work on finite sets, we can relate them to the list versions with the transfer rules shown here.

$$((A \mapsto B) \mapsto \mathsf{LF} \ A \mapsto \mathsf{LF} \ B) \ \mathsf{map} \ \mathsf{map\_fset} \tag{12}$$

$$(\mathsf{LF} \ (\mathsf{LF} \ A) \mapsto \mathsf{LF} \ A) \ \mathsf{concat} \ \mathsf{concat\_fset} \tag{13}$$

These rules allow the Transfer package to work on formerly problematic goals such as map_fset f (concat_fset xss) = concat_fset (map_fset (map_fset f) xss), as long as appropriate transfer rules for equality are also present. The same transfer rules work for all type instances of these constants.

2.4 Transfer Rules with Side Conditions

Some polymorphic functions in Isabelle require side conditions on their parametricity theorems. For example, consider the equality relation =, which has the polymorphic type 'a ⇒ 'a ⇒ bool. Its type would suggest $(A \mapsto A \mapsto op \longleftrightarrow) \ (op =) \ (op =)$, but this does not hold for all relations A—it only holds if A is *bi-unique*, i.e., single-valued and injective.

$$\mathsf{bi\_unique} \ A \Longrightarrow (A \mapsto A \mapsto op \longleftrightarrow) \ (op =) \ (op =) \tag{14}$$

As pointed out by Wadler [14], this restriction on relations is akin to an *eqtype* annotation in ML, or an *Eq* class constraint in Haskell. While Haskell allows users to provide *Eq* instance declarations, the Transfer package allows us to provide additional rules about bi-uniqueness that serve the same purpose, for example: bi_unique ZN, bi_unique $A \Longrightarrow$ bi_unique (set_rel A) and bi_unique $A \Longrightarrow$ bi_unique (list_all2 A).

Using the above rules, the Transfer package is able to relate equality on lists of integers with equality on lists of naturals, using the relation list_all2 ZN. It can similarly relate equality on sets, lists of sets, sets of lists, and so on.

The universal quantifier requires a different side condition on its parametricity rule. While equality requires bi-uniqueness, the universal quantifier requires the relation A to be *bi-total*—i.e., A must be both total and surjective.

$$\mathsf{bi\_total} \ A \Longrightarrow ((A \mapsto op \longleftrightarrow) \mapsto op \longleftrightarrow) \ \mathsf{All} \ \mathsf{All} \tag{15}$$

Universal quantifiers appear in most propositions used with transfer; however, many transfer relations (including ZN) are not bi-total, but only *right-total*, i.e., surjective. The following transfer rule can then be used if no other specialized rule is provided:

$$\mathsf{right\_total} \ A \Longrightarrow ((A \mapsto op \longleftrightarrow) \mapsto op \longleftrightarrow) \ (\mathsf{Ball} \ \{x. \ \mathsf{Domainp} \ A \ x\}) \ \mathsf{All} \tag{16}$$

The predicate Domainp is defined as Domainp $T\,x \equiv \exists y.\ T\,x\,y$. Because it is awkward to work with expressions like Domainp T in the transferred goal, we implemented a post-processing step that can replace Domainp expressions with equivalent but more convenient predicates. This is configured by registering a *transfer domain rule*: Domainp ZN $= (\lambda x.\ x \geq 0)$. We provide transfer domain rules for lists and other types; thus we can replace, for example, Domainp (list_all2 ZN) by list_all $(\lambda x.\ x \geq 0)$. The use of Domainp is not limited to quantifiers—the usual parametricity rules for constants like UNIV, Collect, and set intersection $\cap$ require bi-totality, but we also provide more widely applicable transfer rules using Domainp.

The last condition we can use to restrict relations is being *right-unique*, i.e., single-valued. Bi-totality, right-totality and right-uniqueness are like bi-uniqueness preserved by many relators, including those for lists and sets. We mentioned that ZN is not bi-total but, e.g, total quotients yield bi-total transfer relations; see the overview in Tab. 1.

Handling equality relations. Many propositions contain non-polymorphic constants that remain unchanged by the transfer procedure, e.g., boolean operations. We would like to avoid the necessity for lots of trivial transfer rules like the rule for the boolean conjunction: $(op \longleftrightarrow \Mapsto op \longleftrightarrow \Mapsto op \longleftrightarrow)\ (op \wedge)\ (op \wedge)$. Instead we define a predicate is_equality A, which holds if and only if A is the equality relation on its type, and register a single reflexivity transfer rule is_equality $A \Longrightarrow A\,x\,x$. The is_equality predicate is preserved by all of the standard relators, including lists, sets, pairs, and function space.

2.5 Proving Implications Instead of Equivalences

The transfer proof method can replace a universal with an equivalent bounded quantifier: e.g., $(\forall n{::}\mathsf{nat}.\ n < n + 1)$ is transferred to $(\forall x{::}\mathsf{int} \in \{0..\}.\ x < x + 1)$. This yields a useful extra assumption in the new subgoal. With the transferred attribute, however, it may be preferable to start with a stronger theorem $(\forall x{::}\mathsf{int}.\ x < x + 1)$, without the bounded quantifier. In this case, the Transfer package can prove an implication:

$$(\forall x{::}\mathsf{int}.\ x < x + 1) \longrightarrow (\forall n{::}\mathsf{nat}.\ n < n + 1) \tag{17}$$

The Transfer algorithm works exactly the same; we just need some new transfer rules that encode monotonicity. We provide rules for quantifiers and implication, using various combinations of $\longrightarrow$, $\longleftarrow$, and $\longleftrightarrow$; a few are shown here.

$$\mathsf{right\_total}\ A \Longrightarrow ((A \Mapsto op \longrightarrow) \Mapsto op \longrightarrow)\ \mathsf{All}\ \mathsf{All} \tag{18}$$

$$\mathsf{right\_total}\ A \Longrightarrow ((A \Mapsto op \longleftrightarrow) \Mapsto op \longrightarrow)\ \mathsf{All}\ \mathsf{All} \tag{19}$$

$$(op \longleftarrow \Mapsto op \longrightarrow \Mapsto op \longrightarrow)\ (op \longrightarrow)\ (op \longrightarrow) \tag{20}$$

The derivation of Eq. (17) uses transfer rule (19); rule (18) comes into play when quantifiers are nested. These rules are applicable to relation ZN because it is right-total. Further variants of these rules (involving reverse implication) are used to transfer induction and case analysis rules, which have many nested implications and quantifiers.

Having many different transfer rules for the same constants would tend to introduce a large amount of backtracking search in step two of the transfer algorithm. To counter this, we pre-instantiate some of the relation variables to $\longrightarrow$, $\longleftarrow$, or $\longleftrightarrow$, guided by a simple monotonicity analysis.

3 Lifting Package

The Lifting package allows users to lift terms of the raw type to the abstract type, which is a necessary step in building a library for an abstract type. Lifting defines a new constant by combining coercion functions (Abs and Rep) with the raw term. It also proves an appropriate transfer rule for the Transfer package and, if possible, an equation for the code generator. Doing this lifting manually is mostly tedious and uninteresting; our goal is to automate as much as possible, so users can focus on the interesting bits.

The Lifting package provides two commands: **setup_lifting** for initializing the package to work with a new type, and **lift_definition** for lifting constants. The Lifting package works with four kinds of type abstraction: type copies, subtypes, total quotients and partial quotients. See Tab. 1 for an overview of these.

Example: finite sets. Let us define a type of finite sets as a quotient of lists, where two lists are in the same equivalence class if they represent the same set:

quotient_type 'a fset = 'a list / ($\lambda xs\ ys.$ set xs = set ys)

Now we can define the union of two finite sets as a lifted function of concatenation of two lists append :: 'a list $\Rightarrow$ 'a list $\Rightarrow$ 'a list, which has infix syntax _ @ _:

lift_definition funion :: 'a fset $\Rightarrow$ 'a fset $\Rightarrow$ 'a fset" **is** append

The command opens a proof environment with the following obligation:

$\bigwedge l_1\ l_2\ l_3\ l_4.$ set l_1 = set $l_2 \implies$ set l_3 = set $l_4 \implies$ set $(l_1\ @\ l_3)$ = set $(l_2\ @\ l_4)$

The obligation is called a *respectfulness theorem* and says that append respects the equivalence relation that defines 'a fset. When the user proves the obligation, the new function funion is defined as follows:

funion $A\ B \equiv$ abs_fset $((\text{rep\_fset}\ A)\ @\ (\text{rep\_fset}\ B))$

The package also generates a code equation for the code generator:

funion (abs_fset A) (abs_fset B) = abs_fset $(A\ @\ B)$

And finally, because the package proved internally a corresponding transfer rule, we can prove, e.g., that funion commutes: **lemma** funion $A\ B$ = funion $B\ A$. If we apply the method transfer, we get $\bigwedge A\ B.$ set $(A\ @\ B)$ = set $(B\ @\ A)$, which is easily provable.

If we defined 'a fset by **typedef** 'a fset = $\{A ::$ 'a set. finite $A\}$, i.e., as a subtype of sets, and funion as a lifted function of the set union $\cup$, we would get the proof obligation $\bigwedge s_1\ s_2.$ finite $s_1 \implies$ finite $s_2 \implies$ finite $(s_1 \cup s_2)$ and the code equation rep_fset (funion $A\ B$) = rep_fset $A \cup$ rep_fset B.[1]

3.1 General Case

We abstract from the presented example now and give a description that covers the general case of what the Lifting package does. The input of the lifting is a term $t :: \tau_1$ on

---

[1] See §3.3 for more about what the code equations are and how they are derived.

the concrete level, an abstract type τ_2 and a name f of the new constant. In our example, $t = $ append, $\tau_1 = $ 'a list $\Rightarrow$ 'a list $\Rightarrow$ 'a list, $\tau_2 = $ 'a fset $\Rightarrow$ 'a fset $\Rightarrow$ 'a fset and $f = $ funion. We work generally with types τ which are composed from type constructors κ and other types $\overline{\vartheta}$. Then we write $\tau = \overline{\vartheta}\,\kappa$. Each type parameter of κ can be either co-variant (we write $+$) or contra-variant $(-)$. E.g., in the function type $\alpha \rightarrow \beta$, α is contra-variant whereas β is co-variant.

In this section, we define three functions Morph^p, Relat and Trans. Morph^p is a combination of abstraction and representation functions and gives us the definition of f. The polarity superscript p $(+$ or $-)$ encodes if an abstraction or a representation function should be generated. Relat is a combination of equivalence relations and allows us to describe that t behaves correctly—respects the equivalence classes. Finally, Trans is a composed transfer relation and describes how t and f are related. More formally, if the user proves the respectfulness theorem $\text{Relat}(\tau_1, \tau_2)\ t\ t$, the Lifting package will define the new constant f as $f = \text{Morph}^+(\tau_1, \tau_2)\ t$ and proves the transfer rule $\text{Trans}(\tau_1, \tau_2)\ t\ f$.

For now we will not distinguish between quotients, subtypes, etc. Instead we unify all four kinds of type abstraction with a general notion of an abstract type.

Definition 1. *We say that κ_2 is an* abstract type *of κ_1 if there is a transfer relation $T_{\kappa_1,\kappa_2} ::$ $(\overline{\vartheta})\,\kappa_1 \rightarrow (\overline{\alpha})\,\kappa_2 \rightarrow$ bool *associated with κ_1 and κ_2 (see also Fig. 3a) such that*

1. *T_{κ_1,κ_2} is right-total and right-unique,*
2. *all type variables in $\overline{\vartheta}$ are in $\overline{\alpha}$, which contains only distinct type variables.*

We say that $\tau_2 = (\overline{\rho})\,\kappa_2$ is an instance of an abstract type *of $\tau_1 = (\overline{\sigma})\,\kappa_1$ if*

1. *κ_2 is an abstract type of κ_1 certified by $T_{\kappa_1,\kappa_2} :: (\overline{\vartheta})\,\kappa_1 \rightarrow (\overline{\alpha})\,\kappa_2 \rightarrow$ bool,*
2. *$\overline{\sigma} = \theta\,\overline{\vartheta}$, where $\theta = \text{match}(\overline{\rho}, \overline{\alpha})$ [2].*

In our finite sets example, the **quotient_type** command internally generates a transfer relation between the concrete type 'a list and the abstract type 'a fset. In principle, such a transfer relation alone is sufficient to characterize all four kinds of type abstraction: type copies, subtypes, total and partial quotients. We could build compound transfer relations for compound types and get other components (e.g., the morphisms Morph^p for the definition) from this relation using the choice operator. But it turns out that it is useful to have these other components explicitly, e.g. for generating code equations. The other components that we can derive from each transfer relation T_{κ_1,κ_2} and associate with each abstract type are ($\circ\circ$ is the relation composition):

– Partial equivalence relation R_{κ_1,κ_2} (Fig. 3b), specified by $R_{\kappa_1,\kappa_2} = T_{\kappa_1,\kappa_2} \circ\circ T_{\kappa_1,\kappa_2}^{-1}$.
– Abstraction function Abs_{κ_1,κ_2} (Fig. 3c), specified by $T_{\kappa_1,\kappa_2}\ a\ b \longrightarrow Abs_{\kappa_1,\kappa_2}\ a = b$.
– Representation function Rep_{κ_1,κ_2} (Fig. 3d), specified by $T_{\kappa_1,\kappa_2}\ (Rep_{\kappa_1,\kappa_2}\ a)\ a$.

Since T_{κ_1,κ_2} is right-total and right-unique, there always exist some Abs and Rep functions that meet the above given specification. On the other hand, given R, Abs and Rep, the specification allows only right-total and right-unique T.

The reflexive part of the partial equivalence relation R_{κ_1,κ_2} implicitly specifies which values of the concrete type are used for the construction of the abstract type.[3] The

---

[2] Match is a usual matching algorithm, i.e., $\text{match}(\overline{\beta}, \overline{\alpha})$ yields a substitution θ such that $\beta = \theta\,\alpha$.
[3] We omitted reflexive edges of R_{κ_1,κ_2} in Fig. 3b.

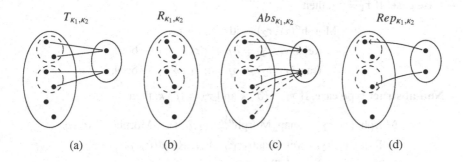

T_{κ_1,κ_2} R_{κ_1,κ_2} Abs_{κ_1,κ_2} Rep_{κ_1,κ_2}

(a)   (b)   (c)   (d)

Fig. 3. Components of an abstract type

representation and abstraction functions map abstract values to concrete values and vice versa. Abs_{κ_1,κ_2} is underspecified outside of a range of T_{κ_1,κ_2} (dashed lines in Fig. 3c) and Rep_{κ_1,κ_2} can select only one of the values in the corresponding class.

Now we come to the key definition of this section. We derived R, Abs and Rep only for transfer relations that are associated with a type constructor. But later on, we build compound transfer relations for general types. What are R, Abs and Rep in this case? Again any functions meeting the above given specification. The following quotient predicate captures this idea and bundles all the components together.

Definition 2. *We define a* quotient predicate *with the syntax* $\langle .,.,.,. \rangle$ *and we say that* $\langle R, Abs, Rep, T \rangle$ *if 1.* $R = T \circ\circ T^{-1}$, *2.* $T\ a\ b \longrightarrow Abs\ a = b$ *and 3.* $T\ (Rep\ a)\ a$.

The following definition requires that $\langle .,.,.,. \rangle$ is preserved by going through the type universe using map functions and relators.

Definition 3. *We say that* map_κ *is a* map function *for* κ *and* rel_κ *is a* relator *for* κ, *where* κ *has arity* n, *if the assumptions* $\langle R_1, m_1^+, m_1^-, T_1 \rangle, \ldots, \langle R_n, m_n^+, m_n^-, T_n \rangle$ *implies* $\langle \mathrm{rel}_\kappa R_1 \ldots R_n, \mathrm{map}_\kappa\ m^{p_\kappa^1} \ldots m^{p_\kappa^{2n}}, \mathrm{map}_\kappa\ m^{-p_\kappa^1} \ldots m^{-p_\kappa^{2n}}, \mathrm{rel}_\kappa T_1 \ldots T_n \rangle$.

Indexes p_κ^i encode which arguments of the map function are co-variant ($+$) or contra-variant ($-$). The map function can in general take $2n$ arguments because each type parameter of κ can be co-variant and contra-variant at the same time, e.g., $\alpha\ \kappa \equiv \alpha \to \alpha$.

Now we finally define Morph^p, Relat and Trans, as we promised to be the main goal of this section. First, let us define auxiliary functions morph^p, relat and trans, which are going to be used as the single step in the main definition of Morph^p, Relat and Trans. Functions morph^p, relat and trans are defined for all types $\tau_1 = (\overline{\sigma})\ \kappa_1$ and $\tau_2 = (\overline{\rho})\ \kappa_2$, where τ_2 is an instance of an abstract type of τ_1, as follows:

- $\mathrm{morph}^+(\tau_1, \tau_2) = Abs_{\kappa_1,\kappa_2} :: \tau_1 \to \tau_2$ – $\mathrm{relat}(\tau_1, \tau_2) = R_{\kappa_1,\kappa_2} :: \tau_1 \to \tau_1 \to \mathrm{bool}$
- $\mathrm{morph}^-(\tau_1, \tau_2) = Rep_{\kappa_1,\kappa_2} :: \tau_2 \to \tau_1$ – $\mathrm{trans}(\tau_1, \tau_2) = T_{\kappa_1,\kappa_2} :: \tau_1 \to \tau_2 \to \mathrm{bool}$

Now we extend the simple step functions morph^p, relat and trans defined only for abstract types to functions Morph^p, Relat and Trans, which take general types τ_1 and τ_2, by doing induction and case split on the type structure:

- **Base case.** If $\tau_1 = \tau_2$, then

$$\mathrm{Morph}^p(\tau_1,\tau_2) = \mathrm{id} :: \tau_1 \to \tau_1,$$
$$\mathrm{Relat}(\tau_1,\tau_2) = op =:: \tau_1 \to \tau_1 \to \mathrm{bool},$$
$$\mathrm{Trans}(\tau_1,\tau_2) = op =:: \tau_1 \to \tau_1 \to \mathrm{bool}.$$

- **Non-abstract type case.** If $\tau_1 = (\overline{\sigma})\,\kappa$ and $\tau_2 = (\overline{\rho})\,\kappa$, then

$$\mathrm{Morph}^p(\tau_1,\tau_2) = \mathrm{map}_\kappa\,\mathrm{Morph}^{p_\kappa^1 p}(\sigma_1,\rho_1)\,\ldots\,\mathrm{Morph}^{p_\kappa^{2n} p}(\sigma_n,\rho_n),$$
$$\mathrm{Relat}(\tau_1,\tau_2) = \mathrm{rel}_\kappa\,\mathrm{Relat}(\sigma_1,\rho_1)\,\ldots\,\mathrm{Relat}(\sigma_n,\rho_n),$$
$$\mathrm{Trans}(\tau_1,\tau_2) = \mathrm{rel}_\kappa\,\mathrm{Trans}(\sigma_1,\rho_1)\,\ldots\,\mathrm{Trans}(\sigma_n,\rho_n),$$

where map_κ is a map function for κ and $p_\kappa^i p$ is a usual multiplication of polarities: $+\cdot- = -\cdot+ = -$ and $+\cdot+ = -\cdot- = +$. The function rel_κ is a relator for type κ.
- **Abstract type case.** If $\tau_1 = (\overline{\sigma})\,\kappa_1$, $\tau_2 = (\overline{\rho})\,\kappa_2$, $\kappa_1 \neq \kappa_2$, and κ_2 is an abstract type of κ_1 certified by $T_{\kappa_1,\kappa_2} :: (\overline{\vartheta})\,\kappa_1 \to (\overline{\alpha})\,\kappa_2 \to \mathrm{bool}$, let us define $\overline{\sigma'} = \theta\,\overline{\vartheta}$, where $\theta = \mathrm{match}(\overline{\rho},\overline{\alpha})$. [4] Then we define these equations

$$\mathrm{Morph}^+(\tau_1,\tau_2) = \mathrm{morph}^+((\overline{\sigma'})\,\kappa_1,\tau_2) \circ \mathrm{Morph}^+(\tau_1,(\overline{\sigma'})\,\kappa_1),$$
$$\mathrm{Morph}^-(\tau_1,\tau_2) = \mathrm{Morph}^-(\tau_1,(\overline{\sigma'})\,\kappa_1) \circ \mathrm{morph}^-((\overline{\sigma'})\,\kappa_1,\tau_2),$$
$$\mathrm{Relat}(\tau_1,\tau_2) = \mathrm{Trans}(\tau_1,(\overline{\sigma'})\,\kappa_1) \circ\circ \mathrm{relat}((\overline{\sigma'})\,\kappa_1,\tau_2) \circ\circ \mathrm{Trans}(\tau_1,(\overline{\sigma'})\,\kappa_1)^{-1},$$
$$\mathrm{Trans}(\tau_1,\tau_2) = \mathrm{Trans}(\tau_1,(\overline{\sigma'})\,\kappa_1) \circ\circ \mathrm{trans}((\overline{\sigma'})\,\kappa_1,\tau_2).$$

The functions Morph^p, Relat and Trans are undefined if κ_2 is not an abstract type for κ_1 in the abstract type case. In such a case the Lifting package reports an error. Let us assume for the rest that we work only with such τ_1 and τ_2 that this does not happen.

Theorem 1. *Morph*^{p}, *Relat and Trans have the following types:* $\mathrm{Morph}^+(\tau_1,\tau_2) :: \tau_1 \to \tau_2$, $\mathrm{Morph}^-(\tau_1,\tau_2) :: \tau_2 \to \tau_1$, $\mathrm{Relat}(\tau_1,\tau_2) :: \tau_1 \to \tau_1 \to \mathrm{bool}$ *and* $\mathrm{Trans}(\tau_1,\tau_2) :: \tau_1 \to \tau_2 \to \mathrm{bool}$.

Proof. By induction on defining equations of Morph^p, Relat and Trans.   □

Thus in our context, where $t :: \tau_1$, the terms $\mathrm{Relat}(\tau_1,\tau_2)\,t\,t$, $f = \mathrm{Morph}^+(\tau_1,\tau_2)\,t$ and $\mathrm{Trans}(\tau_1,\tau_2)\,t\,f$ are well-typed terms and f has indeed type τ_2. The respectfulness theorem $\mathrm{Relat}(\tau_1,\tau_2)\,t\,t$ has to be proven by the user. The definitional theorem $f = \mathrm{Morph}^+(\tau_1,\tau_2)\,t$ is proven by Isabelle. The remaining question is how we get the transfer rule $\mathrm{Trans}(\tau_1,\tau_2)\,t\,f$. Two following theorems give us the desired transfer rule.

Theorem 2. *If* $\langle R, Abs, Rep, T \rangle$, $f = Abs\,t$ *and* $R\,t\,t$, *then* $T\,t\,f$.

Proof. Because $R = T \circ\circ T^{-1}$, and $R\,t\,t$, we have $\exists x.\ T\,t\,x$. Let us denote this x as g. Thus $Abs\,t = g$ follows from $T\,t\,g$. But from $f = Abs\,t$ we have $f = g$ and thus $T\,t\,f$.   □

The following theorem is the key theorem of this section: it proves that our definitions of Morph^p, Relat and Trans are legal, i.e., they have the desired property that they still form a (compound) abstract type, i.e., they meet the quotient predicate.

---

[4] Definition 1 guarantees that all type variables in $\overline{\vartheta}$ are in $\overline{\alpha}$ and thus $\overline{\rho}$ uniquely determines $\overline{\sigma'}$.

Theorem 3. $\langle \text{Relat}(\tau_1, \tau_2), \text{Morph}^+(\tau_1, \tau_2), \text{Morph}^-(\tau_1, \tau_2), \text{Trans}(\tau_1, \tau_2) \rangle$

Proof. By induction on defining equations of Morph^p, Relat and Trans: Base case: $\langle op =, \text{id}, \text{id}, op = \rangle$ holds. Non-abstract type case: $\langle ., ., ., . \rangle$ is preserved as it is required in Definition 3. Abstract type case: for all τ_2, an instance of the abstract type of τ_1, $\langle \text{relat}(\tau_1, \tau_2), \text{morph}^+(\tau_1, \tau_2), \text{morph}^-(\tau_1, \tau_2), \text{trans}(\tau_1, \tau_2) \rangle$ holds by construction. And finally, the key fact that we need: $\langle R_1, Abs_1, Rep_1, T_1 \rangle$ and $\langle R_2, Abs_2, Rep_2, T_2 \rangle$ implies $\langle T_1 \circ\circ R_2 \circ\circ T_1^{-1}, Abs_2 \circ Abs_1, Rep_1 \circ Rep_2, T_1 \circ\circ T_2 \rangle$, i.e., $\langle ., ., ., . \rangle$ is preserved through the composition of abstract types. We proved this key fact in Isabelle/HOL. □

If we compose Theorem 3 with the Theorem 2 and use $f = \text{Morph}^+(\tau_1, \tau_2)$ t and $\text{Relat}(\tau_1, \tau_2)$ $t\,t$, we get the desired transfer rule $\text{Trans}(\tau_1, \tau_2)$ $t\,f$.

In the original Quotient package, $\text{Relat}(\tau_1, \tau_2)$ in the abstract type case was defined as $\text{Relat}(\tau_1, (\overline{\sigma'})\,\kappa_1) \circ\circ \text{relat}((\overline{\sigma'})\,\kappa_1, \tau_2) \circ\circ \text{Relat}(\tau_1, (\overline{\sigma'})\,\kappa_1)$. Although Theorem 1 still holds, Theorem 3 cannot be proven and thus the Quotient package does not cover the whole possible type universe. As a consequence, transferring of theorems did not work for general case of composed abstract types, but the package was still a great progress.

3.2 Implementation

In Isabelle/HOL, $\langle ., ., ., . \rangle$ is defined as the Quotient predicate, whose definition is equivalent to Definition 2. A new abstract type can be registered by a command **setup_lifting** by providing such a Quotient R Abs Rep T theorem, which certifies that the given components R, Abs, Rep and T constitute an abstract type. Quotient theorems for relators and map functions are registered by the attribute quot_map. Such a theorem for the function type is the most prominent one; another example is a theorem for the list type:

lemma fun_quotient: Quotient R_1 abs_1 rep_1 $T_1 \implies$ Quotient R_2 abs_2 rep_2 $T_2 \implies$
   Quotient $(R_1 \Rrightarrow R_2)$ $(rep_1 \mapsto abs_2)$ $(abs_1 \mapsto rep_2)$ $(T_1 \Rrightarrow T_2)$
lemma Quotient_list: Quotient R Abs Rep $T \implies$
   Quotient (list_all2 R) (map Abs) (map Rep) (list_all2 T)

We implemented a syntax-driven procedure that proves a Quotient theorem for a given pair of types τ_1 and τ_2. This procedure recursively descends τ_1 and τ_2 to prove Theorem 3 for τ_1 and τ_2 and the implementation basically follows our induction definition of Morph^p, Relat and Trans in the previous section. The advantage is that this procedure not only proves the compound Quotient theorem in order to derive the transfer theorem, but it also synthesizes the terms $\text{Morph}^p(\tau_1, \tau_2)$, $\text{Relat}(\tau_1, \tau_2)$ and $\text{Trans}(\tau_1, \tau_2)$ as a side effect. This approach was not used in the Quotient package; thanks to it, we got a simpler implementation and managed to remove many technical limitations of the original Quotient package with surprising ease.

Users generally will not prove the Quotient theorem manually for new types, as special commands exist to automate the process. The command **quotient_type** defines a new quotient type, internally proves the corresponding Quotient theorem and registers it with **setup_lifting**. We also support types defined by the command **typedef** The theorem type_definition Rep Abs $\{x. \; P \; x\}$, which axiomatizes the newly defined subtype, can be supplied to **setup_lifting**. It then internally proves the quotient theorem Quotient (invariant P) Abs Rep T, where the transfer relation T is defined as $T \; x \; y \equiv Rep \; y = x$ and the equivalence relation invariant $P \equiv (\lambda x \; y. \; x = y \wedge P \; x)$.

Since the respectfulness theorem is the only proof obligation presented to the user, we also implemented a procedure that does some preprocessing to present this obligation in a user-friendly, readable form in **lift_definition**. The procedure also simplifies the goal if the involved relations come from a subtype. Then the user gets predicates and predicators (e.g., list_all for 'a list) instead of relations and relators. If the relation comes only from type copies, the respectfulness theorem is fully proven by our procedure.

We also implemented a procedure that automatically proves a parameterized transfer rule, which is a stronger transfer rule (see §2.3), in **lift_definition** if the user provides a theorem certifying that the concrete term used in the definition is parametric.

3.3   Code Equations

The code generator is a central component in Isabelle/HOL and is used in a lot of projects for algorithm verification. That is why when we define a new constant by **lift_definition**, we are concerned with how to execute the new constant provided the concrete term is also executable. This can be done by providing *code equations*. See [2] for more about the code equations and how Lifting and Transfer provide efficient code for operations on abstract types. The code generator accepts two types of equations:

- *Representation function equation* has form $Rep\ f = t$, where Rep is not in t and $Abs\ (Rep\ x) = x$ holds, which is provable for any abstract type in our context.
- *Abstract function equation* has form $f\ (Abs_1\ x_1)\ldots(Abs_n\ x_n) = t$.

Here we give only a glimpse how the equations are proven. We assume the usual map function $\mapsto ((f \mapsto g)\ h = g \circ h \circ f)$ and relator $\Mapsto$ (as in Fig. 2) for function type. Let us have the definition $f = Abs\ t$, $f :: \sigma_1 \to \cdots \to \sigma_n \to \sigma_{n+1}$ and the proven respectfulness theorem $R\ t\ t$. Then from the Quotient theorem for the function type and from the construction described in §3.1 it easily follows that $R = R_1 \Mapsto \ldots \Mapsto R_n \Mapsto R_{n+1}$, $Abs = Rep_1 \mapsto \ldots \mapsto Rep_n \mapsto Abs_{n+1}$ and $Rep = Abs_1 \mapsto \ldots \mapsto Abs_n \mapsto Rep_{n+1}$.

Representation function equation. By unfolding the map function $\mapsto$ in the definition of f and using simple facts we get $Rep_{n+1}\ (f\ x_1\ldots x_n) = Rep_{n+1}\ (Abs_{n+1}\ T)$, where $T = t\ (Rep_1\ x_1)\ldots(Rep_n\ x_n)$. If R_{n+1}, Rep_{n+1} and Abs_{n+1} represent a subtype or a type copy, the relation R_{n+1} is a subset of the equality and thus $Rep_{n+1}\ (Abs_{n+1}\ T) = T$ and finally $Rep_{n+1}\ (f\ x_1\ldots x_n) = t\ (Rep_1\ x_1)\ldots(Rep_n\ x_n)$.

Abstraction function equation. By unfolding $\Mapsto$ in R and $\mapsto$ in Rep and using simple facts we get this equation $R_1\ x_1\ x_1 \longrightarrow \ldots \longrightarrow R_n\ x_n\ x_n \longrightarrow f\ (Abs_1\ x_1)\ldots(Abs_n\ x_n) = Abs_{n+1}\ (t\ x_1\ldots x_n)$. If R_1 to R_n are relations that are composed from relators that preserve reflexivity (e.g., holds for any datatype relator) and the abstract types that are involved are total (i.e., a type copy or a total quotient), the procedure that we implemented discharges automatically each of these assumptions and gives us a plain equation.[5]

---

[5] The function relator $\Mapsto$ does not preserve reflexivity in the negative position. But this is not a limitation in practice, because there is hardly a function with a functional parameter that would have an abstract type in the negative position. Because $\Mapsto$ preserves equality, we can still discharge functional parameters using type copies or non-abstract functional parameters.

Table 1. Categorization of abstract types and respective equations

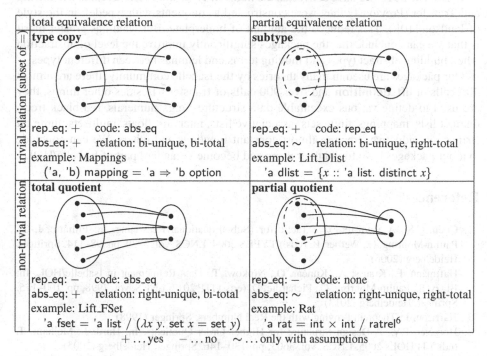

	total equivalence relation	partial equivalence relation
trivial relation (subset of =)	**type copy** rep_eq: + code: abs_eq abs_eq: + relation: bi-unique, bi-total example: Mappings ('a, 'b) mapping = 'a $\Rightarrow$ 'b option	**subtype** rep_eq: + code: rep_eq abs_eq: $\sim$ relation: bi-unique, right-total example: Lift_Dlist 'a dlist = $\{x :: $ 'a list. distinct $x\}$
non-trivial relation	**total quotient** rep_eq: − code: abs_eq abs_eq: + relation: right-unique, bi-total example: Lift_FSet 'a fset = 'a list / $(\lambda x\ y.\ \text{set } x = \text{set } y)$	**partial quotient** rep_eq: − code: none abs_eq: $\sim$ relation: right-unique, right-total example: Rat 'a rat = int $\times$ int / ratrel[6]

+ ... yes   − ... no $\sim$ ... only with assumptions

Thus we can generate representation function equations for type copies and subtypes and abstraction function equations for any abstract type, but we can discharge the extra assumptions only for total types. See Tab. 1 for an overview.

4   Conclusion

We have presented a new design for automation of abstract types in Isabelle/HOL. The distinctive features and the main contributions are:

- The modular design of cleanly separated components with well-defined interfaces yields flexibility, i.e., Lifting is not limited to types defined by **quotient_type** and similarly Transfer to constants defined by **lift_definition**.
- Only one transfer rule is needed for different instances of polymorphic constants like fmap :: ('a $\Rightarrow$ 'b) $\Rightarrow$ 'a fset $\Rightarrow$ 'b fset.
- The Lifting package supports arbitrary type constructors, rather than only co-variant ones plus hard-coded function type.
- The Lifting package can handle a composition of abstract types in all cases.
- The Lifting package generates code equations for the code generator.
- The package generates the statement of the respectfulness theorem, discharges it automatically for type copies, and simplifies it to user-friendly form in other cases.

---

[6] ratrel $x\ y \equiv$ snd $x \neq 0 \wedge$ snd $y \neq 0 \wedge$ fst $x *$ snd $y =$ fst $y *$ snd x.

In Isabelle 2013, we have converted the numeric types int, rat, and real to use Lifting and Transfer (Previously they were constructed as quotients with typedef, in the style of Paulson [10].). This reduced the amount of boilerplate. But what has greater merit is that we can conclude that the packages singificantly improve the level of abstraction when building abstract types and moving terms and lemmas between different types.

Our packages are used in many theories by the Isabelle community: there are almost 400 calls of **lift_definition** and over 900 calls of transfer.[7] Besides other things, they are used to define various executable data structures: e.g., numerals, red-black trees, distinct lists, mappings, finite sets, associative lists, intervals, floats, multisets, finite bit strings, co-inductive streams, almost constant functions and others. We can conclude that our packages have found their users and become a standard part of Isabelle/HOL.

References

1. Coen, C.S.: A Semi-reflexive Tactic for (Sub-)Equational Reasoning. In: Filliâtre, J.-C., Paulin-Mohring, C., Werner, B. (eds.) TYPES 2004. LNCS, vol. 3839, pp. 98–114. Springer, Heidelberg (2006)
2. Haftmann, F., Krauss, A., Kunčar, O., Nipkow, T.: Data Refinement in Isabelle/HOL. In: Blazy, S., Paulin-Mohring, C., Pichardie, D. (eds.) ITP 2013. LNCS, vol. 7998, pp. 100–115. Springer, Heidelberg (2013)
3. Harrison, J.: Theorem Proving with the Real Numbers. Springer (1998)
4. Homeier, P.V.: A Design Structure for Higher Order Quotients. In: Hurd, J., Melham, T. (eds.) TPHOLs 2005. LNCS, vol. 3603, pp. 130–146. Springer, Heidelberg (2005)
5. Kaliszyk, C., Urban, C.: Quotients revisited for Isabelle/HOL. In: Proc. of the 26th ACM Symposium on Applied Computing (SAC 2011), pp. 1639–1644. ACM (2011)
6. Krauss, A.: Simplifying Automated Data Refinement via Quotients. Tech. rep., TU München (2011), http://www21.in.tum.de/~krauss/papers/refinement.pdf
7. Lammich, P.: Automatic data refinement. In: Blazy, S., Paulin-Mohring, C., Pichardie, D. (eds.) ITP 2013. LNCS, vol. 7998, pp. 84–99. Springer, Heidelberg (2013)
8. Magaud, N.: Changing data representation within the Coq system. In: Basin, D., Wolff, B. (eds.) TPHOLs 2003. LNCS, vol. 2758, pp. 87–102. Springer, Heidelberg (2003)
9. Mitchell, J.C.: Representation Independence and Data Abstraction. In: POPL, pp. 263–276. ACM Press (January 1986)
10. Paulson, L.C.: Defining functions on equivalence classes. ACM Trans. Comput. Logic 7(4), 658–675 (2006)
11. Reynolds, J.C.: Types, Abstraction and Parametric Polymorphism. In: IFIP Congress, pp. 513–523 (1983)
12. Slotosch, O.: Higher Order Quotients and their Implementation in Isabelle HOL. In: Gunter, E.L., Felty, A.P. (eds.) TPHOLs 1997. LNCS, vol. 1275, pp. 291–306. Springer, Heidelberg (1997)
13. Sozeau, M.: A New Look at Generalized Rewriting in Type Theory. In: 1st Coq Workshop Proceedings (2009)
14. Wadler, P.: Theorems for free! In: Functional Programming Languages and Computer Architecture, pp. 347–359. ACM Press (1989)

---

[7] Isabelle distribution and *Archive of Formal Proofs* (http://afp.sf.net), release 2013-1.

Refinements for Free!*

Cyril Cohen[1], Maxime Dénès[2], and Anders Mörtberg[1]

[1] Department of Computer Science and Engineering
Chalmers University of Technology and University of Gothenburg
{cyril.cohen,anders.mortberg}@gu.se
[2] INRIA Sophia Antipolis – Méditerranée
mail@maximedenes.fr

Abstract. Formal verification of algorithms often requires a choice between definitions that are easy to reason about and definitions that are computationally efficient. One way to reconcile both consists in adopting a high-level view when proving correctness and then refining stepwise down to an efficient low-level implementation. Some refinement steps are interesting, in the sense that they improve the algorithms involved, while others only express a switch from data representations geared towards proofs to more efficient ones geared towards computations. We relieve the user of these tedious refinements by introducing a framework where correctness is established in a proof-oriented context and automatically transported to computation-oriented data structures. Our design is general enough to encompass a variety of mathematical objects, such as rational numbers, polynomials and matrices over refinable structures. Moreover, the rich formalism of the CoQ proof assistant enables us to develop this within CoQ, without having to maintain an external tool.

Keywords: CoQ, Data refinements, Formal proofs, Efficient algorithms and data structures, Parametricity.

1 Introduction

It is commonly conceived that computationally well-behaved programs and data structures are more difficult to study formally than naive ones. Rich formalisms like the Calculus of Inductive Constructions, on which the CoQ [6] proof assistant relies, allow for several different representations of the same mathematical object so that users can choose the one suiting their needs.

Even simple objects like natural numbers may have both a unary representation which features a very straightforward induction scheme and a binary one which is exponentially more compact, but usually entails more involved proofs. Their respective incarnations in the standard library of CoQ are the two inductive types nat and N along with two isomorphisms N.of_nat : nat -> N and N.to_nat : N -> nat. Recent versions of the library make use of ML-like modules and functors [4] to factor programs and proofs over these two types.

---

* The research leading to these results has received funding from the European Union's 7th Framework Programme under grant agreement nr. 243847 (ForMath).

G. Gonthier and M. Norrish (Eds.): CPP 2013, LNCS 8307, pp. 147–162, 2013.
© Springer International Publishing Switzerland 2013

The traditional approach to abstraction is to first define an interface specifying operators and their properties, then instantiate it with concrete implementations of the operators with proofs that they satisfy the properties. However, this has at least two drawbacks in our context. First, it is not always obvious how to define the correct interface, and it is not clear if a suitable one even exists. Second, having abstract axioms goes against the type-theoretic view of objects with computational content, which means in practice that proof techniques like small scale reflection, as advocated by the SSREFLECT extension [9], are not applicable.

Instead, the approach we describe here consists in proving the correctness of programs on data structures designed for proofs — as opposed to an abstract signature — and then transporting them to more efficient implementations. We distinguish two notions: *program refinements* and *data refinements*. The first of these consists in transforming a program into a more efficient one computing the same thing using a different algorithm, but preserving the involved types. For example, standard matrix multiplication can be refined to a more efficient implementation like Strassen's fast matrix product [25]. The correctness of this kind of refinements is often straightforward to state. In many cases, it suffices to prove that the two algorithms are extensionally equal. The second notion of refinement consists in changing the data representation on which programs operate while preserving the algorithm, for example a multiplication algorithm on dense polynomials may be refined to an algorithm on sparse polynomials. This kind of refinement is more subtle to express as it involves transporting both programs and their correctness proofs to the new data representation.

The two kinds of refinements can be treated independently and in the following, we focus on data refinements. A key feature of these should be compositionality, meaning that we can combine multiple data refinements. For instance, given both a refinement from dense to sparse polynomials and a refinement from unary to binary integers we get a refinement from dense polynomials over unary integers to sparse polynomials over binary integers.

In a previous work [8], two of the authors defined a framework for refining *algebraic structures* in a comparable way, while allowing a step-by-step approach to prove the correctness of algorithms. The present work[1] improves several aspects by considering the following methodology:

1. *relate* a proof-oriented data representation with a more efficient one (Sect. 2),
2. *parametrize* algorithms and the data on which they operate by an abstract type and its basic operations (Sect. 3),
3. *instantiate* these algorithms with proof-oriented data types and their basic operations, and prove the correctness of that instance,
4. use *parametricity* of the algorithm (with respect to the data representation on which it operates), together with points 2 and 3, to deduce that the algorithm instantiated with the more efficient data representation is also correct (Sect. 4).

---

[1] The formal development is available at http://www.maximedenes.fr/coqeal/

Further, this paper also contains a detailed example application of this new framework to Strassen's algorithm for efficient matrix multiplication (Sect. 5). Section 6 provides an overview of related work.

2   Data Refinements

In this section we will study various data refinements by considering some examples. All of these fit in a general framework of data refinements based on heterogeneous relations which relate *proof-oriented* types for convenient proofs with *computation-oriented* types for efficient computation.

2.1   Refinement Relations

In some cases we can define (possibly partial) functions from proof-oriented to computation-oriented types and *vice versa*. We call a function from proof-oriented to computation-oriented types an *implementation* function, and a function going the other way around a *specification* function.

Note that a specification function alone suffices to define a refinement relation between the two data types: a proof-oriented term p refines to a computation-oriented term c if the specification of c is p. We write the following helper functions to map respectively total and partial specification functions to the corresponding refinement relations:

```
Definition fun_hrel A B (f : B -> A) : A -> B -> Prop :=
  fun a b => f b = a.
```

```
Definition ofun_hrel A B (f : B -> option A) : A -> B -> Prop :=
  fun a b => f b = Some a.
```

Isomorphic Types. Isomorphic types correspond to the simple case where the implementation and specification functions are inverse of each other.

The introduction mentions the two types nat and N which represent unary and binary natural numbers. These are isomorphic, which is witnessed by the implementation function N.of_nat : nat -> N and the specification function N.to_nat : N -> nat. Here, the proof-oriented type is nat and the computation-oriented type is N. Another example of isomorphic types is the efficient binary representation Z of integers in the COQ standard library that can be declared as a refinement of the unary, nat-based, representation int of integers in the SSRE-FLECT library.

Quotients. Quotients correspond to the case where the specification and implementation functions are total and where the specification is a left inverse of the implementation. This means that the computation-oriented type may have "more elements" and that the implementation function is not necessarily surjective (unless the quotient is trivial). In this case the proof-oriented type can be seen as a quotient of the computation-oriented type by an equivalence relation defined by the specification function, i.e. two computation-oriented objects are related if their specifications are equal. This way of relating types by quotients is linked to the general notion of quotient types in type theory [5]. The specification corresponds to the canonical surjection in the quotient, while the implementation corresponds to the choice of a canonical representative. However, here we are not interested in studying the proof-oriented type, which is the quotient type. Instead, we are interested in studying the computation-oriented type, which is the type being quotiented.

An important example of quotients is the type of polynomials. These are represented in SSREFLECT as a record type with a list and a proof that the last element is nonzero, however this proof is only interesting when developing theory about polynomials and not for computation. Hence a computation-oriented type can be just the list of coefficients and the specification function would normalize polynomials by removing zeros in the end.

A better representation of polynomials is sparse Horner normal form [10] which can be implemented as:

```
Inductive hpoly := Pc : A -> hpoly
                 | PX : A -> pos -> hpoly -> hpoly.
```

Here A is an arbitrary type and pos is the type of positive numbers, the first constructor represents a constant polynomial and PX a n p should be interpreted as $a + X^n p$ where a is a constant, n is a positive number and p is another polynomial in sparse Horner normal form. However, with this representation not only polynomials with zeros in the end can be represented but there are also multiple ways to represent polynomials like X^2 as it can be represented by either $0 + X^2 \cdot 1$ or $0 + X^1(0 + X^1 \cdot 1)$. To remedy this we implement a specification function that normalize polynomials and translate them to SSREFLECT polynomials.

Partial Quotients. Quotient based refinement relations cover a larger class of data refinements than the relations defined by isomorphisms, but there are still interesting examples that are not covered, for example when the specification function is partial. To illustrate this, let us consider rational numbers. The SSREFLECT library contains a definition where they are defined as pairs of coprime integers with nonzero denominator:

```
Record rat : Set := Rat {
  valq : int * int;
  _ : (0 < valq.2) && coprime '|valq.1| '|valq.2|
}.
```

Here '|valq.1| denotes the absolute value of the first projection of the valq pair. This definition is well-suited for proofs, notably because elements of type rat can be compared using Leibniz equality since they are normalized. But maintaining this invariant during computations is often too costly since it requires multiple gcd computations. Besides, the structure also contains a proof which is not interesting for computations but only for developing the theory of rational numbers.

In order to be able to compute efficiently we would like to refine this to pairs of integers (int * int) that are not necessarily normalized and perform all operations on the subset of pairs with nonzero second component. The link between the two representations is depicted in Fig. 1:

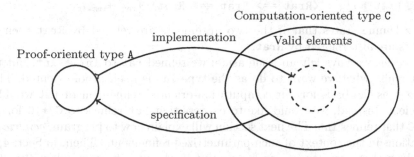

Fig. 1. Partial quotients

In the example of rational numbers the proof-oriented type is rat while the computation-oriented type is int * int and computations should be performed on the subset of valid elements of the computation-oriented type, i.e. pairs with nonzero second component. In order to conveniently implement this, the output type of the specification function has been extended to option A in order to make it total. The key property of the implementation and specification functions is still that the specification is a left inverse of the implementation. This means that the proof-oriented type can be seen as a quotient of the set of valid elements, i.e. elements that are not sent to None by the specification function. For rational numbers the implementation and specification functions and their correctness looks like:

```
Definition rat_to_Qint (r : rat) : int * int := valq r.
Definition Qint_to_rat (r : int * int) : option rat :=
  if r.2 != 0 then Some (r.1%:Q / r.2%:Q) else None.
```

```
Lemma Qrat_to_intK :
  forall (x : rat), Qint_to_rat (rat_to_Qint x) = Some x.
```

The notation %:Q is the cast from int to rat. Here the lemma says that the composition of the implementation with the specification is the identity. Using this, we get a relation between rat and int * int by using ofun_hrel defined at the beginning of this section:

```
Definition Rrat : rat -> int * int -> Prop := ofun_hrel Qint_to_rat.
```

Functional Relations. Partial quotients often work for the data types we define, but fails to describe refinement relations on functions. Given two relations R : A -> B -> Prop and R' : A' -> B' -> Prop we build a relation on the function space: `R ==> R' : (A -> A') -> (B -> B') -> Prop`. It is a heterogeneous generalization of the `respectful` functions defined for generalized rewriting [22].

This definition is such that two functions are related by `R ==> R'` if they send related inputs to related outputs. We can now use this to define the correctness of addition on rational numbers:

Lemma Rrat_addq : (Rrat ==> Rrat ==> Rrat) $+_{rat}$ $+_{int*int}$.

The lemma states that if the two arguments are related by `Rrat` then the outputs are also related by `Rrat`.

However, we have left an issue aside: we refined `rat` to `int * int`, but this is not really what we want to do as the type `int` is itself proof-oriented. Thus, taking it as the basis for our computation-oriented refinement of `rat` would be inefficient. Instead, we would like to express that `rat` refines to `C * C` for *any* type `C` that refines `int`. The next section will explain how to program, *generically*, operations in the context of such parametrized refinements. Then, in Sect. 4, we will show that correctness can be proved in the specific case when `C` is `int`, and automatically transported to any other refinement by taking advantage of parametricity.

2.2   Comparison with the Previous Approach

We gain in generality with regard to the previous work on refinements [8] in several ways. The previous work assumed a total injective implementation function, which intuitively corresponds to a partial isomorphism: the proof-oriented type is isomorphic to a subtype of the computation-oriented type. Since we do not rely on those translation functions anymore, we can now express refinement relations on functions. Moreover, we take advantage of (possibly partial) specification functions, rather than implementation functions.

Another important improvement is that we do not need any notion of equality on the computation-oriented type anymore. Indeed, the development used to rely on Leibniz equality, which prevented us from using setoids [2] as computation-oriented types. In Sect. 2.1, we use the setoid `int * int` of rational numbers, but the setoid equality is left implicit. This is in accordance with our principle never to do proofs on computation-oriented types. We often implement algorithms to decide equality, but these are treated as any other operation (Sect. 3).

2.3   Indexing and Using Refinements

We use the COQ type class mechanism [23] to maintain a database of lemmas establishing refinement relations between proof-oriented and computation-oriented terms. The way this database is used is detailed in Sect. 4.

In order to achieve this, we define a heterogeneous generalization of the `Proper` relations from generalized rewriting [22]. We call this class of relations `param` and define it by:

```
Class param (R : A -> B -> Prop) (a : A) (b : B) := param_rel : R a b.
```

Here R is meant to be a refinement relation from A to B, and we can register an instance of this class whenever we have two elements a and b and a proof that R a b. For example, we register the lemma Rrat_addq from Sect. 2.1 using the following instance:

```
Instance Rrat_addq : param (Rrat ==> Rrat ==> Rrat) +rat +int*int.
```

Given a term x, type class resolution searches for y and a proof of param R x y. If R was obtained from a specification function, then $x = \text{spec } y$ and we can always substitute x by spec y and compute y, thus taking advantage of our framework to do efficient computation steps within proofs.

3 Generic Programming

We may want to provide operations on the computation-oriented type corresponding to operations on the proof-oriented type. For example, we want to define an addition addQ on computation-oriented rationals C * C, corresponding to the addition ($+_\text{rat}$) on rat. However this computation-oriented operation relies on both addition ($+_C$) and multiplication ($*_C$) on C, so we parametrize addQ by ($+_C$) and ($*_C$):

```
Definition addQ C (+c) (*c) : (C * C) -> (C * C) -> (C * C) :=
  fun x y => (x.1 *c y.2 +c y.1 *c x.2, x.2 *c y.2).
```

This operation is correct if ($+_\text{rat}$) refines to (addQ C ($+_C$) ($*_C$)) whenever ($+_\text{int}$) refines to ($+_C$) and ($*_\text{int}$) refines to ($*_C$). The refinement from ($+_\text{rat}$) to (addQ C ($+_C$) ($*_C$)) is explained in Sect. 4.1.

Since we abstracted over operations of the underlying data type, only one implementation of each algorithm suffices, the same code can be used for doing both correctness proofs and efficient computations as it can be instantiated by both proof-oriented and computation-oriented types and programs. This means that the programs need only be written once and code is never duplicated, which is an improvement compared to the previous development.

In order to ease the writing of this kind of programs and refinement statements in the code, we use operational type classes [24] for standard operations like addition and multiplication together with appropriate notations. This means we define a class for each operator and a generic notation referring to the corresponding operation. For example, in the code of addQ we can always write (+) and (*) and let the system infer the operations,

```
Instance addQ C '{add C, mul C} : add (C * C) :=
  fun x y => (x.1 * y.2 + y.1 * x.2, x.2 * y.2).
```

Here '{add C, mul C} means that C comes with type classes for addition and multiplication operators. Declaring addQ as an instance of addition on C * C enables the use of the generic (+) notation to denote addQ.

4   Parametricity

The approach presented in the above section is incomplete though: once we have proven that the instantiation of a generic algorithm to proof-oriented structures is correct, how can we guarantee that other instances will be correct as well? Proving correctness directly on computation-oriented types is precisely what we are trying to avoid.

Informally, since our generic algorithms are polymorphic in their types and operators, their behavior has to be uniform across all instances. Hence, a correctness proof should be portable from one instance to another, so long as the operators instances are themselves correct.

The exact same idea is behind the interpretation of polymorphism in relational models of pure type systems [3]. The present section builds on this analogy to formalize the automated transport of a correctness proof from a proof-oriented instance to other instances of the same generic algorithm.

4.1   Splitting Refinement Relations

Let us illustrate the parametrization process by an example on rational numbers. For simplicity, we consider negation which is implemented by:

```
Instance oppQ C '{opp C} : opp (C * C) :=
  fun (a : Q C) => (-_c a.1, a.2).
```

The function takes a negation operation in the underlying type C and define negation on C * C by negating the first projection of the pair (the numerator). Now let us assume that C is a refinement of int for a relation Rint : int -> C -> Prop and that we have:

```
(Rint ==> Rint) (-_int) (-_c)
(Rrat ==> Rrat) (-_rat) (oppQ int (-_int))
```

The first of these states that the $(-_c)$ parameter of oppQ is correctly instantiated, while the second one expresses that the proof-oriented instance of oppQ is correct. Assuming this, we want to show that $(-_{rat})$ refines all the way to oppQ, but instantiated with C and $(-_c)$ instead of their proof-oriented counterparts (int and $(-_{int})$).

In order to write this formally, we define the product and composition of relations as R * S := fun x y => R x.1 y.1 /\ S x.2 y.2 and R \o S := fun x y => exists z, R x z /\ S z y. Using this we can define the relation RratC : rat -> C * C -> Prop as RratC := Rrat \o (Rint * Rint). We want to show:

```
(RratC ==> RratC) (-_rat) (oppQ C (-_c))
```

A small automated procedure, relying on type class instance resolution, first splits this goal in two, following the composition \o in the definition of RratC:

```
(Rrat ==> Rrat) (-_rat) (oppQ int (-_int))
(Rint * Rint ==> Rint * Rint) (oppQ int (-_int)) (oppQ C (-_c))
```

The first of these is one of the assumptions while the second relates the results of the proof-oriented instance of oppQ to another instance. This is precisely where parametricity comes into play, as we will show in the next section.

4.2 Parametricity for Refinements

While studying the semantics of polymorphism, Reynolds introduced a relational interpretation of types [19]. Parametricity [27] is a reformulation based on the fact that if a type has no free variable, its relational interpretation expresses a property shared by all terms of this type. This result extends to pure type systems [3] and provides a meta-level transformation $\llbracket \cdot \rrbracket$ defined inductively on terms and contexts. In the closed case, this transformation is such that if $\vdash A : B$, then $\vdash \llbracket A \rrbracket : \llbracket B \rrbracket \, A \, A$. That is, for any term A of type B, it gives a procedure to build a proof that A is related to itself for the relation interpreting the type B.

The observation we make is that the last statement of Sect. 4.1 is an instance of such a *free theorem*. More precisely, we know that $\llbracket \text{oppQ} \rrbracket$ is a proof of

$$\llbracket \forall Z, (Z \to Z) \to Z * Z \to Z * Z \rrbracket \text{ oppQ oppQ}$$

which expands to

$$\forall Z : \text{Type}, \qquad \forall Z' : \text{Type}, \qquad \forall Z_R : Z \to Z' \to \text{Prop},$$
$$\forall \text{oppZ} : Z \to Z, \quad \forall \text{oppZ}' : Z' \to Z', \quad \llbracket Z \to Z \rrbracket \text{ oppZ oppZ}' \to$$
$$\llbracket Z * Z \to Z * Z \rrbracket \ (\text{oppQ } Z \text{ oppZ}) \ (\text{oppQ } Z' \text{ oppZ}').$$

Then, instantiating Z to int, Z' to C and Z_R to Rint gives us the exact statement we wanted to prove, since $\llbracket Z \to Z \rrbracket$ is what we denoted $Z_R \Longrightarrow Z_R$.

Following the term transformation $\llbracket \cdot \rrbracket$, we design a logic program in order to derive proofs of closed instances of the parametricity theorem. Indeed, it should be possible in practice to establish the parametric relation between two terms like oppQ and itself, since oppQ is closed.

For now, we can only express and infer parametricity on polymorphic expressions (no dependent types allowed), by putting the polymorphic types outside the relation. Hence we do not need to introduce a quantification over relations.

4.3 Generating the Parametricity Lemma

Rather than giving the details of how we programmed the proof search using type classes and hints in the CoQ system, we instead show an execution of this logic program on our simple example, starting from:

```
(Rrat ==> Rrat) (-rat) (oppQ C (-c))
```

Let us first introduce the variables and their relations, and we get to prove

```
(Rint * Rint) (oppQ int (-int) a) (oppQ C (-c) b)
```

knowing that `((Rint ==> Rint) (-int) (-c))` and `((Rint * Rint) a b)`. By unfolding oppQ, it suffices to show:

```
(Rint * Rint) (-int a.1, a.2) (-C b.1, b.2)
```

To show that, we use parametricity theorems for the pair constructor `pair` and eliminators `_.1` and `_.2`. In our context, we have to give manual proofs for them. Indeed, we lack automation for the axioms, but the number of combinators to treat by hand is negligible compared to the number of occurrences in user-defined operations. These lemmas look like:

```
param_pair := forall RA RB, (RA ==> RB ==> RA * RB) pair pair
param_fst := forall RA RB, (RA * RB ==> RA) _.1 _.1
param_snd := forall RA RB, (RA * RB ==> RB) _.2 _.2
```

Unfolding the last of these gives:

```
forall (RA : A -> A' -> Prop) (RB : B -> B' -> Prop)
       (a : A) (a' : A') (b : B) (b' : B'),
  RA a a' -> RB b b' -> (RA * RB) (a, b) (a', b')
```

This can be applied to the initial goal, giving two subgoals:

```
Rint (-int a.1) (-C b.1)
Rint a.2 b.2
```

The second of these follow directly from `param_snd` and to show the first it suffices to prove:

```
(Rint ==> Rint) (-int) (-C)
Rint a.1 b.1
```

The first of these is one of the assumptions we started with and the second follows directly from `param_fst`.

5   Example: Strassen's Matrix Product

In the previous development an important application of the refinement framework was Strassen's algorithm for the product of two square matrices of size n with time complexity $\mathcal{O}(n^{2.81})$ [25]. We show here how we adapted it to the new framework described in this paper.

Let us begin with one step of Strassen's algorithm: given a function f which computes the product of two matrices of size `p`, we define, generically, a function `Strassen_step f` which multiplies two matrices of size `p + p`:

```
Variable mxA : nat -> nat -> Type.

Context '{hadd mxA, hsub mxA, hmul mxA, hcast mxA, block mxA}.
Context '{ulsub mxA, ursub mxA, dlsub mxA, drsub mxA}.

Definition Strassen_step {p : positive} (A B : mxA (p+p) (p+p))
  (f : mxA p p -> mxA p p -> mxA p p) : mxA (p+p) (p+p) :=
  let A11 := ulsubmx A in let A12 := ursubmx A in
```

```
let A21 := dlsubmx A in let A22 := drsubmx A in
let B11 := ulsubmx B in let B12 := ursubmx B in
let B21 := dlsubmx B in let B22 := drsubmx B in
let X := A11 - A21 in let Y := B22 - B12 in
let C21 := f X Y in let X := A21 + A22 in
let Y := B12 - B11 in let C22 := f X Y in
let X := X - A11 in let Y := B22 - Y in
let C12 := f X Y in let X := A12 - X in
let C11 := f X B22 in let X := f A11 B11 in
let C12 := X + C12 in let C21 := C12 + C21 in
let C12 := C12 + C22 in let C22 := C21 + C22 in
let C12 := C12 + C11 in let Y := Y - B21 in
let C11 := f A22 Y in let C21 := C21 - C11 in
let C11 := f A12 B21 in let C11 := X + C11 in
block_mx C11 C12 C21 C22.
```

The `mxA` variable represents the type of matrices indexed by their sizes. The various operations on this type are abstracted over by operational type classes, as shown in Sect. 3. Playing with notations and scopes allows us to make this generic implementation look much like an equivalent one involving SSREFLECT matrices.

Note that `Strassen_step` expresses matrix sizes by the `positive` type. These are positive binary numbers, whose recursion scheme matches the one of Strassen's algorithm through matrix block decomposition. This is made compatible with the `nat`-indexed `mxA` type thanks to a hidden coercion `nat_of_pos`.

The full algorithm is expressed by induction over `positive`. However, in order to be able to state parametricity lemmas, we do not use the primitive `Fixpoint` construction. Instead, we use the recursion scheme attached to `positive`:

```
positive_rect : forall P : positive -> Type,
(forall p : positive, P p -> P (p~1)%positive) ->
(forall p : positive, P p -> P (p~0)%positive) ->
P 1%positive -> forall p : positive, P p
```

We thus implement three functions corresponding to the three cases given by the constructor of the `positive` inductive type: `Strassen_xI` for odd-sized matrices, `Strassen_x0` for even-sized ones and `Strassen_xH` for matrices of size 1. Strassen's algorithm is then defined as:

```
Definition Strassen :=
  (positive_rect (fun p => (mxA p p -> mxA p p -> mxA p p))
                  Strassen_xI Strassen_x0 Strassen_xH).
```

Then we instantiate the `mxA` type and all the associated operational type classes to SSREFLECT proof-oriented matrix type and operators. In this context, we prove the program refinement from the naive matrix product `mulmx` to Strassen's algorithm:

```
Lemma StrassenP p : param (eq ==> eq ==> eq) mulmx (@Strassen p).
```

The proof is essentially unchanged from [8], the present work improving only the data refinement part. The last step consists in stating and proving the parametricity lemmas. This is done in a context abstracted over both a representation type for matrices and a refinement relation:

```
Context (A : ringType) (mxC : nat -> nat -> Type).
Context (RmxA : forall {m n}, 'M[A]_(m, n) -> mxC m n -> Prop).
```

Operations on matrices are also abstracted, but we require them to have an associated refinement lemma with respect to the corresponding operation on proof-oriented matrices. For instance, for addition we write as follows:

```
Context '{hadd mxC, forall m n, param (RmxA ==> RmxA ==> RmxA)
          (@addmx A m n) (@hadd_op _ _ _ m n)}.
```

We also have to prove the parametricity lemma associated to our recursion scheme on **positive**:

```
Instance param_elim_positive P P'
  (R : forall p, P p -> P' p -> Prop) txI txI' txO txO' txH txH' :
  (forall p, param (R p ==> R (p~1)) (txI p) (txI' p)) ->
  (forall p, param (R p ==> R (p~0)) (txO p) (txO' p)) ->
  (param (R 1) txH txH') ->
  forall p, param (R p) (positive_rect P txI txO txH p)
                        (positive_rect P' txI' txO' txH' p).
```

We declare this lemma as an Instance of the **param** type class. This allows to automate data refinement proofs requiring induction over **positive**. Finally, we prove parametricity lemmas for **Strassen_step** and **Strassen**:

```
Instance param_Strassen_step p :
  param (RmxA ==> RmxA ==> (RmxA ==> RmxA ==> RmxA) ==> RmxA)
        (@Strassen_step (@matrix A) p) (@Strassen_step mxC p).
```

```
Instance param_Strassen p :
  param (RmxA ==> RmxA ==> RmxA)
        (@Strassen (@matrix A) p) (@Strassen mxC p).
```

Here, the improvement over [8] is twofold: only one generic implementation of the algorithm is now required and refinement proofs are now mostly automated, including induction steps.

A possible drawback is that our generic description of the algorithms requires all the operators to take the sizes of the matrices involved as arguments, which are sometimes not required for computation-oriented operators. However, some preliminary benchmarks seem to indicate that this does not entail a significant performance penalty.

6   Related Work

Our work addresses a fundamental problem: how to change data representations in a compositional way. As such, it is no surprise that it shares aims with other work. We already mentioned ML-like modules and functors, that are available in COQ, but forbid proof methods to have a computational content.

The most general example of refinement relations we consider are partial quotients, which are often represented in type theory by setoids over partial equivalence relations [2] and manipulated using generalized rewriting [22]. The techniques we are using are very close to a kind of heterogeneous version of the latter. Indeed, it usually involves a relation R : A -> A -> Prop for a given type A, whereas our refinement relations have the shape R : A -> B -> Prop where A and B can be two different types.

Some years ago, a plugin was developed for COQ for changing data representations and converting proofs from a type to another [16]. However, this approach was limited to isomorphic types, and does not provide a way to achieve generic programming (only proofs are ported). Our design is thus more general, and we do not rely on an external plugin which can be costly to maintain.

In [15], a methodology for modular specification and development of programs in type theory is presented. The key idea is to express algebraic specifications using sigma-types which can be refined using refinement maps, and realized by concrete programs. This approach is close to the use of ML-like modules, since objects are abstracted and their behavior is represented by a set of equational properties. A key difference to our work is that these equational properties are stated using an abstract congruence relation, while we aim at proving correctness on objects that can be compared with Leibniz equality, making reasoning more convenient. This is made possible by our more relaxed relation between proof-oriented and computation-oriented representations.

Another way to reconcile data abstraction and computational content is the use of *views* [17,26]. In particular, it allows to derive induction schemes independently of concrete representations of data. This can be used in our setting to write generic programs utilizing these induction schemes for defining recursive programs and proving properties for generic types, in particular `param_elim_positive` (Sect. 5) is an example of a view.

The closest work to ours is probably the automatic data refinement tool AU-TOREF implemented independently for ISABELLE [14]. While many ideas, like the use of parametricity, are close to ours, the choice is made to rely on an external tool to synthesize executable instances of generic algorithms and refinement proofs. The richer formalism that we have at our disposal, in particular full polymorphism and dependent types makes it easier to internalize the instantiation of generic programs.

Another recent work that is related to this paper is [11] in which the authors explain how the ISABELLE/HOL code generator uses data refinements to generate executable versions of abstract programs on abstract types like sets. In the paper they use a refinement relation that is very similar to our partial quotients (they use a domain predicate instead of an option type to denote what values are valid and which are not). The main difference though is that they are applying data refinements for code generation while in our case this is not necessary since all programs written in CoQ can be executed as they are and data refinements are only useful to perform more efficient computations.

7   Conclusions and Future Work

In this paper an approach to data refinements has been presented where the user only needs to supply the minimum amount of necessary information and both programs and their correctness proofs gets transported to new data representation. The three main parts of the approach are:

1. a lightweight and general refinement interface to support any heterogeneous relation between two types,
2. operational type classes to increase generality of implementations and
3. parametricity to automatically transport correctness proofs.

As mentioned in the introduction of this paper, this work is an improvement of a previous work [8]. More precisely it improves the approach presented in Sect. 5 of [8] in the following aspects.

1. Generality: it extends to previously unsupported data types, like the type of non-normalized rationals (Sect. 2.2).
2. Modularity: each operator is refined in isolation instead of refining whole algebraic structures (Sect. 2.3), as suggested in the future work section of the previous paper.
3. Genericity: before, every operation had to be implemented both for the proof-oriented and computation-oriented types, now only one generic implementation is sufficient (Sect. 3).
4. Automation: the current approach has a clearer separation between the different steps of data refinements which makes it possible to use parametricity (Sect. 4) in order to automate proofs that previously had to be done by hand.

The implementation of points 2, 3 and 4 relies on the type class mechanism of CoQ in two different ways: in order to support ad-hoc polymorphism of algebraic operations, and in order to do proof and term reconstruction automatically through logic programming. The automation of proof and term search is achieved by the same set of lemmas as in the previous paper, but now these do not impact the interesting proofs anymore.

The use of operational type classes is very convenient for generic programming. But the more complicated programs get, the more arguments they need. In particular, we may want to bundle operators in order to reduce the size of contexts that users need to write when defining generic algorithms.

The handling of parametricity is currently done by meta-programming but requires some user input and deals only with polymorphic constructions. We should address these two issues by providing a systematic way of producing parametricity lemmas for inductive types [3] and extending relation constructions with dependent types. We may adopt Keller and Lasson's [13] way of producing parametricity theorems and their proofs for closed terms.

Currently all formalizations have been done using standard COQ, but it would be interesting to see how the univalent foundations [18] can be used for simplifying our approach to data refinements. Indeed, in the presence of the univalence axiom, isomorphic structures are equal [1,7] which should be useful when refining isomorphic types. Also in the univalent foundations there are ways to represent quotient types (see for example [20]). This could be used to refine types that are related by quotients or even partial quotients.

The work presented in this paper is currently being used as a new basis for COQEAL — The COQ Effective Algebra Library — which is a library, currently in development, containing many formally verified program refinements, for instance: Strassen's fast matrix product [25], Karatsuba's fast polynomial product [12], the Sasaki-Murao algorithm for efficiently computing the characteristic polynomial of a matrix [21] and an algorithm for computing the Smith normal form of matrices over Euclidean rings.

Acknowledgments. The authors are grateful to the anonymous reviewers for their useful comments and feedback. We also thank Bassel Mannaa and Dan Rosén for proof reading the final version of this paper.

References

1. Ahrens, B., Kapulkin, C., Shulman, M.: Univalent categories and the Rezk completion (2013) (Preprint), http://arxiv.org/abs/1303.0584
2. Barthe, G., Capretta, V., Pons, O.: Setoids in type theory. Journal of Functional Programming 13(2), 261–293 (2003)
3. Bernardy, J.-P., Jansson, P., Paterson, R.: Proofs for free. Journal of Functional Programming 22(2), 107–152 (2012)
4. Chrząszcz, J.: Implementing Modules in the Coq System. In: Basin, D., Wolff, B. (eds.) TPHOLs 2003. LNCS, vol. 2758, pp. 270–286. Springer, Heidelberg (2003)
5. Cohen, C.: Pragmatic Quotient Types in COQ. In: Blazy, S., Paulin-Mohring, C., Pichardie, D. (eds.) ITP 2013. LNCS, vol. 7998, pp. 213–228. Springer, Heidelberg (2013)
6. COQ development team. The COQ Proof Assistant Reference Manual, version 8.4. Technical report, Inria (2012)
7. Danielsson, N.A., Coquand, T.: Isomorphism is Equality (2013) (Preprint), http://www.cse.chalmers.se/~nad/publications/coquand-danielsson-isomorphism-is-equality.html
8. Dénès, M., Mörtberg, A., Siles, V.: A Refinement-Based Approach to Computational Algebra in COQ. In: Beringer, L., Felty, A. (eds.) ITP 2012. LNCS, vol. 7406, pp. 83–98. Springer, Heidelberg (2012)
9. Gonthier, G., Mahboubi, A.: A Small Scale Reflection Extension for the Coq system. Technical report, Microsoft Research INRIA (2009)
10. Grégoire, B., Mahboubi, A.: Proving Equalities in a Commutative Ring Done Right in Coq. In: Hurd, J., Melham, T. (eds.) TPHOLs 2005. LNCS, vol. 3603, pp. 98–113. Springer, Heidelberg (2005)
11. Haftmann, F., Krauss, A., Kunčar, O., Nipkow, T.: Data Refinement in Isabelle/HOL. In: Blazy, S., Paulin-Mohring, C., Pichardie, D. (eds.) ITP 2013. LNCS, vol. 7998, pp. 100–115. Springer, Heidelberg (2013)

12. Karatsuba, A., Ofman, Y.: Multiplication of many-digital numbers by automatic computers. USSR Academy of Sciences 145, 293–294 (1962)
13. Keller, C., Lasson, M.: Parametricity in an Impredicative Sort. In: CSL, vol. 16, pp. 381–395. Schloss Dagstuhl - Leibniz-Zentrum fuer Informatik (2012)
14. Lammich, P.: Automatic Data Refinement. In: Blazy, S., Paulin-Mohring, C., Pichardie, D. (eds.) ITP 2013. LNCS, vol. 7998, pp. 84–99. Springer, Heidelberg (2013)
15. Luo, Z.: Computation and reasoning: a type theory for computer science. Oxford University Press, Inc., New York (1994)
16. Magaud, N.: Changing Data Representation within the Coq System. In: Basin, D., Wolff, B. (eds.) TPHOLs 2003. LNCS, vol. 2758, pp. 87–102. Springer, Heidelberg (2003)
17. McBride, C., McKinna, J.: The view from the left. Journal of Functional Programming 14(1), 69–111 (2004)
18. T. U. F. Program: Homotopy Type Theory: Univalent Foundations of Mathematics. Institute for Advanced Study (2013), http://homotopytypetheory.org/book/
19. Reynolds, J.C.: Types, abstraction and parametric polymorphism. In: IFIP Congress, pp. 513–523 (1983)
20. Rijke, E., Spitters, B.: Sets in homotopy type theory (2013) (Preprint), http://arxiv.org/abs/1305.3835
21. Sasaki, T., Murao, H.: Efficient Gaussian Elimination Method for Symbolic Determinants and Linear Systems. ACM Trans. Math. Softw. 8(3), 277–289 (1982)
22. Sozeau, M.: A new look at generalized rewriting in type theory. Journal of Formalized Reasoning 2(1), 41–62 (2009)
23. Sozeau, M., Oury, N.: First-Class Type Classes. In: Mohamed, O.A., Muñoz, C., Tahar, S. (eds.) TPHOLs 2008. LNCS, vol. 5170, pp. 278–293. Springer, Heidelberg (2008)
24. Spitters, B., van der Weegen, E.: Type Classes for Mathematics in Type Theory. MSCS, Special Issue on 'Interactive Theorem Proving and the Formalization of Mathematics' 21, 1–31 (2011)
25. Strassen, V.: Gaussian elimination is not optimal. Numerische Mathematik 13(4), 354–356 (1969)
26. Wadler, P.: Views: A way for pattern matching to cohabit with data abstraction. In: POPL, pp. 307–313. ACM Press (1987)
27. Wadler, P.: Theorems for free? In: Functional Programming Languages and Computer Architecture, pp. 347–359. ACM Press (1989)

A Formal Proof of Borodin-Trakhtenbrot's Gap Theorem

Andrea Asperti

Department of Computer Science and Engineering – DISI
University of Bologna
asperti@cs.unibo.it

Abstract. In this paper, we discuss the formalization of the well known Gap Theorem of Complexity Theory, asserting the existence of arbitrarily large gaps between complexity classes. The proof is done at an abstract, machine independent level, and is particularly aimed to identify the minimal set of assumptions required to prove the result (smaller than expected, actually). The work is part of a long term *reverse complexity* program, whose goal is to obtain, via a reverse methodological approach, a formal treatment of Complexity Theory at a comfortable level of abstraction and logical rigor.

1 Introduction

The Gap Theorem, first proved by Boris Trakhtenbrot in 1964 [33] and independently rediscovered eight years later by Allan Borodin [12], is a major theorem of Complexity Theory stating the existence of arbitrarily large gaps in the hierarchy of complexity classes. More explicitly, given a computable function g representing an increase in computational resources, one can effectively find a recursive function t such that the complexity classes with boundary functions t and $g \circ t$ are identical. In Borodin's words [12] "no matter how much better one computer may seem compared to the other, there will be a t such that the set of functions computable in time t is the same for both computers".

The Gap Theorem is a typical example of an "abstract" complexity result, that is a fact that can be proved without any reference to concrete computational models. Actually, our main motivation for addressing the formalization of this theorem was to derive, along a reverse methodological approach, a minimal set of logical assumptions sufficient to entail the result. The work is part of a larger "reverse complexity" program, outlined in [2], that applies the methodology of reverse mathematics [20,30] to Complexity Theory, reconstructing from proofs the basic notions and assumptions underlying the major results of this field. The final, long term goal would be to obtain a formal, axiomatic treatment of Complexity Theory at a *comfortable* level of abstraction and mathematical rigor, reviving under a new perspective and through an innovative methodological approach the old quest for a machine-independent theory of complexity; we refer the reader to [2] for a short historical survey and a more exhaustive discussion of the Reverse Complexity program.

G. Gonthier and M. Norrish (Eds.): CPP 2013, LNCS 8307, pp. 163–177, 2013.
© Springer International Publishing Switzerland 2013

Mechanical devices such as proof assistants and interactive theorem provers play a major role in our program, not only to check the formal correctness of the resulting theory, but as actual *drivers* of the research. In fact, the reverse methodology presupposes a deep and frequent refactoring of the formalization, playing with different axiomatizations, improving the readability and maintainability of the code, or reducing its complexity: it is natural to expect to be supported by automatic devices along this process. As we already observed in [4], the situation is similar to the role of type checkers in software development, that are not simply meant to discriminate good programs from bad ones: type checkers are essential drivers of the development phase, and major tools for the deployment of lightweight, adaptive software methodologies, requiring frequent modifications and refactoring. This interactive exploitment of proof checkers, more than their batch usage as oracles to discriminate between correct and wrong arguments is, in our opinion, the new and challenging frontier of interactive provers.

The formalization of the Gap Theorem discussed in this paper was done with the assistance of the Matita Interactive Theorem Prover [7]. Matita is a light implementation of the Calculus of Inductive Construction developed and maintained at the University of Bologna. We do not have enough space to describe here the syntax of Matita's script language, so we shall omit formal proofs. We only wish to remark that Matita is a constructive system, and all proofs in this paper are constructive.

The development itself is accessible (and executable!) through the web interface of Matita [6] at the following url: `http://matita.cs.unibo.it/matitaweb.shtml`. An offline version can be downloaded at `http://www.cs.unibo.it/~asperti/gap.tar`.

The structure of the paper is the following. In the next section we shall start giving a rigorous formulation of the gap theorem and the original proof of Borodin [12], discussing it as well as other, later versions of the proof [15,35,17]. Section 3 is devoted to a brief review of the main modules of the Matita library that will be required for the formalization of the result, and in particular: bounded quantification, big operators and minimization, iteration, and a bit of combinatorics. In Section 4, we introduce the axiomatic setting that we shall use for the proof, essentially based on the existence of a suitable function (intuitively) playing the role of Kleene's T-predicate. Section 5 contains the formal proof, as well as the computation of an interesting and apparently original upper bound for the gap operator. Conclusions are discussed in Section 6.

2   Borodin's Proof of the Gap Theorem

The gap theorem can be stated and proved without any reference to a concrete computational model. The typical setting adopted for expressing and proving it is Blum's abstract complexity framework [11], that applies to time, space, and many other reasonable complexity measures.

We write $f(x)\downarrow$ to express that the *partial* function f is defined for input x.

Definition 1. *(Blum [11]) A pair $\langle \varphi, \Phi \rangle$ is a computational complexity measure if φ is a principal effective enumeration of all partial recursive functions and Φ satisfies the following axioms:*

$$(a)\ \varphi_i(n) \downarrow \leftrightarrow \Phi_i(n) \downarrow$$
$$(b)\ \text{the predicate } \Phi_i(n) = m \text{ is decidable}$$

We adopt the convention that $\Phi_i(n) = \infty$ if $\Phi_i(n) \uparrow$; in particular, the relation $\Phi_i(n) > n$ also holds when $\Phi_i(n)$ is undefined.

Blum's axioms are very weak and very general, and nevertheless they are sufficient to prove a large number of interesting results in Complexity Theory. In particular, they proved to be a convenient setting to investigate the order structure of complexity classes under set theoretic inclusion [12,26], their recursive presentability and the computational quality of such a presentation [25,34,24]. It is important to observe that, from a strictly formal point of view, Blum's "axioms" do not provide a real axiomatization, since they rely on the delicate notion of *computable function*. The fact that φ is a *principal effective enumeration* (see e.g.[28]) of all partial recursive functions is used in an essential way in most proofs based on Blum's axioms, usually by an invocation of Church Thesis. For this reason, Blum's axioms are not easy to use in a strictly formal framework, urging us to look for a more convenient and possibly more primitive axiomatisation.

Borodin's proof of the gap theorem is very concise and elegant, so we report his original argument here; we just slightly rephrased it for notational reasons, and retouched some bounds in order to get a more elegant formalization.

Theorem (Gap Theorem). Let $\langle \varphi, \Phi \rangle$ be a complexity measure, g a nondecreasing recursive function such that $\forall x. x \leq g(x)$. Then there exists a nondecreasing recursive function t such that, for sufficiently large n,

$$\Phi_i(n) \leq t(n) \qquad \text{or} \qquad \Phi_i(n) > g \circ t(n)$$

PROOF. Define t as follows:

- $t(0) = 1$,
- $t(n+1) = \mu k \geq t(n)\{\forall i \leq n.[\Phi_i(n) \leq k \text{ or } \Phi_i(n) > g(k)]\}$

Then:

1. for any n, k exists, since forall $i \leq n$ if $\Phi_i(n) \uparrow$ then $\forall k. \Phi_i(n) > g(k)$, and if $\Phi_i(n) \downarrow$ then $\exists k. \Phi_i(n) \leq k$.
2. k can be found recursively, since Φ is a complexity measure and thus $\Phi_i(n) \leq k$ and $\Phi_i(n) > g(k)$ are recursive predicates.
3. t satisfies the theorem, since $n \geq i$ implies that either $\Phi_i(n) \leq t(n)$ or $\Phi_i(n) > g \circ t(n)$.

QED.

An arbitrarily large t can be found to satisfy the conditions of the gap theorem, by taking k larger than $\max\{r(n), t(n)\}$ (for a suitable function r) in the definition of $t(n+1)$.

2.1   Discussion

The first problem in formalizing the previous proof in a proof system like Matita is due to the definition of t, that is formulated by means of general (unbounded) minimization. In general, this kind of recursive functions cannot be directly expressed in the Calculus of Constructions, and you should resort to an indirect encoding, by means of a suitable predicate, that is not particularly elegant. A second problem is point 1. in the proof, that seems to use *tertium non datur* on a semidecidable predicate, namely if $\Phi_i(n) \downarrow$ or not.

Luckily, as already pointed out by [35] (see also [17]) the existence of k can be proved in a more constructive way, and this will also induce a more constrained (primitive recursive) definition of t.

The general idea is relatively simple. Suppose we wish to find a k larger than a base value b, such that (for given i and n)

$$\Phi_i(n) \leq k \text{ or } \Phi_i(n) > g(k) \tag{1}$$

If $\Phi_i(n) \leq b$ then we take $k = b$ (note that the test $\Phi_i(n) \leq b$ is decidable!); otherwise we check if $\Phi_i(n) < g(b)$: if the answer is yes, we take $k = g(b)$ and otherwise we again take $k = b$ (the interesting point is not the decidability of equation (1), that is obvious, but the fact that we can put an upper bound to the search for a k solving the equation).

The previous reasoning can be iterated over all $i \leq n$: in particular, in the interval between b and $g^{n+1}(b)$ there must exist at least one k such that

$$\forall i \leq n.\Phi_i(n) \leq k \text{ or } \Phi_i(n) > g(k)$$

Suppose that at least j functions terminate within $b_j \leq g^j(b)$; if no other function terminates within $g(b_j)$ we are done; otherwise we take $b_{j+1} = g(b_j) \leq g^{j+1}(b)$ and go on. Since the number of terminating functions increases at each iteration, we shall eventually stop after $n+1$ steps.

Stated in a different way, let us consider the intervals $[g^i(b), g^{i+1}(b)[$ for $0 \leq i \leq n$ and all functions with index $j < n$ such that $\Phi_j(n) \leq g^{n+1}(b)$. We have at most n functions to distribute over $n+1$ intervals, so at least one interval must remain empty.

An interesting consequence of the previous reasoning, that apparently has never been emphasized by any author, is that we can compute an explicit upper bound u for t. In particular, let $\sigma(n) = \sum_{i \leq n} i = n \cdot (n+1)/2$; then, for any n,

$$t(n) \leq g^{\sigma(n)}(1) \leq g^{n^2}(1)$$

(see Section 5.2 for the simple proof).

3   Preliminaries

In this section, we shall discuss some of the background material we need for our development: bounded quantification 3.1, big operators and minimization 3.2, iteration 3.3 and a few combinatorial results 3.4.

Most of the results in this section are absolutely standard; we present them for the sake of completeness, in order to provide a self-contained description of the formalization, fixing names and notations.

In the rest of the article, all parts inside round boxes are Matita code; all proofs are skipped, but they are really simple.

3.1   Bounded Quantification

We need to exploit a small library of results about bounded quantification. A proposition P is decidable if $P \vee \neg P$ is provable:

> **definition** decidable : Prop $\rightarrow$ Prop := λA:Prop. A $\vee \neg$ A.

It is trivial to prove that decidable propositions are closed with respect to logical connectives and bounded quantification:

> **lemma** decidable_not: $\forall$P. decidable P $\rightarrow$ decidable ($\neg$ P).
>
> **lemma** decidable_or: $\forall$P,Q. decidable P $\rightarrow$ decidable Q $\rightarrow$ decidable (P$\vee$ Q).
>
> **lemma** decidable_forall: $\forall$P. ($\forall$i.decidable (P i)) $\rightarrow$ $\forall$n.decidable ($\forall$i. i < n $\rightarrow$ P i).
>
> **lemma** decidable_exists: $\forall$P. ($\forall$i.decidable (P i)) $\rightarrow$ $\forall$n.decidable ($\exists$i. i < n $\wedge$ P i).

On a decidable predicate we have the usual duality properties we know from classical logic, and in particular:

> **lemma** not_exists_to_forall: $\forall$P,n.
> $\neg$($\exists$i. i < n $\wedge$ P i) $\rightarrow$ $\forall$i. i < n $\rightarrow$ $\neg$P i.
>
> **lemma** not_forall_to_exists: $\forall$P,n. ($\forall$i.decidable (P i)) $\rightarrow$
> $\neg$($\forall$i. i < n $\rightarrow$ P i) $\rightarrow$ ($\exists$i. i < n $\wedge \neg$(P i)).

3.2   Big Operators and Minimization

Matita's library offers a well developed module on big operators, that has been described in some detail in [5].

A big operator is a higher-order construction that is supposed to iterate a function F over all elements in a given range, combining the results with an operator op; a nil value is returned when the range is empty. The range, the function F, the operator op and the value nil are all explicit parameters of the big operator.

Matita's notation is relatively standard ([10]), and has the following shape:

$$\big[\mathrm{op}, \mathrm{nil}]\_\{ \text{ range description } \} \ \mathrm{F}$$

The range description gives a name to the iteration variable and fixes the domain over which this variable is supposed to range. The elements in the range are supposed to be enumerated (that is not a limitation, considering that the range must be finite), hence the range is specified as an interval $i \in [a, b]$ where a is the lower bound and b is the upper bound (both included in the range). In case the lower bound is 0, the simpler notation $i \leq b$ can also be used. The variable i whose name can obviously be chosen by the user, is bound by the notation, and it usually occurs free in F.

The range can be further restricted specifying an additional boolean predicate, acting as a filter. For instance, the following notation represents the product of all primes less or equal to n

$$\big[\mathrm{times}, 1]\_\{\mathrm{p} \ \leq \mathrm{n} \ | \ \mathrm{primeb} \ \mathrm{p}\} \ \mathrm{p}$$

In this paper, we shall use big operators to iterate boolean functions over finite domains; for instance, the notation

$$\big[\mathrm{andb}, \mathrm{true}]\_\{\mathrm{i} < \mathrm{n}\} \ (\mathrm{b} \ \mathrm{i}).$$

expresses the boolean conjunction of all $(b\ i)$ for all i less than n.

Minimization is essentially a big operator where we iterate the binary minimum function **min** on all elements in a given range enjoying a suitable predicate; the only problem is the definition of a default **nil** element. A relatively natural choice, in case we found no element in the range $[a, b[$ matching the test, is to return b:

$$\textbf{definition Min} := \lambda\mathrm{a}, \mathrm{b}, \mathrm{f}. \big[\min, \mathrm{b}]\_\{\mathrm{i} \in [\mathrm{a}, \mathrm{b}] \ | \ \mathrm{f} \ \mathrm{i}\} \ \mathrm{i}.$$

Although the definition is elegant, the possibility to exploit results on big operators for proving properties of Min poses some problems, in this case. The point is that the lemmas on big operations are hierarchically organized according to the algebraic structure associated with the operator. In the case of the minimum, we have associativity and commutativity, but we do not have a (generic) neutral element (see also [10] for the discussion of a similar problem relative to maximization on real numbers), so we have only access to very basic results.

For minimization we shall use the following ad hoc notation:

$$\mu\_\{ \ \mathrm{i} \in [\mathrm{a}, \mathrm{b}] \ \} \ \mathrm{p}$$

to express the minimum element in the range $[a, b]$ that satisfies p (and returns the successor of b is no such element is found).

The main results about minimization that we shall exploit are the following: under the assumption that there exists an element in the range $[a, b]$ that satisfies f, then the minimum m satisfies f and moreover it is not greater than b (as a matter of fact, the definition of the function t of the gap theorem does not exploit minimality, but only existence).

lemma f_min_true: ∀f,a,b.
  (∃i. a ≤ i ∧ i ≤ b ∧ f i = true) → f (μ_{i ∈[a,b]} (f i)) = true.

lemma min_up: ∀f,a,b.
  (∃i. a ≤ i ∧ i ≤ b ∧ f i = true) → μ_{i ∈[a,b]}(f i) ≤ b.

3.3   Iteration

We shall need to consider progressive intervals of the kind $[g^i(b), g^{i+1}(b)[$, that requires a simple higher-order iterator:

```
let rec iter  (A:Type[0]) (g:A→ A) n a on n :=
match n with
 [O ⇒ a
 |S m ⇒ g (iter A g m a)].
```

The notation $g^i(b)$ is hence a shorthand for (iter nat g i b).
For the proof of the gap theorem we only need the following result:

lemma le_iter: ∀g,a. (∀x. x ≤ g x) → ∀i. a ≤ g^i a.

A few more simple lemmas about composition and monotonicity are used for computing an upper bound of the gap operator:

lemma iter_iter: ∀A.∀g:A→ A.∀a,b,c. g^c (g^b a) = g^(b+c) a.

lemma monotonic_iter: ∀g,a,b,i. (monotonic ? le g) → a ≤ b →
  g^i a ≤ g^i b.

lemma monotonic_iter2: ∀g,a,i,j. (∀x. x ≤ g x) → i ≤ j → g^i a ≤ g^j a.

The question mark in monotonic_iter is an *implicit parameter*, that is an argument automatically filled in by the type inference algorithm (in this case, nat).

3.4   A Bit of Combinatorics

The final ingredient we need for the proof of the gap theorem is a bit of combinatorics. The only delicate point in the definition of t is the termination of the minimization. The general idea is to consider a succession of $n + 1$ disjoint intervals $[r_i, r_{i+1}[$ for $0 ≤ i ≤ n$; then, we consider a set of at most n values to distribute over them (expressing the resources required by a machine with index $i < n$ to terminate on a specific input). Since we have strictly less items than intervals, one of the interval $[r_k, r_{k+1}[$ must remain empty, that gives the desired k. This is essentially a variant (an inverse form) of the *Pigeonhole principle* (also know as Dirichlet's drawer principle), that states that if n items are put into m

pigeonholes where $n > m$, then at least one pigeonhole must contain more than one item.

A simple way to formalize the principle is by considering lists of natural numbers. Given a list l we shall denote with $|l|$ the length of l, and we shall write $x \in l$ to express that x is an element f the list. Let us consider a list l of *distinct* numbers in the interval $[0, n[$; then, obviously, $|l| \leq n$. The interesting point is that

$$|l| = n \leftrightarrow \forall i.i < n \to i \in l$$

This is expressed by the following notions and results in the library of Matita. The **unique** predicate express the fact that the list has no duplicates:

```
let rec unique A (l: list  A) on l :=
  match l with
  [ nil  ⇒ True
  |cons a  tl  ⇒ ¬a ∈tl ∧ unique A tl].
```

Then, we can prove the following results (the proofs are not entirely straightforward, but these basic combinatorial principles belong by now to the folklore of interactive proving, so we do not discuss them).

```
lemma length_unique_le: ∀n,l.
   unique ? l   → (∀x. x ∈ l → x < n) → |l| ≤ n.

lemma eq_length_to_mem_all: ∀n,l.
   |l| = n → unique ? l → (∀x. x ∈ l → x < n) → ∀i. i < n → i ∈ l.

lemma lt_length_to_not_mem: ∀n,l.
   unique ? l   → (∀x. x ∈ l → x < n) → |l| < n → ∃i. i < n ∧ ¬(i ∈ l).
```

4   Kleene's Predicate

The starting point of our axiomatization is Kleene's predicate, that we shall represent with a function U with the following type:

```
axiom U: nat → nat → nat → option nat.
```

The intuitive idea is that

$$U \, i \, x \, r = \begin{cases} \text{Some } y & \text{if program } i \text{ on input } x \text{ returns } y \text{ with resource bound } r \\ \text{None} & \text{otherwise} \end{cases}$$

You should think of U as some agent performing the execution of the program, and checking that it respects the given resource bounds. The only assumption we make about U is about its "monotonicity" with respect to the amount of resources at our disposal:

```
axiom monotonic_U: ∀i,n,m,y. n ≤ m →
   U i x n = Some ? y → U i x m = Some ? y.
```

From the previous axiom we easily conclude that U is single valued:

> **lemma** unique_U: $\forall$i,x,n,m,yn,ym.
>   U i x n = Some ? yn $\rightarrow$ U i x m = Some ? ym $\rightarrow$ yn = ym.

We say that the computation of program x on input y terminates with resource bound r (notation: $\langle i, x \rangle \downarrow r$) if there exists y such that $U\,i\,x\,r =$ Some y:

> **definition** terminate $:=\lambda$i,x,r. $\exists$y. U i x r = Some ? y.

It is straightforward to prove that the previous notion of (bounded) termination is decidable:

> **lemma** terminate_dec: $\forall$x,i,n. $\langle$x,i$\rangle$ $\downarrow$ n $\vee \neg \langle$x,i$\rangle$ $\downarrow$ n.

In order to define the gap operator, we need a boolean version of the termination test:

> **definition** termb $:=\lambda$i,x,t.
>   **match** U i x t **with** [None $\Rightarrow$ false |Some y $\Rightarrow$ true].

It is easy to prove that **termb** *reflects* **terminate** in the sense of [21]:

> **lemma** termb_true_to_term: $\forall$i,x,t. termb i x t = true $\rightarrow \langle$i,x$\rangle$ $\downarrow$ t.
>
> **lemma** term_to_termb_true: $\forall$i,x,t. $\langle$i,x$\rangle$ $\downarrow$ t $\rightarrow$ termb i x t = true.

Exploiting the decidability of termination and the closure properties of section 3.1 it is easy to prove that that the test used in the definition of the gap function is decidable too:

> **lemma** decidable_test : $\forall$n,x,r,r1.
>   ($\forall$i. i $<$ n $\rightarrow \langle$i,x$\rangle$ $\downarrow$ r $\vee \neg \langle$i,x$\rangle$ $\downarrow$ r1) $\vee$
>   ($\exists$i. i $<$ n $\wedge (\neg \langle$i,x$\rangle$ $\downarrow$ r $\wedge \langle$i,x$\rangle$ $\downarrow$ r1)).

5 The Proof of the Gap Theorem

Let us define the following predicate **gapP** n x g r expressing that for all programs up to n, there is a gap between r and $g\,r$ on input x:

> **definition** gapP $:=\lambda$n,x,g,r. $\forall$i. i $<$ n $\rightarrow \langle$i,x$\rangle$ $\downarrow$ r $\vee \neg \langle$i,x$\rangle$ $\downarrow$ g r.

The important fact is that, for any b, g, n, x we can always find a r in the interval between b and $g^n b$ such that (**gapP** n x g r):

> **lemma** upper_bound: $\forall$g,b,n,x. ($\forall$x. x $\leq$ g x) $\rightarrow$
> $\exists$r.b $\leq$ r $\wedge$ r $\leq$ g^n b $\wedge$ gapP n x g r.

For the proof, we pass through the following auxiliary lemma

> **lemma** upper_bound_aux: $\forall$g,b,n,x. ($\forall$x. x $\leq$ g x) $\rightarrow$ $\forall$k.
> ($\exists$j.j $<$ k $\wedge$
>   ($\forall$i. i $<$ n $\rightarrow$ $\langle$i,x$\rangle$ $\downarrow$ g^j b $\vee$ $\neg$ $\langle$i,x$\rangle$ $\downarrow$ g^(S j) b)) $\vee$
> $\exists$l. |l| = k $\wedge$ unique ? l $\wedge$ $\forall$i. i $\in$ l $\rightarrow$ i $<$ n $\wedge$ $\langle$i,x$\rangle$ $\downarrow$ g^k b .

This is proved by induction on k. At the inductive step $k0$ we reason by cases on the inductive hypothesis: we already found our j or we have a list of programs terminating with bound $g^{k0} b$ on input x. In the first case, we are done. In the other case we reason by cases on `decidable_test n x (g^k0 b) (g^(S k0) b))`. In the first case, we can take $j = k0$, and otherwise we have a program i that does not terminate in $g^{k0} b$ but terminates in $g^{k0+1} b$ on input x, and we add i to the list l.

Starting from `upper_bound_aux` it is now easy to prove `upper_bound`. The idea is to proceed by cases on (`upper_bound_aux g b n x Hg n`), where `Hg` is the hypothesis that g is increasing. In case we have a j, we take $r = g^j b$ and we conclude easily. Otherwise, we have a list of programs terminating with bound $g^n b$ on input x. Since the list has length n, by property `eq_length_to_mem_all` all programs up to n must appear in this list, and we can just take $r = g^n b$.

5.1 The Gap Operator

The first step for defining the gap operator is to express the gap predicate `gapP` as a computable boolean function; a simple approach is to use big operators to encode bounded quantification:

> **definition** gapb :=λn,x,g,r.
> \big[andb,true]_{i $<$ n} ((termb i x r) $\vee$ $\neg$ (termb i x (g r))).

It is straightforward to prove that `gapb` reflects the gap predicate `gapP`, and in particular:

> **lemma** gapb_true_to_gapP : $\forall$n,x,g,t.
>   gapb n x g t = true $\rightarrow$ $\forall$i. i $<$ n $\rightarrow$ $\langle$i,x$\rangle$$\downarrow$t $\vee$ $\neg$ ($\langle$i,x$\rangle$ $\downarrow$ (g t)).
>
> **lemma** gapP_to_gapb_true : $\forall$n,x,g,r.
>   ($\forall$i. i $<$ n $\rightarrow$ $\langle$i,x$\rangle$$\downarrow$r $\vee$ $\neg$ ($\langle$i,x$\rangle$ $\downarrow$ (g r))) $\rightarrow$ gapb n x g r = true.

It is now easy to define the gap operator as a higher-order function parametric in g:

> **let rec** gap g n on n :=
>   **match** n **with**
>   [O $\Rightarrow$ 1
> | S m $\Rightarrow$ **let** b :=gap g m **in** μ_{k $\in$[b,g^n b]} (gapb n n g k)
>   ].

From `upper_bound` it is easy to derive an analogous upper bound for `gapb`:

> **lemma** upper_bound_gapb: $\forall$g,m. ($\forall$x. x $\leq$ g x) $\rightarrow$
> $\exists$r.gap g m $\leq$ r $\wedge$ r $\leq$ g^(S m) (gap g m) $\wedge$ gapb (S m) (S m) g r = true.

Then, using property `f_min_true` we easily conclude:

> **lemma** gapS_true: $\forall$g,m. ($\forall$x. x $\leq$ g x) $\rightarrow$ gapb (S m) (S m) g (gap g (S m)) = true.

and from the previous result we derive the expected behaviour of gap operator, in the general case:

> **theorem** gap_theorem: $\forall$g,i.($\forall$x. x $\leq$ g x)$\rightarrow \exists$k.$\forall$n.k < n $\rightarrow$
> $\langle$i,n$\rangle$ $\downarrow$ (gap g n) $\vee \neg \langle$i,n$\rangle$ $\downarrow$ (g (gap g n)).

We just instantiate k with i and proceed by cases on i.

5.2 An Upper Bound

We conclude this section providing a simple upper bound for $gap\,g$, namely, for any n

$$\text{gap } g\,n \leq g^{\sigma(n)}(1) \leq g^{n^2}(1)$$

where $\sigma(n) = \sum_{i \leq n} i = n \cdot (n+1)/2$.

> **let rec** sigma n :=
>   **match** n **with**
>   [O $\Rightarrow$ 0 | S m $\Rightarrow$ n + sigma m].
>
> **lemma** gap_bound: $\forall$g. ($\forall$x. x $\leq$ g x) $\rightarrow$ (monotonic ? le g) $\rightarrow$
> $\forall$n.gap g n $\leq$ g^(sigma n) 1.

The proof is a simple induction on n. If $n = 0$ both sides are equal to 1. In the inductive case:

$$
\begin{aligned}
\text{gap } g\,(S\,n) &\leq g^{(S\,n)}(\text{gap } g\,n) && \text{by min\_up using upper\_bound\_gapb} \\
&\leq g^{(S\,n)}(g^{\sigma(n)}1) && \text{by induction hypothesis} \\
&= g^{(S\,n+\sigma(n)}1 && \text{by iter\_iter} \\
&= g^{\sigma(S\,n)}1 && \text{by definition of sigma}
\end{aligned}
$$

It is worth observing that if g is primitive recursive, than (gap g) is too, and not too far away from g in the elementary hierarchy.

Many authors (see e.g Papadimitriou [27]) note the "fantastically fast growth" of the gap function (without providing an explicit bound), but after all it is no *so* scary (at least, compared to the enormous complexity of other logical problems [19]). Of course, the growth-rate of the function has little to do with its ability to create a gap: its upper bound $g^{\sigma(n)}1$ is a (space and time) constructible function, hence the hierarchy theorems apply and it does not define any gap. The really surprising fact is that in a relatively small interval as that comprised between $g(n)$ and $g^{\sigma(n)}1$ we can find a function with such a strange behaviour as (gap g).

6   Conclusions

In this paper, we presented a formalization in the Matita Interactive Theorem Prover of Borodin-Trakhtenbrot's Gap Theorem of Computational Complexity. The work is part of a huge program of formal revisitation of Complexity Theory, that we call *reverse complexity*, based on the application of methodologies typical of *reverse mathematics* [20,30], consisting in a backward reconstruction from proofs of the basic notions and assumptions underlying the main results of the field.

The final goal is to understand, at a suitable level of abstraction and logical rigor, what *really matters* for a foundational investigation of Complexity, since we know that the details of the different, specific computational models are largely uninfluential.

The need for a better understanding of the logical grounds of complexity theory is testified by a long series of works aimed to provide machine-independent characterizations, spanning from the old works of Blum [11], to the recent field of Implicit Computation Complexity (see [8], and the bibliography therein), passing through a multitude of systems defined by controlling different aspects of the computation: explicit bounds on the growth rate of functions [14,13], the logical power required for proving termination [18], the use and replication of computational resources [9]. See also [16] for a modern treatment of bounded arithmetical systems and an investigation of proof complexity from the point of view of computational complexity.

Even in the relatively simple case of the Gap Theorem, the reverse methodology was instructive, allowing us to clarify that the full power of Blums' abstract framework is not required for this proof. In particular, there is no need to refer to a *principal enumeration* of partial recursive functions, that would be a difficult notion to characterize at an abstract level.

We only postulated the existence of a function U, intuitively playing the role of Kleene's T'-predicate, but avoiding any explicit reference to a system of computable functions; we just assumed U to be monotonic:

axiom U: nat $\rightarrow$ nat $\rightarrow$ nat $\rightarrow$ option nat.

axiom monotonic_U: $\forall$i,n,m,y. n $\leq$ m $\rightarrow$
 U i x n = Some ? y $\rightarrow$ U i x m = Some ? y.

The U function seems to provide an interesting starting point for many different investigations. For instance, exploiting the idea embodied in Kleene's normal form, we can easily axiomatize the existence of an interpreter (universal machine):

axiom universal: $\exists$u.$\forall$i,x,y.
 $\exists$n. U u $\langle$i,x$\rangle$ n = Some y $\leftrightarrow$$\exists$m.U i x m = Some y.

In [3], we proved that any indexed set of partial functions that is closed under composition, contains all projections, an interpreter, and satisfies the s-m-n theorem of Recursion Theory is algorithmically complete, that is, it enumerates all computable functions. So, adding a few more axioms, we get a natural, abstract theory of computable functions. Morevoer, following the ideas outlined in [1], we can integrate the closure conditions on the class of computable functions by suitable complexity conditions, obtaining an interesting formal framework to address complexity theory.

Even more interestingly, we can investigate weaker logical frameworks, corresponding to system of subrecursive functions. For instance, for many interesting results of Complexity Theory, you do not need the existence of a full interpreter, but just the possibility to perform a restricted form of *bounded* interpretation. This is for instance the case of the well known hierarchy theorems of computational complexity [22,31], whose formalization was investigated in [2]. The relation between full and bound interpretation from the point of view of Complexity Theory seems to be an argument worth to be further investigated too.

The new, major milestone in our program is however to provide a suitable, abstract axiomatization of the so called "reachability method". The general idea is to consider the graph of all possible configurations of the computational device, reducing the existence of a computation to a reachability problem in such a graph. Time bounds the dimension of the graph, and in turn the dimension of each configuration bounds the number of possible distinct nodes in the graph, allowing to establish the main relations between time and space. This is largely indepedent from any specific computational device, and it seems important to identify the right *abstract* setting underlying the previous ideas, paving the way to a reverse investigation of the well known theorems of Savitch [29] and Immerman-Szelepcsényi [23,32].

References

1. Asperti, A.: The intensional content of Rice's theorem. In: Proceedings of the 35th ACM SIGPLAN-SIGACT Symposium on Principles of Programming Languages (POPL), San Francisco, California, USA, January 7-12, pp. 113–119. ACM (2008)
2. Asperti, A.: Reverse complexity. Submitted for publication (2013)
3. Asperti, A., Ciabattoni, A.: Effective applicative structures. In: Johnstone, P.T., Rydeheard, D.E., Pitt, D.H. (eds.) CTCS 1995. LNCS, vol. 953, pp. 81–95. Springer, Heidelberg (1995)
4. Asperti, A., Geuvers, H., Natarajan, R.: Social processes, program verification and all that. Mathematical Structures in Computer Science 19(5), 877–896 (2009)
5. Asperti, A., Ricciotti, W.: A proof of Bertrand's postulate. Journal of Formalized Reasoning 5(1), 37–57 (2012)
6. Asperti, A., Ricciotti, W.: A web interface for Matita. In: Jeuring, J., Campbell, J.A., Carette, J., Dos Reis, G., Sojka, P., Wenzel, M., Sorge, V. (eds.) CICM 2012. LNCS, vol. 7362, pp. 417–421. Springer, Heidelberg (2012)
7. Asperti, A., Ricciotti, W., Sacerdoti Coen, C., Tassi, E.: The Matita interactive theorem prover. In: Bjørner, N., Sofronie-Stokkermans, V. (eds.) CADE 2011. LNCS, vol. 6803, pp. 64–69. Springer, Heidelberg (2011)
8. Baillot, P., Marion, J.-Y., Della Rocca, S.R. (eds.): Special issue on implicit complexity. ACM Transactions on Computational Logic 10(4) (2009)

9. Bellantoni, S., Cook, S.A.: A new recursion-theoretic characterization of the poly-time functions (extended abstract). In: Proceedings of the 24th Annual ACM Symposium on Theory of Computing, Victoria, British Columbia, Canada, May 4-6, pp. 283–293. ACM (1992)
10. Bertot, Y., Gonthier, G., Ould Biha, S., Pasca, I.: Canonical big operators. In: Mohamed, O.A., Muñoz, C., Tahar, S. (eds.) TPHOLs 2008. LNCS, vol. 5170, pp. 86–101. Springer, Heidelberg (2008)
11. Blum, M.: A machine-independent theory of the complexity of recursive functions. J. ACM 14(2), 322–336 (1967)
12. Borodin, A.: Computational complexity and the existence of complexity gaps. J. ACM 19(1), 158–174 (1972)
13. Clote, P., Takeuti, G.: On the computational complexity of algorithms. Annals of Pure and Applied Logic 56(1-3), 73–117 (1992)
14. Cobham, A.: The intrinsic computational difficulty of functions. In: Proceedings of the 1964 International Congress for Logic, Methodology, and Philosophy of Science, pp. 24–30. North-Holland, Amsterdam (1964)
15. Constable, R.L.: The operator gap. J. ACM 19(1), 175–183 (1972)
16. Cook, S., Nguyen, P.: Logical Foundations of Proof Complexity. Cambridge University Press (2010)
17. Cutland, N.J.: Computability: An Introduction to Recursive Function Theory. Cambridge University Press (1980)
18. Leivant, D.: A foundational delineation of computational feasibility. In: Proceedings of the Sixth Annual IEEE Symposium on Logic in Computer Science (LICS), pp. 2–11. IEEE (1991)
19. Friedman, H.: Some decision problems of enormous complexity. In: 14th Annual IEEE Symposium on Logic in Computer Science, Trento, Italy, July 2-5, pp. 2–12. IEEE Computer Society (1999)
20. Friedman, H., Simpson, S.G.: Issues and problems in reverse mathematics. Contemporary Mathematics 257, 127–143 (2000)
21. Gonthier, G., Mahboubi, A.: An introduction to small scale reflection in coq. Journal of Formalized Reasoning 3(2), 95–152 (2010)
22. Hartmanis, J., Stearns, R.E.: On the computational complexity of algorithms. Transaction of the American Mathematical Society 117, 285–306 (1965)
23. Immerman, N.: Nondeterministic space is closed under complementation. SIAM J. Comput. 17(5), 935–938 (1988)
24. Landweber, L.H., Robertson, E.L.: Recursive properties of abstract complexity classes. Journal of ACM 19(2), 296–308 (1972)
25. Lewis, F.D.: Unsolvability considerations in computational complexity. In: Proceedings of the Second Annual ACM Symposium on Theory of Computing (STOC), Northampton, Massachusetts, USA, May 4-6, pp. 296–308. ACM (1970)
26. McCreight, E.M., Meyer, A.R.: Classes of computable functions defined by bounds on computation. In: Proceedings of the 1st Annual ACM Symposium on Theory of Computing (STOC), Victoria, British Columbia, Canada, May 4-6, pp. 79–88. ACM (1969)
27. Papadimitriou, C.H.: Computational Complexity. Addison-Wesley (1994)
28. Rogers, H.: Theory of Recursive Functions and Effective Computability. MIT Press (1987)
29. Savitch, W.J.: Relationships between nondeterministic and deterministic tape complexities. J. Comput. Syst. Sci. 4(2), 177–192 (1970)
30. Simpson, S.G.: Subsystems of second order arithmetic. Cambridge University Press (2009)

31. Stearns, R.E., Hartmanis, J., Lewis, P.M.: Hierachies of memory limited computations. In: Proceedings of the 6th Annual Symposium on Switching Circuit Theory and Logical Design (SWCT 1965), FOCS, pp. 179–190 (1965)
32. Szelepcsényi, R.: The method of forced enumeration for nondeterministic automata. Acta Inf. 26(3), 279–284 (1988)
33. Trakhtenbrot, B.: Turing computations with logarithmic delay. Algebra and Logic 3(4), 33–48 (1964)
34. Young, P.R.: Toward a theory of enumeration. Journal of ACM 16(2), 328–348 (1969)
35. Young, P.R.: Easy constructions in complexity theory: gap and speed-up theorems. Proceedings of A.M.S. 37(2), 555–563 (1973)

Certified Kruskal's Tree Theorem

Christian Sternagel*

JAIST, Japan
c-sterna@jaist.ac.jp

Abstract. This paper gives the first formalization of Kruskal's tree theorem in a proof assistant. More concretely, an Isabelle/HOL development of Nash-Williams' minimal bad sequence argument for proving the tree theorem is presented. Along the way, the proofs of Dickson's lemma and Higman's lemma are discussed.

Keywords: Well-Quasi-Orders, Dickson's Lemma, Minimal Bad Sequences, Higman's Lemma, Kruskal's Tree Theorem.

1 Introduction

Kruskal's tree theorem [1] (sometimes called *the* tree theorem in the following) is a famous result in combinatorics, more precisely well-quasi-order (wqo) theory.

Kruskal's Tree Theorem. *When a set A is wqo'd (by a relation $\preceq$), then so is the set of finite trees over A (by homeomorphic embedding w.r.t. $\preceq$).*

Nash-Williams gave a short and elegant proof of the tree theorem [2], where he first established what is now known as the *minimal bad sequence argument*: assume the existence of a minimal "bad" infinite sequence of elements, construct an even smaller "bad" infinite sequence, thus contradicting minimality and proving wqo'dness (since the definition of wqo requires all infinite sequences of elements to be "good").

Besides the minimal bad sequence argument, Nash-Williams' work [2] contains proofs of Dickson's lemma [3] (*if A and B are wqo'd, then so is the Cartesian product $A \times B$*) and a variant of Higman's lemma [4] (*if A is wqo'd, then so is the set of finite subsets of A*), where the latter also incorporates an instance of the minimal bad sequence argument.

The work at hand constitutes a formalization of Nash-Williams' original proofs in the proof assistant Isabelle [5].[1] As indicated also by others, his argumentation is short (in fact, Nash-Williams' paper consists of only two and a half pages in total) and elegant (which was also the main reason for basing the formalization on his work). However, formalizations using proof assistants typically require us to be more rigorous than with pen and paper. Thus, the formalization is

---

* Supported by the Austrian Science Fund (FWF): J3202.
[1] Available from http://isabelle.in.tum.de (try Isabelle/jEdit for browsing).

G. Gonthier and M. Norrish (Eds.): CPP 2013, LNCS 8307, pp. 178–193, 2013.
© Springer International Publishing Switzerland 2013

more detailed in places, which results in somewhat longer (about three and a half thousand lines of Isabelle/HOL theories) and slightly less elegant proofs. Fortunately the most detailed part could be localized (pun intended), thus not derogating the elegance of the remaining proofs.

The author wants to stress that everything presented in the following, is formalized using the proof assistant Isabelle. In this paper, a high-level overview of this formalization is given. The full development is part of the *Archive of Formal Proofs* [6].

Contributions. To the best of the author's knowledge, the presented work constitutes the first unrestricted formalization of Higman's lemma in Isabelle/HOL as well as the first formalization of Kruskal's tree theorem ever. Both are important combinatorial results with applications in rewriting theory. For example, the theory of simplification orders [7] was formalized as part of IsaFoR,[2] where it is applied to show well-foundedness of the Knuth-Bendix-Order [8].

Moreover, the author believes that besides their high trustworthiness (which is of course very important), formalizations of existing mathematical results are also of archival and educational value. The reason is that a formalization contains *all* non-trivial steps of a proof. No doubt, more often than not, those steps were already conducted in the minds of the original proof authors. However, when the original author writes down a proof in condensed form for publishing, some of the steps may get lost. If, much later, another person tries to understand the proof, there may be some mental gaps (or in the worst case even errors).

Finally, formalizations are often hard to read for non-experts (but note that the Isar language for Isabelle [9] is a huge improvement in that respect). Thus, the author hopes that this high-level overview makes the presented formalization more accessible.

Differences to Nash-Williams' Work. The formalization presented here differs from the original presentation of Nash-Williams in several details: As stated above, Nash-Williams proved a variant of Higman's lemma [2, Lemma 2]. Also his version of the tree theorem [2, Theorem 1] does not mention homeomorphic embedding. In the following, every reference to Higman's lemma, means *"If A is wqo'd, then A* is wqo'd by homeomorphic embedding on lists,"* and every reference to the tree theorem, means *"If A is wqo'd, then the set of finite trees over A is wqo'd by homeomorphic embedding on trees."* The structure of the proofs, stays the same.

Overview. The remainder is structured as follows. In Section 2, necessary preliminaries are covered. Then, in Section 3, the structure of Nash-Williams' original proofs is reviewed. The next four sections present a formalization of Dickson's lemma (featuring a proof of a variant of Dickson's lemma for almost-full relations, i.e., not relying on transitivity), in Section 4, a general construction of minimal bad sequences, in Section 5; a formalization of Higman's lemma, in Section 6; and ultimately, a formalization of Kruskal's tree theorem, in Section 7.

---

[2] http://cl-informatik.uibk.ac.at/software/ceta/

Finally, the paper concludes in Section 8, where also applications are sketched, and future as well as related work is discussed.

2   Preliminaries

Throughout this exposition, standard mathematical notation is used as far as possible. However, additionally some Isabelle specific notation is employed, since Isabelle's document preparation facilities were used for typesetting all lemmas and theorems (in the words of Haftmann et al. [10]: *no typos, no omissions, no sweat*; alas, this does not extend to the regular text). Thus, some explanation might be in order.

Isabelle/HOL is a higher-order logic based on the simply-typed lambda calculus. Thus, every term has a type, where Greek letters α, β, γ, ... are used for *type variables*; and *type constructors* like *nat* for natural numbers, $\alpha \Rightarrow \beta$ for the function space, $\alpha \times \beta$ for ordered pairs, α *set* for sets, and α *list* for finite lists. *Type constraints* are written $t::\tau$ and denote that term t is of type τ. As usual for lambda calculi, function application is denoted by juxtaposition, i.e., $f\,x$ denotes the application of function f to the argument x. The type $\alpha \Rightarrow \alpha \Rightarrow bool$ is used to encode binary relations. (An alternative would have been to use $(\alpha \times \alpha)$ *set*. However, the two representations are mostly equivalent and the former is used for many binary relations of Isabelle/HOL's library.)

Further, the following constants from Isabelle/HOL's library are freely used: $\circ::(\alpha \Rightarrow \beta) \Rightarrow (\gamma \Rightarrow \alpha) \Rightarrow \gamma \Rightarrow \beta$, where $f \circ g$ denotes the functional composition of the two functions f and g, i.e., $f \circ g \stackrel{\text{def}}{=} \lambda x.\, f\,(g\,x)$, and sometimes f_φ is used instead of $f \circ \varphi$ for brevity (especially when f denotes an infinite sequence and φ is an index-mapping); $fst::\alpha \times \beta \Rightarrow \alpha$ and $snd::\alpha \times \beta \Rightarrow \beta$ extract the first and second component of a pair, respectively; $set::\alpha$ *list* $\Rightarrow \alpha$ *set*, where $set\,xs$ is the set of elements occurring in the list xs; $[]::\alpha$ *list*, the empty list; $\cdot::\alpha \Rightarrow \alpha$ *list* $\Rightarrow \alpha$ *list*, where $x \cdot xs$ denotes "consing" the element x in front of the list xs; and $@::\alpha$ *list* $\Rightarrow \alpha$ *list* $\Rightarrow \alpha$ *list*, where $xs\ @\ ys$ denotes the concatenation of the two lists xs and ys. Note that since $\cdot$ and $@$ are both right-associative and have the same priority, $xs\ @\ y \cdot ys$ is the same as $xs\ @\ (y \cdot ys)$ and denotes a list that is constructed by inserting the element y between those of xs and ys.

When stating formulas, sometimes Isabelle specific notation is used. Then, $\bigwedge$ denotes universal quantification and $\Longrightarrow$ (right-associative) implication. Moreover, nested implication, like $A \Longrightarrow B \Longrightarrow C$, is abbreviated to $[\![A;\ B]\!] \Longrightarrow C$.

Let $\preceq$ be a binary relation and A a set. The relation $\preceq$ is *reflexive on* A, written $refl_A(\preceq)$, iff $\forall x \in A.\ x \preceq x$; and *transitive on* A, written $trans_A(\preceq)$, iff $\forall x \in A.\ \forall y \in A.\ \forall z \in A.\ x \preceq y \land y \preceq z \longrightarrow x \preceq z$.

Infinite sequences over elements of type α are represented by functions of type $nat \Rightarrow \alpha$. A binary relation $\preceq$ is *transitive on a sequence* f, written $trans_f(\preceq)$, iff $\forall i\,j.\ i < j \longrightarrow f\,i \preceq f\,j$. A sequence f is *good w.r.t.* $\preceq$, written $good_\preceq(f)$, iff $\exists i\,j.\ i < j \land f\,i \preceq f\,j$. If a sequence is not good, it is called *bad*.

The author follows Veldman [11] and Vytiniotis et al. [12] in basing wqos on *almost-full* relations (which are basically wqos without transitivity). The main reason for doing so, is that all the properties of interest also hold for almost-full relations and are easily extended to wqos.

The relation $\preceq$ is *almost-full on* A, written $af_A(\preceq)$, iff all infinite sequences over elements of A are good, i.e., $af_A(\preceq) \overset{\text{def}}{=} \forall f.\, (\forall i.\, f\, i \in A) \longrightarrow good_{\preceq}(f)$. Note that every almost-full relation is necessarily reflexive (see, e.g., [13, Lemma 1]).

Let $\preceq$ be almost-full on A. If in addition $\preceq$ is transitive on A, then $\preceq$ is a *wqo on* A (equivalently, A is wqo'd by $\preceq$), written $wqo_A(\preceq)$.

3   Nash-Williams' Proof

Before a detailed account of the formalization is given (in the sections to come), let us review Nash-Williams' original proofs. The purpose of this section is to familiarize the reader with the overall structure of those proofs, highlight differences to the approach of this paper, and indicate places where the formalization requires additional work – marked by (D1)–(D3). Since the full proofs are not reproduced, a copy of Nash-Williams' paper [2] might be useful for reference.

Nash-Williams starts by giving a proof of Dickson's lemma: *If A and B are wqo'd, then so is $A \times B$* [2, Lemma 1]. Assume there are two infinite sequences a (over elements of A) and b (over elements of B) that are both known to be good. Then, witnesses i and j such that $i < j$ and $(a\, i, b\, i) \preceq (a\, j, b\, j)$ have to be constructed. First, construct a subsequence a_φ of a, such that $a_\varphi\, i \preceq_1 a_\varphi\, (i{+}1)$ for all i. Then, since b is good, indices $i < j$ with $b_\varphi\, i \preceq_2 b_\varphi\, j$ are obtained. At this point, in order to obtain $a_\varphi\, i \preceq_1 a_\varphi\, j$ and thus $(a_\varphi\, i, b_\varphi\, i) \preceq (a_\varphi\, j, b_\varphi\, j)$, transitivity of $\preceq_1$ is essential (refer to theory *Dickson-with-Transitivity* for a formalization of Nash-Williams' original proof, where *TRANS* marks the step in which transitivity is applied). In contrast, the presented formalization proves Dickson's lemma for almost-full relations based on an existing formalization of Ramsey's theorem, thus avoiding the transitivity requirement on $\preceq_1$.

Next comes a proof of Higman's lemma: *If A is wqo'd, then A^* is wqo'd by homeomorphic embedding on lists* [2, Lemma 2]. Assume that the statement is false. Then construct a bad sequence in which every element is as small as possible, i.e., a bad sequence such that replacing any given element by a smaller one, the resulting sequence would be good. The construction is described roughly as follows (where A° is used to denote the set of "objects" built over elements of A; which might refer to the set of finite subsets, the set of finite lists, the set of finite trees, ... in a concrete case):

> Select an $x_1 \in A^\circ$ such that x_1 is the first term of a bad sequence of members of A° and $|x_1|$ is as small as possible. Then select an x_2 such that x_1, x_2 (in that order) are the first two terms of a bad sequence of members of A° and $|x_2|$ is as small as possible [...]. Assuming the axiom of choice, this process yields a bad sequence [...]

In the formalization, this construction is realized by a recursive definition (assuming the existence of an appropriate choice-function). But the definition alone is not enough. It has to be shown that

  the definition is well-defined and results in a minimal bad sequence   (D1)

where well-definedness relies on the existence of the mentioned choice-function. This is the first place where the formalization requires drastically more details than the original proof. Moreover, it constitutes the most technical part of the formalization.

For now, assume that there is a minimal bad sequence m (which is a sequence of finite lists). Let h be the sequence of heads of m and t the sequence of corresponding tails. It is then shown that

  there is no φ such that t_φ is bad and $\varphi\ 0 \leq \varphi\ i$ for all i,   (S1)

since otherwise m would not be minimal. Furthermore, let $T = \{t\ i \mid i \geq 0\}$. Then it is stated, without proof, that

  a bad sequence over T indicates a sequence of shape (S1).   (D2)

In the formalization the corresponding proof is mandatory. From the above it follows that T is wqo'd, since there are no bad sequences. Let $H = \{h\ i \mid i \geq 0\}$, which is wqo'd since A is. Then, by Dickson's lemma, $H \times T$ is wqo'd. Hence, there are i and j such that $i < j$ and $(h\ i, t\ i) \preceq (h\ j, t\ j)$, which implies $m\ i \preceq m\ j$ and thus contradicts the badness of m.

Finally, for the tree theorem the proof structure is very similar to the previous one (only using finite trees instead of finite lists and homeomorphic embedding on trees instead of homeomorphic embedding on lists). Assume that the statement is false. Again a minimal bad sequence m has to be constructed. Instead of heads and tails of lists, now roots and direct subtrees (which are also called successors) of trees are considered. Let r and s denote the sequences of roots and successors of m and $S\ i$ be the set of successors of the i-th tree, i.e., $S\ i = \{x \mid x \in set\ (s\ i)\}$ (note that s is a sequence of finite lists). Then it is shown that

  there is no bad sequence t, such that $t\ i \in S_\varphi\ i$ and $\varphi\ 0 \leq \varphi\ i$ for all i, (S2)

since otherwise m would not be minimal. Let $S' = \{t \mid \exists i.\ t \in set\ (s\ i)\}$. Then it is stated, without proof, that

  a bad sequence over S' indicates a sequence of shape (S2).   (D3)

Again the formalization needs to provide the corresponding proof.

4   Dickson's Lemma

In essence, the formalization is about preservation of wqo'dness by certain type constructors (Dickson's lemma for pairs, Higman's lemma for lists, and the tree theorem for trees). For each of these constructors, a way to extend the orders on the base types to an order on the newly constructed type is required. For Dickson's lemma the following is used: Given two orders $\preceq_1$ and $\preceq_2$, the pointwise order on pairs is defined by $(a_1,\ a_2) \preceq (b_1,\ b_2) \overset{\text{def}}{=} a_1 \preceq_1 b_1 \wedge a_2 \preceq_2 b_2$.

Before proving Dickson's lemma (i.e., that the pointwise combination of orders preserves wqo'dness when forming Cartesian products), let us have a look at how Ramsey's theorem allows us to disregard transitivity (and hence prove a similar lemma already for almost-full relations rather than wqos; see theory *Almost-Full-Relations* for the formal proof development).

The following variant of Ramsey's theorem (which is part of Isabelle/HOL's library; `~~/src/HOL/Library/Ramsey.thy`) is used:

$[\![\textit{infinite } Z; \forall i \in Z.\ \forall j \in Z.\ i \neq j \longrightarrow h\ \{i,\ j\} < n]\!]$
$\Longrightarrow \exists I\ c.\ I \subseteq Z \wedge \textit{infinite } I \wedge c < n \wedge (\forall i \in I.\ \forall j \in I.\ i \neq j \longrightarrow h\ \{i,\ j\} = c)$

In words: Let Z be an infinite set and let h be a function that, given a two-element subset of Z, returns a natural number smaller than n. Then there is an infinite subset I of Z and a natural number c smaller than n such that h encodes all two-element subsets of I by c. More abstractly, assume there is an infinite graph with nodes from Z such that every edge has exactly one of n colors. Then there is an infinite subgraph with nodes from I and all edges of color c.

Using Ramsey's theorem, the auxiliary fact that whenever the union of two binary relations is transitive on an infinite sequence, then there is an infinite subsequence on which either the first or the second relation is transitive, is shown.

Lemma 1. $\textit{trans}_f(\preceq_1 \cup \preceq_2) \Longrightarrow \exists \varphi.\ \textit{trans}_\varphi(<) \wedge (\textit{trans}_{f_\varphi}(\preceq_1) \vee \textit{trans}_{f_\varphi}(\preceq_2))$

Here φ is a strictly monotone (since $<$ is transitive on it) mapping from natural numbers to natural numbers. Hence, f_φ is a subsequence of f whose elements are in the same relative order.

Proof (of Lemma 1). Assume $\textit{trans}_f(\preceq_1 \cup \preceq_2)$, which means that

$$\text{for all } i < j, \text{ either } f\ i \preceq_1 f\ j \text{ or } f\ i \preceq_2 f\ j. \tag{$\star$}$$

Then colorize the set of two-element subsets $\{i,\ j\}$ of the natural numbers using h, defined by, if $i < j$ and $f\ i \preceq_1 f\ j$, then $h\ \{i,\ j\}$ is *0* (*white*), otherwise *1* (*black*). Now Ramsey's theorem can be applied (since the set of natural numbers is infinite and there are exactly two colors). Thus, an infinite set I of natural numbers and a color c such that for all $i \neq j$ in I, the corresponding color $h\ \{i,\ j\}$ is c, is obtained. Since I is well-ordered, there is a function $\varphi::\textit{nat} \Rightarrow \textit{nat}$ that enumerates its elements in increasing order, i.e., $\textit{trans}_\varphi(<)$. Consider the two cases (for arbitrary but fixed $i < j$):

- **case** (c is *white*). Since φ is strictly monotone, also $\varphi\ i < \varphi\ j$. Therefore, $h\ \{\varphi\ i,\ \varphi\ j\} = 0$, and thus $f_\varphi\ i \preceq_1 f_\varphi\ j$.
- **case** (c is *black*). Again, $\varphi\ i < \varphi\ j$. Thus $h\ \{\varphi\ i,\ \varphi\ j\} = 1$, which together with ($\star$) implies $f_\varphi\ i \preceq_2 f_\varphi\ j$.   □

Using this auxiliary fact, Dickson's lemma for almost-full relations is shown.

Lemma 2. $[af_{A_1}(\preceq_1);\ af_{A_2}(\preceq_2)] \implies af_{A_1 \times A_2}(\preceq)$

Proof. Assume $af_{A_1}(\preceq_1)$ and $af_{A_2}(\preceq_2)$. Moreover, to derive a contradiction, assume $\neg\ af_{A_1 \times A_2}(\preceq)$. Then there is some sequence f on $A_1 \times A_2$ which is bad. Let $x \lhd y$ and $x \blacktriangleleft y$ denote *fst* $x \npreceq_1$ *fst* y and *snd* $x \npreceq_2$ *snd* y, respectively. Since f is bad, also $\forall i\, j.\ i < j \longrightarrow f\,i \lhd f\,j \vee f\,i \blacktriangleleft f\,j$, i.e., $trans_f(\lhd \cup \blacktriangleleft)$. Then, by Lemma 1, a strictly monotone mapping φ such that $trans_{f_\varphi}(\lhd)$ or $trans_{f_\varphi}(\blacktriangleleft)$ is obtained. In the first case *fst* $\circ\ f_\varphi$ is bad and in the second *snd* $\circ\ f_\varphi$ is bad, both contradicting the assumptions. $\qquad\Box$

The previous lemma trivially extends to wqos.

Dickson's Lemma. $[wqo_{A_1}(\preceq_1);\ wqo_{A_2}(\preceq_2)] \implies wqo_{A_1 \times A_2}(\preceq)$

Proof. Assuming transitivity of $\preceq_1$ on A_1 and $\preceq_2$ on A_2, it is trivial to show transitivity of $\preceq$ on $A_1 \times A_2$. With Lemma 2, this yields Dickson's lemma. $\qquad\Box$

5   Minimal Bad Sequences

Since the minimal bad sequence argument is needed for Higman's lemma as well as the tree theorem, a general construction that is applicable to both cases is provided (see theory *Minimal-Bad-Sequences* for the formal proof development). To this end, Isabelle/HOL's locale mechanism is employed which allows us to define new constants and prove facts using an "interface" of hypothetical constants and assumptions. As long as the assumptions can be discharged, the new constants and proven facts can be instantiated to arbitrary special cases.

Below, the locale *mbs* which captures the construction of a minimal bad sequence over elements from a given set is described (an early version, that could be simplified drastically since, was presented at the *Isabelle Users Workshop* in 2012 [13]). The locale fixes the following constants:

– The set of elements A, and
– a binary relation $\lhd$ that is used to compare the structural size of elements.

Furthermore, it has the assumptions:

$$wf_A(\lhd) \tag{M1}$$
$$[x \lhd y;\ y \lhd z] \implies x \lhd z \tag{M2}$$

That is, the structural comparison is well-founded on A (M1) (thus, it makes sense to talk about *minimal* elements) and transitive (M2). It turns out that these ingredients are enough to construct – under the assumption that there is a bad sequence – a minimal bad sequence. Informally, an infinite bad sequence is a *minimal* bad sequence, when replacing any element by a smaller one, turns it into a good sequence.

Definition 1 (Minimality). More formally, let an infinite sequence f be *minimal at position* n, written $min^n_\preceq(f)$, iff

$$\forall g. \ (\forall i. \ g \ i \in A) \wedge (\forall i{<}n. \ g \ i = f \ i) \wedge g \ n \lhd f \ n \longrightarrow good_\preceq(g)$$

A sequence is *minimal* if it is minimal at every position.

In words, the definition of $min^n_\preceq(f)$ is: for every sequence g whose initial part up to (but not including) position n coincides with f and where the n-th element of g is strictly smaller than the n-th element of f; g is good. This definition facilitates the construction of a minimal bad sequence from a given bad sequence by iterating over its positions: elements before the current position stay fixed and at the current position an element that is as small as possible is inserted.

As indicated above, a given sequence is modified iteratively. To this end the following auxiliary lemma is employed (which shows that from a sequence that is minimal at position n, a sequence that is also minimal at the next position $n{+}1$, can be obtained):

Lemma 3. $[\![\forall i. \ f \ i \in A; \ bad_\preceq(f); \ min^n_\preceq(f)]\!]$
$\implies \exists g. \ (\forall i{\leq}n. \ g \ i = f \ i) \wedge$
$\qquad g \ (n{+}1) \unlhd f \ (n{+}1) \wedge (\forall i. \ g \ i \in A) \wedge bad_\preceq(g) \wedge min^{n+1}_\preceq(g)$

Proof. Since $\lhd$ is well-founded on A, the induction schema

$$[\![x \in A; \ \textstyle\bigwedge x. \ [\![x \in A; \ \textstyle\bigwedge y. \ [\![y \in A; \ y \lhd x]\!] \implies P \ y]\!] \implies P \ x]\!] \implies P \ x$$

is valid. Assume $\forall i. \ f \ i \in A$, $bad_\preceq(f)$, and $min^n_\preceq(f)$. Let $\exists g. \ \mathfrak{C} \ g \ f \ (f \ (n{+}1))$ abbreviate the conclusion of Lemma 3 (parametrized over the sequences g and f and the element on which induction will be applied). In order for the later induction to go through, a slightly stronger statement than Lemma 3 is shown. To this end, let $\mathfrak{J} \ x$ abbreviate

$$\forall f. \ x = f \ (n{+}1) \wedge (\forall i. \ f \ i \in A) \wedge bad_\preceq(f) \wedge min^n_\preceq(f) \longrightarrow (\exists g. \ \mathfrak{C} \ g \ f \ x)$$

(i.e., generalize over f and let x – on which well-founded induction will be applied – equal the $n{+}1$-th element of f).

For an arbitrary but fixed x, let $x = f \ (n{+}1)$. Hence, from the assumption $\forall i. \ f \ i \in A$ it follows that $x \in A$. Now the above induction schema is used to prove (discharging its first assumption by $x \in A$): $\bigwedge x. \ x = f \ (n{+}1) \implies \mathfrak{J} \ x$.

Thus, $x \in A$ for some arbitrary but fixed x, and $\bigwedge y. \ [\![y \in A; \ y \lhd x]\!] \implies \mathfrak{J} \ y$ is the induction hypothesis (IH). Then show $\mathfrak{J} \ x$.

To this end, assume $x = f \ (n{+}1)$, $\forall i. \ f \ i \in A$, $min^n_\preceq(f)$, and $bad_\preceq(f)$ for some arbitrary but fixed f. Now either $min^{n+1}_\preceq(f)$, concluding the proof, or there is a sequence h such that

$$h \ (n{+}1) \lhd f \ (n{+}1) \tag{1}$$
$$\forall i{<}n{+}1. \ h \ i = f \ i \tag{2}$$
$$\forall i. \ h \ i \in A \tag{3}$$
$$bad_\preceq(h) \tag{4}$$

employing Definition 1. From (1), (3), and the IH, obtain $\mathfrak{I}\,(h\,(n{+}1))$. Moreover, from (2) and $min^n_{\preceq}(f)$ it follows that $min^n_{\preceq}(h)$, which together with (4) yields a sequence m that satisfies $\mathfrak{C}\,m\,h\,(h\,(n{+}1))$. Additionally, from (1) and transitivity of $\trianglelefteq$ it follows that $m\,(n{+}1) \trianglelefteq x$. Combining the previous facts, we obtain $\exists\,m.\;\mathfrak{C}\,m\,f\,x$, thus finishing the prove of $\bigwedge x.\;x = f\,(n{+}1) \Longrightarrow \mathfrak{I}\,x$. Choosing $x = f\,(n{+}1)$ (i.e., discharging the first assumption by reflexivity) and using the initial assumptions, yields $\exists\,g.\;\mathfrak{C}\,g\,f\,(f\,(n{+}1))$. $\qquad\square$

For a step-wise construction of a minimal bad sequence it remains to be shown that from an arbitrary bad sequence, one that is minimal at position 0 can be obtained. This is taken care of by the next lemma.

Lemma 4. $[\![\forall\,i.\;f\,i \in A;\;bad_{\prec}(f)]\!] \Longrightarrow \exists\,g.\;(\forall\,i.\;g\,i \in A) \wedge min^0_{\preceq}(g) \wedge bad_{\prec}(g)$

Proof. Similar structure to the proof of Lemma 3 (but much simpler). $\qquad\square$

At this point it can be shown that if a relation is not almost-full, then there is a minimal bad sequence, thereby taking care of (D1).

Theorem 1. $\neg\;af_A(\preceq) \Longrightarrow \exists\,m.\;bad_{\prec}(m) \wedge (\forall\,n.\;min^n_{\preceq}(m)) \wedge (\forall\,i.\;m\,i \in A)$

Proof. Assume $\neg\;af_A(\preceq)$. Then there is a bad sequence f, i.e., $\forall\,i.\;f\,i \in A$ and $bad_{\prec}(f)$. From Lemma 4 a bad sequence g that is minimal at its first position is obtained. Then, with Lemma 3, together with the axiom of choice,[3] a choice function ν such that

$$\forall f\,n.$$
$$(\forall\,i.\;f\,i \in A) \wedge min^n_{\preceq}(f) \wedge bad_{\prec}(f) \longrightarrow$$
$$(\forall\,i.\;\nu\,f\,n\,i \in A) \wedge$$
$$(\forall\,i{\leq}n.\;\nu\,f\,n\,i = f\,i) \wedge$$
$$\nu\,f\,n\,(n{+}1) \trianglelefteq f\,(n{+}1) \wedge bad_{\prec}(\nu\,f\,n) \wedge min^{n+1}_{\preceq}(\nu\,f\,n)$$

is obtained. That is, $\nu\,f\,n$ provides a witness to Lemma 3, provided that f and n satisfy its assumptions.

Then define an auxiliary sequence (of sequences) m' by $m'\,0 = g$ and $m'\,(n{+}1) = \nu\,(m'\,n)\,n$. The desired minimal bad sequence m, is defined to be $\lambda i.\;m'\,i\,i$ (i.e., the "diagonal" of the auxiliary sequence m'). Of course, it has to be *proven* that m actually is a minimal bad sequence. To this end, the following statements are simultaneously shown by induction on n (i.e., they are true for any n):

$$\forall\,i.\;m'\,n\,i \in A \qquad\qquad \forall\,i{\leq}n.\;min^i_{\preceq}(m'\,n)$$
$$\forall\,i{\leq}n.\;m\,i = m'\,n\,i \qquad\qquad bad_{\prec}(m'\,n)$$

From this $bad_{\prec}(m)$ can be shown as follows: Assume that m is not bad, then there are indices i and j, such that $m\,i \preceq m\,j$; but then also $m'\,j\,i \preceq m'\,j\,j$, contradicting $bad_{\prec}(m'\,j)$. Moreover, from $\forall\,n\,i.\;i \leq n \longrightarrow min^i_{\preceq}(m'\,n)$ it is easy to show that $\forall\,n.\;min^n_{\preceq}(m)$. Ultimately, from $\forall\,n\,i.\;m'\,n\,i \in A$, it follows that $\forall\,i.\;m\,i \in A$, concluding the proof. $\qquad\square$

---

[3] In Isabelle/HOL: $\forall x.\;\exists y.\;Q\,x\,y \Longrightarrow \exists f.\;\forall x.\;Q\,x\,(f\,x)$.

6   Higman's Lemma

Before Higman's lemma for almost-full relations is stated formally, a construction that extends a given order on elements to an order on lists is required: *homeomorphic embedding*. Furthermore, a kind of structural comparison between lists as well as the set of lists built over a given set of elements is needed. The set of lists over elements from a set A, written A^*, is defined inductively:

$$\frac{}{[] \in A^*} \qquad \frac{x \in A \qquad xs \in A^*}{x \cdot xs \in A^*}$$

The list xs is a *proper suffix* of the list ys iff $\exists us.\ ys = us \ @\ xs \wedge us \neq []$ (written $xs < ys$). Homeomorphic embedding on lists, for a given base order $\preceq$, is defined inductively by the rules

$$\frac{}{[] \preceq^* ys} \qquad \frac{xs \preceq^* ys}{xs \preceq^* y \cdot ys} \qquad \frac{x \preceq^= y \qquad xs \preceq^* ys}{x \cdot xs \preceq^* y \cdot ys}$$

where $R^=$ denotes the reflexive closure of R. Note that this definition makes $\preceq^*$ reflexive for arbitrary $\preceq$. For reflexive (and thus also for almost-full) $\preceq$, the assumption $x \preceq^= y$ can be replaced by $x \preceq y$. Intuitively, it might be easier to think about homeomorphic embedding on lists as follows: a list xs is embedded in a list ys iff xs can be obtained from ys by dropping elements and replacing elements with arbitrary smaller ones (w.r.t. the base order). An important special case of embedding is $=^*$, which is called the *sublist relation*. Then, $xs =^* ys$ iff the list xs can be obtained from the list ys by dropping elements.

Using the definitions above, the *mbs* locale can be instantiated as follows (for some arbitrary relation $\preceq$): use A^* for A and $<$ for $\lhd$. The assumptions of the *mbs* locale are discharged by the following facts:

$$wf_{A^*}(<) \qquad\qquad [\![xs < ys;\ ys < zs]\!] \implies xs < zs$$

Thus,

$$\neg\ af_{A^*}(\preceq^*) \implies \exists m.\ bad_{\preceq^*}(m) \wedge (\forall n.\ min_{\preceq^*}^n(m)) \wedge (\forall i.\ m\ i \in A^*)$$

which allows us to prove Higman's lemma for almost-full relations.

Lemma 5. $af_A(\preceq) \implies af_{A^*}(\preceq^*)$

Proof. Assume $af_A(\preceq)$ but $\neg\ af_{A^*}(\preceq^*)$, for the sake of a contradiction. Then there is a bad sequence f. This, in turn, implies the existence of a minimal bad sequence m. All lists in m are non-empty (since otherwise m would be good). Hence, there are sequences h and t of heads and tails of m (i.e., $m\ i = h\ i \cdot t\ i$).

First, it is shown that there is no index-mapping φ such that $\varphi\ 0 \leq \varphi\ i$ for all i and the sequence t_φ is bad. Assume, to the contrary, that such a φ exists. Let n abbreviate $\varphi\ 0$ and c be the combination of m with t, defined by $c\ i \stackrel{\text{def}}{=} if\ i < n\ then\ m\ i\ else\ t\ (\varphi\ (i - n))$ (i.e., c is the same as t_φ, but prepended by the first n elements of m). Then c is bad, since otherwise a contradiction is obtained as follows: Assume c is good. Then there are $i < j$ such that $c\ i \preceq^* c\ j$. Now, analyze the following cases:

- **case** $(j < n)$. Then $m\ i \preceq^* m\ j$, contradicting badness of m.
- **case** $(n \leq i)$. Let $i' = i - n$ and $j' = j - n$. Then $i' < j'$ and $t_\varphi\ i' \preceq^* t_\varphi\ j'$, contradicting badness of t_φ.
- **case** $(i < n$ and $n \leq j)$. Let $j' = j - n$. Then $t\ (\varphi\ j') \leq m\ (\varphi\ j')$ (since the tail of a non-empty list is obviously also a suffix) and $m\ i \preceq^* t\ (\varphi\ j')$ (from $c\ i \preceq^* c\ j$). Moreover, $m\ i \preceq^* m\ (\varphi\ j')$ (since the suffix relation is a special case of embedding and embedding is transitive). Together with $i < \varphi\ j'$, this contradicts the badness of m.

Thus, c is bad. Furthermore, $\forall i < n.\ c\ i = m\ i$ and $c\ n < m\ n$, and thus c is good (since m is minimal): A contradiction, concluding the proof of

$$\nexists \varphi.\ (\forall i.\ \varphi\ 0 \leq \varphi\ i) \wedge bad_{\preceq^*}(t_\varphi). \tag{$\star$}$$

Let H and T denote the sets of heads and tails of the lists in m, respectively, i.e., $H = \{h\ i \mid i \geq 0\}$ and $T = \{t\ i \mid i \geq 0\}$. Obviously $\preceq$ is almost-full on H, since $H \subseteq A$ and $\preceq$ is almost-full on A. Moreover, since every bad sequence over T would admit a subsequence of the shape in $(\star)$, the relation $\preceq^*$ is almost-full on T. With Lemma 2, it is shown that the pointwise combination of $\preceq$ and $\preceq^*$ is almost-full on $H \times T$. Thus, there are $i < j$ with $h\ i \preceq^= h\ j$ and $t\ i \preceq^* t\ j$. By definition of $\preceq^*$, this implies $m\ i \preceq^* m\ j$, contradicting the badness of m. □

But wait a moment, "*since every bad sequence over T ...*" above, is exactly (D2), for which a proof has to be provided.

Lemma 6. $[\![refl_{\{t\ i \mid i \geq 0\}}(\preceq);\ \forall i.\ f\ i \in \{t\ i \mid i \geq 0\};\ bad_{\preceq}(f)]\!]$
$\implies \exists \varphi.\ (\forall i.\ \varphi\ 0 \leq \varphi\ i) \wedge bad_{\preceq}(t_\varphi)$

Proof. Assume that $\preceq$ is reflexive (on $\{t\ i \mid i \geq 0\}$), and f is a bad sequence (over $\{t\ i \mid i \geq 0\}$). First note that for every i, there exists a j such that $f\ i = t\ j$. By the axiom of choice, an index-mapping φ' with $f\ i = t_{\varphi'}\ i$ for all i is obtained. Since f is bad, also $t_{\varphi'}$ is bad. Next it is shown that

$$\text{for every } i \text{ there is a } j > i \text{ such that } \varphi'\ 0 \leq \varphi'\ j. \tag{$\star$}$$

Assume otherwise, then there is some i such that for all $j > i$ the index-mapping satisfies $\varphi\ j < \varphi'\ 0$. Thus, the image of φ' under $\{j \mid i < j\}$ is finite, whereas $\{j \mid i < j\}$ itself is infinite. By the pigeonhole principle, a $k > i$ is obtained such that there are infinitely many $j > i$ with $\varphi'\ j = \varphi'\ k$. But then, there is some $l > k$ for which $\varphi'\ l = \varphi'\ k$. Since $\preceq$ is reflexive and $k < l$, this implies that $t_{\varphi'}$ is good; a contradiction. Using $(\star)$ and the axiom of choice, an index-mapping ψ' such that $i < \psi'\ i$ and $\varphi'\ 0 \leq \varphi'\ (\psi'\ i)$ for all i, is obtained. Now, let ψ abbreviate $\lambda i.\ \psi'^i\ 0$ (the i-fold application of ψ' to 0) and φ abbreviate $\varphi' \circ \psi$. Then, ψ is strictly monotone and $\varphi\ 0 \leq \varphi\ i$ for all i. Moreover, since $t_{\varphi'}$ is bad and ψ is monotone, also t_φ is bad. This concludes the proof. □

Higman's Lemma. $wqo_A(\preceq) \implies wqo_{A^*}(\preceq^*)$

Proof. For transitivity of $\preceq^*$ (under the assumption that $\preceq$ is transitive), refer to lemma *list-hembeq-trans* in theory *Sublist*. Together with Lemma 5, this yields Higman's lemma. □

7   The Tree Theorem

The tree theorem is for finite trees, what Higman's lemma is for finite lists. However, whereas for finite lists, their representation inside Isabelle/HOL is quite unambiguous and the existing data type is generally applicable; this is not so much the case for finite trees. Consider the following two data types

> **datatype** $\alpha\ t = Node\ \alpha\ (\alpha\ t\ list)$
> **datatype** $\alpha\ t' = Empty\ |\ Node\ \alpha\ (\alpha\ t'\ list)$

or the type of first-order terms

> **datatype** $(\alpha,\ \beta)\ term = Var\ \beta\ |\ Fun\ \alpha\ ((\alpha,\ \beta)\ term\ list)$

also a kind of finite tree (and more importantly, one of the types to which the tree theorem is applied, in order to formalize the fact that the Knuth-Bendix order is a simplification order [8]). Restricting the tree theorem to a specific data type would strongly restrict its applicability. Therefore, again Isabelle/HOL's locale mechanism is employed. This time, for a locale *finite-tree* that fixes the following constants (see theory *Finite-Tree* for details):

- A function $mk::\beta \Rightarrow \alpha\ list \Rightarrow \alpha$ that is used to construct a finite tree from a given node and a given list of finite trees.
- A function $root::\alpha \Rightarrow \beta$ that extracts the root node from a given tree.
- As well as a function $succs::\alpha \Rightarrow \alpha\ list$ that extracts the list of direct subtrees (successors) from a given tree.

These constants are required to satisfy the following assumptions (thereby turning mk into kind of a data type constructor with extractors $root$ and $succs$):

$$root\ (mk\ f\ ts) = f \tag{F1}$$
$$succs\ (mk\ f\ ts) = ts \tag{F2}$$
$$(mk\ f\ ss = mk\ g\ ts) = (f = g \wedge ss = ts) \tag{F3}$$

As opposed to a real data type, the above assumptions do not guarantee that *all* finite trees are built from a finite number of applications of mk. Thus, the set of finite trees over nodes from A, written $\mathcal{T}(A)$, is defined inductively by:

$$\frac{f \in A \qquad \forall t \in set\ ts.\ t \in \mathcal{T}(A)}{mk\ f\ ts \in \mathcal{T}(A)}$$

The notion of structural decrease, as needed to instantiate the *mbs* locale, is provided by the *subtree* relation:

$$\frac{t \in set\ ts}{t \lhd mk\ f\ ts} \qquad \frac{s \lhd t \qquad t \in set\ ts}{s \lhd mk\ f\ ts}$$

Where a tree s is a proper subtree of another tree t, if it is either a direct subtree of t itself or a proper subtree of one of the direct subtrees of t.

Homeomorphic embedding on finite trees is also defined inductively by:

$$\frac{t \in set\ ts}{t \preceq_{\text{emb}} mk\ f\ ts} \qquad \frac{s \preceq_{\text{emb}} t \qquad t \preceq_{\text{emb}} u}{s \preceq_{\text{emb}} u}$$

$$\frac{s \preceq_{\text{emb}} t}{mk\ f\ (ss_1\ @\ s\cdot ss_2) \preceq_{\text{emb}} mk\ f\ (ss_1\ @\ t\cdot ss_2)} \qquad \frac{f \preceq^= g \qquad ss =^* ts}{mk\ f\ ss \preceq_{\text{emb}} mk\ g\ ts}$$

The first three rules are easy: homeomorphic embedding extends the subtree relation, is transitive, and is closed under contexts. The last rule states that the nodes of a tree may be replaced by smaller ones (w.r.t. $\preceq$) and that arbitrary successors may be dropped. From this definition, the following property can be shown:

Lemma 7. $[\![f \preceq^= g;\ ss \preceq_{\text{emb}}^* ts]\!] \implies mk\ f\ ss \preceq_{\text{emb}} mk\ g\ ts$

Proof. This property seems obvious, as $\preceq_{\text{emb}}$ is reflexive, transitive, and closed under contexts. However, it turns out to be surprisingly tedious to formalize (or at least *the author* did not find an elegant way). To spare the reader some tedium the details (to be found in lemma *tree-hembeq-list-hembeq* of theory *Finite-Tree*) are skipped.   □

To instantiate the *mbs* locale, the following facts (see [6] for proofs) are shown:

$$wf_{\mathcal{T}(A)}(\lhd) \qquad\qquad [\![s \lhd t;\ t \lhd u]\!] \implies s \lhd u$$

Thus,

$$\neg\ af_{\mathcal{T}(A)}(\preceq_{\text{emb}}) \implies \exists m.\ bad_{\preceq_{\text{emb}}}(m) \wedge (\forall n.\ min^n_{\preceq_{\text{emb}}}(m)) \wedge (\forall i.\ m\ i \in \mathcal{T}(A))$$

Finally, the tree theorem for almost-full relations can be stated and proved (see theory *Kruskal* for details).

Theorem 2. $af_A(\preceq) \implies af_{\mathcal{T}(A)}(\preceq_{\text{emb}})$

Proof. Assume $af_A(\preceq)$ but $\neg\ af_{\mathcal{T}(A)}(\preceq_{\text{emb}})$ for the sake of a contradiction. Then there is a bad sequence and thus a minimal bad sequence m. All trees in m are in the set $\mathcal{T}(A)$ (and thus non-empty). Hence, there are sequences r and s of roots and successor lists of the trees in m (i.e., $m\ i = mk\ (r\ i)\ (s\ i)$).

First it is shown that there is no sequence of trees t and index-mapping φ such that $t\ i \in set\ (s_\varphi\ i)$ (i.e., the sequence t selects an arbitrary successor of $m_\varphi\ i$ as its i-th element) and $\varphi\ 0 \leq \varphi\ i$ for all i, and t is bad. Assume, to the contrary, that such t and φ exist. Let n abbreviate $\varphi\ 0$ and c be the sequence defined by $c\ i \stackrel{\text{def}}{=} if\ i < n\ then\ m\ i\ else\ t\ (i - n)$. Then c is bad, since assuming that it was good results in a contradiction by a similar case analysis conducted in the proof of Lemma 5 above. Furthermore, $\forall i<n.\ c\ i = m\ i$ and $c\ n \lhd m\ n$, and thus c is good (since m is minimal). This contradiction concludes the proof of

$$\nexists t\ \varphi.\ (\forall i.\ t\ i \in set\ (s_\varphi\ i) \wedge \varphi\ 0 \le \varphi\ i) \wedge bad_{\preceq_{emb}}(t). \qquad\qquad (\star)$$

Let R and S denote the sets of roots and successor lists of trees in m, respectively, i.e., $R = \{r\ i \mid i \ge 0\}$ and $S = \{s\ i \mid i \ge 0\}$). Clearly, $\preceq$ is almost-full on R (since $R \subseteq A$). Let S' abbreviate $\{t \mid \exists i.\ t \in set\ (s\ i)\}$. Every bad sequence over S' would admit a sequence of the shape in $(\star)$, thus $\preceq_{emb}$ is almost-full on S'. From Lemma 5, together with $S \subseteq S'^*$, it follows that $\preceq_{emb}{}^*$ is almost-full on S. With Lemma 2, it follows that the pointwise combination of $\preceq$ and $\preceq_{emb}{}^*$ is almost-full on $R \times S$. Thus, there are $i < j$ such that $r\ i \preceq^= r\ j$ and $s\ i \preceq_{emb}{}^* s\ j$, which, employing Lemma 7, implies that $m\ i \preceq_{emb} m\ j$ and thus contradicts the badness of m. □

Note that "*Every bad sequence over S' ...*" above, corresponds to (D3). The corresponding proof is required.

Lemma 8. *Let $\preceq$ be a binary relation and X be the set $\{t \mid \exists i.\ t \in set\ (s\ i)\}$ for a sequence of lists s. Then,*

$$[\![refl_X(\preceq);\ \forall i.\ f\ i \in X;\ bad_\preceq(f)]\!]$$
$$\implies \exists t\ \varphi.\ (\forall i.\ t\ i \in set\ (s_\varphi\ i) \wedge \varphi\ 0 \le \varphi\ i) \wedge bad_\preceq(t)$$

Proof. The proof is structured similarly to the proof of Lemma 6 but slightly more involved, due to the extra indirection via list elements. For details, refer to lemma *bad-of-special-shape'* in theory *Kruskal-Auxiliaries* of [6]. □

Kruskal's Tree Theorem. $wqo_A(\preceq) \implies wqo_{T(A)}(\preceq_{emb})$

Proof. Theorem 2 and transitivity of $\preceq_{emb}$ yield the tree theorem. □

8   Conclusions and Related Work

An Isabelle/HOL formalization of three important results from combinatorics was presented: Dickson's lemma, Higman's lemma, and Kruskal's tree theorem.

Parts of the presented formalization were used by Wu et al. [14] to formalize a proof of: *For every language A, the languages of sub- and superstrings of A are regular.* (Details are presented in a submitted journal version of [15].)

Moreover, the presented formalization of the tree theorem is employed for a proof that the Knuth-Bendix order is a simplification order [8]. To this end, actually a variant of the tree theorem as presented in this paper is needed – which might be called *the term theorem*. The reason is that in the above mentioned proof it is essential to consider arities of function symbols, whereas in Section 7, a node in a tree is allowed to have an arbitrary (finite) number of successors.

It is left as future work to investigate whether the tedious induction in the proof of Theorem 1 can be replaced by an invocation of Zorn's lemma (and this in turn, by an application of open induction [16,17], thereby hopefully giving also insight into the computational content of the minimal bad sequence argument).

There are formalizations of Higman's lemma in Isabelle/HOL by Berghofer [18] and using other proof assistants by Murthy [19], Fridlender [20], Herbelin [21], Seisenberger [22], and Martín-Mateos et al. [23].

Since Berghofer's work was also conducted using Isabelle/HOL, some comments on the relation to the presented work are in order. First note that Berghofer's formalization is constructive (based on an earlier proof by Coquand and Fridlender in an unpublished manuscript entitled *A Proof of Higman's Lemma by Structural Induction*). Furthermore, it is restricted to a two letter alphabet (and Berghofer notes that *"the extension of the proof to an arbitrary finite alphabet is not at all trivial"*). Also noteworthy is that the focus of Berghofer's work is on program extraction and the computational behavior of the resulting program. In contrast, the presented work constitutes a formalization of Higman's lemma without restricting the alphabet, i.e., the alphabet may be infinite as long as it is equipped with a wqo (which is always the case for finite alphabets).

An intuitionistic proof of Kruskal's tree theorem is presented in [11]. However, to the best of the author's knowledge the presented work constitutes the first formalization of the tree theorem in a proof assistant ever.

Acknowledgments. I thank Mizuhito Ogawa for helpful discussions on everything related to the tree theorem, as well as enabling (together with the Austrian Science Fund) my stay in Japan.

References

1. Kruskal, J.B.: Well-quasi-ordering, the tree theorem, and Vazsonyi's conjecture. Trans. Amer. Math. Soc. 95(2), 210–225 (1960), doi:10.2307/1993287
2. Nash-Williams, C.S.J.A.: On well-quasi-ordering finite trees. Proc. Cambridge Philos. Soc. 59(4), 833–835 (1963), doi:10.1017/S0305004100003844
3. Dickson, L.E.: Finiteness of the odd perfect and primitive abundant numbers with n distinct prime factors. Amer. J. Math. 35(4), 413–422 (1913), doi:10.2307/2370405
4. Higman, G.: Ordering by divisibility in abstract algebras. Proc. London Math. Soc. s3-2(1), 326–336 (1952), doi:10.1112/plms/s3-2.1.326
5. Nipkow, T., Paulson, L.C., Wenzel, M.: Isabelle/HOL - A Proof Assistant for Higher-Order Logic. LNCS, vol. 2283. Springer, Heidelberg (2002), doi:10.1007/3-540-45949-9
6. Sternagel, C.: Well-Quasi-Orders. In: Klein, G., Nipkow, T., Paulson, L.C. (eds.) AFP (2012), http://afp.sf.net/devel-entries/Well_Quasi_Orders.shtml
7. Middeldorp, A., Zantema, H.: Simple termination of rewrite systems. Theor. Comput. Sci. 175(1), 127–158 (1997), doi:10.1016/S0304-3975(96)00172-7
8. Sternagel, C., Thiemann, R.: Formalizing Knuth-Bendix orders and Knuth-Bendix completion. In: van Raamsdonk, F. (ed.) RTA 2013. LIPIcs, vol. 21, pp. 286–301, Schloss Dagstuhl (2013), doi:10.4230/LIPIcs.RTA.2013287
9. Wenzel, M.: Isabelle/Isar – A Versatile Environment for Human-readable Formal Proof Documents. PhD thesis, Technische Universität München (2002), http://tumb1.biblio.tu-muenchen.de/publ/diss/in/2002/wenzel.pdf
10. Haftmann, F., Klein, G., Nipkow, T., Schirmer, N.: LaTeX sugar for Isabelle documents (2013), http://isabelle.in.tum.de/dist/Isabelle2013/doc/sugar.pdf

11. Veldman, W.: An intuitionistic proof of Kruskal's theorem. Arch. Math. Logic 43(2), 215–264 (2004), doi:10.1007/s00153-003-0207-x
12. Vytiniotis, D., Coquand, T., Wahlstedt, D.: Stop when you are almost-full - adventures in constructive termination. In: Beringer, L., Felty, A. (eds.) ITP 2012. LNCS, vol. 7406, pp. 250–265. Springer, Heidelberg (2012), doi:10.1007/978-3-642-32347-8_17
13. Sternagel, C.: A locale for minimal bad sequences. In: IUW 2012, arxiv 1208.1366 (2012)
14. Wu, C., Zhang, X., Urban, C.: The Myhill-Nerode theorem based on regular expressions. In: Klein, G., Nipkow, T., Paulson, L.C. (eds.) AFP (2011), http://afp.sf.net/entries/Myhill-Nerode.shtml
15. Wu, C., Zhang, X., Urban, C.: A formalisation of the Myhill-Nerode theorem based on regular expressions (proof pearl). In: van Eekelen, M., Geuvers, H., Schmaltz, J., Wiedijk, F. (eds.) ITP 2011. LNCS, vol. 6898, pp. 341–356. Springer, Heidelberg (2011), doi:10.1007/978-3-642-22863-6_25
16. Raoult, J.C.: Proving open properties by induction. Inform. Process. Lett. 29(1), 19–23 (1988), doi:10.1016/0020-0190(88)90126-3
17. Ogawa, M., Sternagel, C.: Open Induction. In: Klein, G., Nipkow, T., Paulson, L.C. (eds.) AFP (2012), http://afp.sf.net/devel-entries/Open_Induction.shtml
18. Berghofer, S.: A constructive proof of Higman's lemma in Isabelle. In: Berardi, S., Coppo, M., Damiani, F. (eds.) TYPES 2003. LNCS, vol. 3085, pp. 66–82. Springer, Heidelberg (2004), doi:10.1007/978-3-540-24849-1_5
19. Murthy, C.R.: Extracting Constructive Content from Classical Proofs. PhD thesis, Cornell University (1990), http://hdl.handle.net/1813/6991
20. Fridlender, D.: Higman's lemma in type theory. In: Giménez, E., Paulin-Mohring, C. (eds.) TYPES 1996. LNCS, vol. 1512, pp. 112–133. Springer, Heidelberg (1996), doi:10.1007/BFb0097789
21. Herbelin, H.: A program from an A-translated impredicative proof of Higman's lemma (1994), http://coq.inria.fr/pylons/contribs/view/HigmanNW/v8.3
22. Seisenberger, M.: On the Constructive Content of Proofs. PhD thesis, LMU Munich (2003), http://nbn-resolving.de/urn:nbn:de:bvb:19-16190
23. Martín-Mateos, F.J., Ruiz-Reina, J.L., Alonso, J.A., Hidalgo, M.J.: Proof pearl: A formal proof of Higman's lemma in ACL2. J. Autom. Reason. 47(3), 229–250 (2011), doi:10.1007/s10817-010-9178-x

Extracting Proofs from Tabled Proof Search[*]

Dale Miller[1] and Alwen Tiu[2]

[1] INRIA-Saclay & LIX/École Polytechnique
[2] Research School of Computer Science, The Australian National University
& School of Computer Engineering, Nanyang Technological University

Abstract. We consider the problem of model checking specifications involving co-inductive definitions such as are available for bisimulation. A proof search approach to model checking with such specifications often involves state exploration. We consider four different tabling strategies that can minimize such exploration significantly. In general, tabling involves storing previously proved subgoals and reusing (instead of reproving) them in proof search. In the case of co-inductive proof search, tables allow a limited form of loop checking, which is often necessary for, say, checking bisimulation of non-terminating processes. We enhance the notion of tabled proof search by allowing a limited deduction from tabled entries when performing table lookup. The main problem with this enhanced tabling method is that it is generally unsound when co-inductive definitions are involved and when tabled entries contain unproved entries. We design a proof system with tables and show that by managing tabled entries carefully, one would still be able to obtain a sound proof system. That is, we show how one can extract a post-fixed point from a tabled proof for a co-inductive goal. We then apply this idea to the technique of bisimulation "up-to" commonly used in process algebra.

1   Introduction

Model checking and theorem proving are usually considered two distinct techniques in formal verification: the former is concerned mainly with satisfiability in a given model while the latter is concerned mainly with provability (e.g., validity in all models). Viewed algorithmically, model checking can be loosely characterized as a model exploration technique (e.g., explorations of states in a transition systems, worlds in a Kripke structure, etc). We adopt this view here. When inference and proof are enriched to contain flexible treatments of (least and greatest) fixed points, model checking can be seen as deduction. As such, the model checkers can be expected to output *proof certificates* justifying their completed state explorations in a manner similar to what one might expect to have output from automatic or interactive theorem provers.

[*] We thank the anonymous referees for their helpful comments. The first author has been supported by the ERC Advanced Grant ProofCert and the second author has been supported by the ARC Discovery Grant DP110103173.

G. Gonthier and M. Norrish (Eds.): CPP 2013, LNCS 8307, pp. 194–210, 2013.
© Springer International Publishing Switzerland 2013

In this paper, formal proofs will be based on the Linc sequent calculus [15,17] (see Section 2), which generalizes Gentzen's sequent calculus LJ for intuitionistic logic with induction and co-induction. We shall also focus on the Bedwyr model checking implementation of part of Linc [1], particularly the form of *tabled deduction* that is implemented in that system. Bedwyr has been most successfully applied to domains where model checking is performed on syntactically rich domains (involving expressions taken from process calculi and programming languages) instead of more simple state-like domains comprised of tuples of booleans, small integers, etc.

1.1 Model Checking as Proof Search

In this paper, we address the problem of integrating (co-)inductively proved theorems with model checking and we will use bisimulation as a specific and important example. In the setting of Linc, bisimulation is defined as the greatest fixed point of the following recursive definition.

$$\text{bisim}(P,Q) \overset{\nu}{=} [\forall P'\forall A.\ P \xrightarrow{A} P' \supset \exists Q'.\ Q \xrightarrow{A} Q' \wedge \text{bisim}(P',Q')] \wedge$$
$$[\forall Q'\forall A.\ Q \xrightarrow{A} Q' \supset \exists P'.\ P \xrightarrow{A} P' \wedge \text{bisim}(Q',P')]$$

Bedwyr's proof search mechanism will turn this definition into a state exploration procedure. Such a direct and intimate connection between bisimulation defined as a logical formula and an algorithmic state exploration algorithm provides at least two important novelties. First, logical encodings may clarify some aspects of the theories being encoded: for example, the difference between late bisimulation and open bisimulation for the π-calculus can be explained as the distinction between intuitionistic and classical logic, i.e., the presence (or absence) of the excluded middle principle applied to the equality of names [16]. Second, since model checking can be seen as building Linc proofs, a checker should be able to output a formal *proof certificate*: for example, successful proof search in Bedwyr for a query concerning bisimilarity of two processes or satisfiability of a modal formulas by a process yields a Linc proof which can be extracted and checked independently.

Although such logical encoding of state exploration techniques is in principle straightforward, naive proof search techniques can yield inefficient algorithmic search and proof certificates that are unacceptably large. One way to address these problems is to use *tabled deduction* so that proved subgoals can be shared and not reproved. For example, Bedwyr stores certain (sub)goals that have been proved and attempts to reuse them when proving other (sub)goals. The tabling of proved subgoals is not, however, sufficient to deal with model checking of potentially non-terminating systems. For example, to prove bisimilarity of simple processes such as $!a$ and $!(a+a)$, a naive unfolding of the processes will not terminate (of course, checking bisimilarity is undecidable so no fixed strategy will yield bounded search in all cases). A more clever approach to showing bisimulation is the *bisimulation up-to* technique [11], which can employ additional information about bisimulation in order to reduce the size of the relation (the table) needed to demonstrate bisimilarity.

1.2 Four Tabling Strategies

In this paper, we examine how tabled deduction can be used to build a bisimulation as well as a bisimulation-up-to. In particular, we examine four tabling strategies in model checking that allow building smaller witnesses (ultimately, proof certificates) of relationships on possibly non-terminating processes. In each case, the main technical difficulties involves extracting an independently checkable proof certificate: obviously, such extraction guarantees soundness of the tabling method. As a case study, we show how bisimulation up-to techniques for process calculi can be encoded in proof search in one of the tabling strategies, and show how proof certificates can be generated.

Since we view model checking as a certain process for building a proof, at any particular moment, the state of that process can be abstracted to be roughly two items: the *partial proof* and the *table*. The first of these is a tree structure of nodes that is labeled by atomic formulas. Nodes are either *leaf* nodes or *interior* nodes and both of these classes can be further divided between *open* and *closed*. A *closed leaf node* is one that has been proved and an *open leaf node* is one for which no proof has yet been found. A *closed interior node* is one all of whose descendant leaf nodes are closed and an *open interior node* is one with some descendant leaf node that is open. The second component of the model checker's state, the table, is a set of node occurrences. We shall always allow a table to contain closed leaf occurrences from the associated partial proof. We shall also use the term *history atom* to describe a formula that labels an interior node.

Two independent choices are available in describing a tabling strategy: the first choice is between allowing or not allowing history atoms into the table and the second is allowing the table to infer an atom by simply checking its membership in the table or by allowing a deduction from tabled atoms and some assumed set of theories. As an example of this latter choice, consider a table that contains the atomic statements $(\text{bisim}(p_1, p_2))$ and $(\text{bisim}(p_2, p_3))$. If the table is only used to infer its members, we can infer these two atoms. If we have proved elsewhere (using a proof assistant that understands (co-)induction) that the $(\text{bisim}(\cdot, \cdot))$ relationship is transitive, then a table that incorporates that theorem could also conclude $(\text{bisim}(p_1, p_3))$. More formally, if R is the set of atoms in a table and T is a set of theories, then we are allowed to infer the atomic formula G from this table if formula $(R \wedge T) \rightarrow G$ is provable. In this paper, we shall assume that T is a set of hereditary Harrop (hH) formulas (formulas containing only conjunction, implication, and universal quantifiers: these formulas subsume Horn clauses and are basis of λProlog [7]). While in most of our examples, such hH formulas will form a simple, decidable theory, we shall not assume a priori that theories are, in fact, decidable.

We identify the following four tabling strategies.

I History atoms are not tabled; the table only infers its members.
II History atoms are not tabled; the table uses theories to infer additional atoms.
III History atoms can be tabled; the table only infers its members.
IV History atoms can be tabled; the table uses theories to infer additional atoms.

The first two strategies yield proof certificates that simply use the cut rule: these two strategies are always sound as long as the theory (in Strategy II) is known to be valid, i.e., proved elsewhere using (co-)inductive techniques. Actually, Strategy I collapses into Strategy II if the empty set is an allowed theory. Soundness of these two strategies is not difficult to establish and it follows the work presented in [8]. Strategy III is sound only when the tabled entries are co-inductive predicates: furthermore, a proof certificate can always be constructed and it will be essentially a post-fixed point found within the table. When tabled entries are restricted to co-inductive predicates, the last strategy corresponds to the bisimulation up-to technique, as described in, say [10]. In this case, the theory $\mathcal{T}$ that is used to expand the table no longer corresponds to a lemma in the meta logic. They instead encode functions on relations, and soundness of a tabled proof in this case depends on the soundness of these functions, i.e., whether they allow one to construct a post-fixed point of a co-inductive definition. We shall focus on strategy III and IV in this paper, but the soundness results for Strategy I and II can be found in [9, Appendix B].

There is significant precedent in the literature related to the use of history atoms to capture aspects of co-inductive proofs, notably works on *cyclic proofs* for logics with induction and co-induction [14,4]. In particular, proof search strategies similar to strategy III above have also been used in cyclic theorem provers [3] and tabling methods in co-inductive logic programming (see e.g. [5,13]). Soundness of cyclic proofs (inductive or co-inductive) is not difficult to establish semantically and there are well known syntactic criteria for cyclic proof systems to be sound, e.g., the notion of a *progressing trace* that dates back to work on modal μ-calculus [18] and its first-order extensions [14]. However, there are two main distinguishing features of our work compared to these related work:

First, we do not justify the soundness of cyclic proofs via semantics but instead we translate cyclic proofs into a more standard proof system that uses explicit (co-)induction rules, e.g., the logic Linc or higher-order logic, for which the issue of soundness has been well established and for which there is a well developed proof theory. Such translation is in general difficult: Sprenger and Dam in [14] provide such a general translation but it requires annotations of fixed point operators with ordinals. For annotation-free cyclic proof systems such as that of Brotherston [4], the translation from cyclic proofs to proofs with explicit (co-)induction rules remains an open problem. While our cyclic proof system (for strategy III) does not introduce explicit ordinal annotations, the kind of cyclic structures allowed in that proof system is much simpler than in [14,4] and forbids cross-branch cycles and mutually recursive definitions. We are thus able to give simple constructions of proofs with explicit (co-)induction rules.

Second, our strategy IV has no counterpart in literature of cyclic proofs. The interpretation of such a cyclic proof is not a straightforward construction of post fixed points since the circularity induced by applications of the theory component in this strategy does not obey the notion of progressing traces underlying existing cyclic proof systems mentioned above. Of course semantic soundness for such applications is known in the literature of bisimulation up-to [10]; our work can be seen as a formal logical formulation of the soundness criteria in [10].

We note that strategies I to III have been implemented in the current development version of Bedwyr, and a preliminary version of strategy IV is being developed at the Parsifal team at INRIA. An example in [9, Section A] illustrates the use of strategy IV to prove bisimilarity of two non-terminating processes, something which is not possible with other strategies.

In Section 2, we present the proof system for intuitionistic logic that we use in the rest of this paper. In Section 3, we present a proof system which uses tables. The four tabling strategies outlined above are differentiated in this tabled system by a function that filters appropriate elements of the tables and the theories that are assumed in the proof. Soundness of strategy III is proved in Section 4, where we show how to construct a post fixed point from tabled entries. In Section 5 we show how to interpret theories as up-to functions and tabled entries as a post fixed point "up-to". In Section 6, we show how compositions of up-to functions can be encoded as compositions of logical theories. We then show, via a permutation argument, that up-to functions can be freely and soundly composed, provided certain conditions related to how these theories permute over each other hold. In Section 7, we discuss further work. The appendix of the companion paper [9] contains several proofs that are omitted in the main text.

2   Backgrounds

We give an overview of the logical framework used as the foundation of this work, i.e., the logic Linc [15], and the bisimulation up-to techniques [11,10].

The Linc logic is essentially a version of Church's Simple Theory of Types with the following differences. (i) Linc is based on intuitionistic provability (described here using a two-sided sequent calculus similar to Gentzen's LJ proof system). (ii) The type of quantified variables are restricted to those not containing the type of propositions (i.e., the type o in Church's notation): thus, Linc does not allow predicate quantification. (iii) Linc also contains *free equality*, i.e., equality in the term model, and *inductive* and *co-inductive* definitions as *logical connectives* and these will be given introduction rules in the sequent calculus. (iv) Finally, Linc also contains the ∇-quantifier (see, for example, [16]) but we can safely ignore it in this paper.

Each predicate symbol in Linc is given a designation as either *undefined*, *inductive* or *co-inductive*. An undefined predicate is the usual one in first-order intuitionistic logic, i.e., its interpretation in a model is allowed to be an arbitrary subset of the domain of interpretation. To each (co-)inductive predicate

$$\frac{\{\Gamma[\rho] \longrightarrow C[\rho]\}_{\rho \in \mathbb{U}(s,t)}}{s = t, \Gamma \longrightarrow C} \ \text{eq}\mathcal{L} \qquad\qquad \frac{}{\Gamma \longrightarrow t = t} \ \text{eq}\mathcal{R}$$

$$\frac{B S \vec{y} \longrightarrow S \vec{y} \quad \Gamma, S \vec{t} \longrightarrow C}{\Gamma, p \vec{t} \longrightarrow C} \ \text{I}\mathcal{L}, \ p \vec{x} \overset{\mu}{=} B p \vec{x} \qquad\qquad \frac{\Gamma \longrightarrow B p \vec{t}}{\Gamma \longrightarrow p \vec{t}} \ \text{I}\mathcal{R}, \ p \vec{x} \overset{\mu}{=} B p \vec{x}$$

$$\frac{B p \vec{t}, \Gamma \longrightarrow C}{p \vec{t}, \Gamma \longrightarrow C} \ \text{CI}\mathcal{L}, \ p \vec{x} \overset{\nu}{=} B p \vec{x} \qquad\qquad \frac{\Gamma \longrightarrow S \vec{t} \quad S \vec{x} \longrightarrow B S \vec{x}}{\Gamma \longrightarrow p \vec{t}} \ \text{CI}\mathcal{R}, \ p \vec{x} \overset{\nu}{=} B p \vec{x}$$

Fig. 1. The Linc inference rules for equality and the least and greatest fixed points

p, we associate a *definition*, i.e., a formula possibly containing occurrences of p. Formally, we write $p\vec{x} \overset{\mu}{=} D p \vec{x}$ to denote an inductive definition of p. Here D is an abstraction, containing no occurrences of p, that is applied to p and variables $\vec{x}$. We shall require that p occurs strictly positively in $D p \vec{x}$. A co-inductive definition is similarly defined, with $\overset{\nu}{=}$ replacing $\overset{\mu}{=}$. We write $p\vec{x} \overset{\triangle}{=} D p \vec{x}$ to denote either an inductive or a co-inductive definition.

In Section 1.1, the definition of bisimulation illustrates this scheme by setting the schema variable D to be the λ-term with abstractions $\lambda\text{bisim}\lambda P\lambda Q$ and with its body being the entire right-hand-side of the definition. Further restrictions are needed, e.g., restrictions on mutual recursions between inductive and co-inductive definitions, to guarantee cut-elimination; see [15,17] for details.

We consider terms as equal modulo α-conversion and assume the usual notion of capture-avoiding substitutions for λ-calculus. The application of a substitution θ to a term t is written $t[\theta]$. This notation extends to application of substitutions to multisets of formulas, i.e., $\Gamma[\theta] = \{B[\theta] \mid B \in \Gamma\}$. The inference rules of Linc are those for LJ plus the rules for equality and fixed points that are given in Figure 1. In eq$\mathcal{L}$, the expression $\mathbb{U}(s,t)$ is used to denote a complete set of unifiers for s and t. Since equality has introduction rules, it is a logical connective and not a predicate. The rules for the introduction of inductive predicates on the right or co-inductive predicates on the left are given by familiar unfolding rules while the introduction of inductive predicates on the left or co-inductive predicates on the right are given by the corresponding induction or co-induction principles. In this latter case, the predicate variable S in those inference rules correspond to the invariant (pre-fixed point) or co-inductive invariant (post-fixed point). Notice that unfolding inductive predicates on the left and co-inductive predicates on the right are admissible (sound) inference rules.

We shall often need to restrict ourselves to the "level 0/1 fragment" [1] of Linc. To define this fragment, we assume that every predicate symbol is either inductive or co-inductive and is assigned a *level* of 0 or 1. A formula is *level-0* if it contains no predicates of level 1 and contains no occurrences of implication or universal quantifier. Level-1 formulas satisfy the following grammar:

$$F ::= \bot \mid \top \mid t = s \mid p\vec{t} \mid \exists x.F \mid \forall x.F \mid G \supset F \mid F \wedge F \mid F \vee F.$$

where G ranges over level-0 formulas and p ranges over level-0 or level-1 predicates. A definition $p\, \vec{x} \triangleq B$ is a level-0 (level-1) definition if both p and B are level-0 (resp. level-1) formulas.

Bisimulation up-to [11] refers to a technique for proving bisimilarity of processes that aims at reducing the size of the relation one needs to construct to prove bisimilarity. Bisimulation is a binary relation $\mathcal{R}$ that satisfies some closure properties w.r.t. the transition system generated by processes, as shown in the diagram on the left below. The up-to technique modifies this definition to allow P' and Q' to be related by a larger relation $\mathcal{F}(\mathcal{R})$, defined via an *up-to function* $\mathcal{F}$, as shown in the diagram on the right below.

$$
\begin{array}{ccccccccc}
P & \mathcal{R} & Q & & & P & \mathcal{R} & Q \\
\alpha\downarrow & & \downarrow\alpha & & & \alpha\downarrow & & \downarrow\alpha \\
P' & \mathcal{R} & Q' & & & P' & \mathcal{F}(\mathcal{R}) & Q'
\end{array}
$$

Let $\mathbb{B}$ be the function on binary relations defined by

$$
\mathbb{B}(R) = \{\langle P, Q\rangle \mid [\forall P' \forall A.\ P \xrightarrow{A} P' \supset \exists Q'.\ Q \xrightarrow{A} Q' \wedge R(P', Q')] \wedge
$$
$$
[\forall Q' \forall A.\ Q \xrightarrow{A} Q' \supset \exists P'.\ P \xrightarrow{A} P' \wedge R(Q', P')]\}
$$

Then bisimilarity, denoted by $\sim$, is defined as the greatest fixed point of $\mathbb{B}$. The left-diagram above shows that $\mathcal{R} \subseteq \mathbb{B}(\mathcal{R})$, i.e., that $\mathcal{R}$ is a post-fixed point of $\mathbb{B}$. Since $\mathbb{B}$ is monotone, the Knaster-Tarski fixed point theorem implies that $\mathcal{R}$ is included in $\sim$. The right-diagram, on the other hand, only proves that $\mathcal{R} \subseteq \mathbb{B}(\mathcal{F}(\mathcal{R}))$ and in general this does not establish $\mathcal{R}$ as a post-fixed point of $\mathbb{B}$, so one needs to prove that the function $\mathcal{F}$ is *sound*, i.e., for every $\mathcal{R}$, if $\mathcal{R} \subseteq \mathbb{B}(\mathcal{F}(\mathcal{R}))$ then $\mathcal{R} \subseteq \sim$. This up-to technique is not limited to bisimulation and it can be used with other co-inductive definitions [10].

3   Tabled Deduction Presented as a Proof System

When inductive and co-inductive predicates are not used, tabled deduction is easily justified using the cut inference rules of sequent calculus [8]. For example, proving $A \wedge B$ from assumptions Γ can proceed as follows:

$$
\cfrac{\Xi_A \quad \cfrac{\cfrac{}{A, \Gamma \longrightarrow A}\ init \quad \cfrac{\Xi_B}{A, \Gamma \longrightarrow B}}{A, \Gamma \longrightarrow A \wedge B}\ \wedge R}{\Gamma \longrightarrow A \wedge B}\ cut
$$

where $\Xi_A : \Gamma \longrightarrow A$.

Here, A is both proved by the subproof Ξ_A and is an assumption in the subproof Ξ_B of B from Γ.

When co-inductive predicates are present, one way to establish a co-inductive goal, say bisimulation, is to allow a form of *circular proofs*. In a circular proof, a branch in the proof tree is allowed to close when there is a 'loop', i.e., the sequent at the leaf of the branch matches another sequent lower in the tree. This is a familiar notion in fixed point logics and conditions that guarantee soundness for such circular proofs are known: e.g., the notion of a *progressing trace* in [4]. Such conditions include forbidding loops across minor premises of an inference rule, and every loop must be 'guarded', i.e., there must be an unfolding of a co-inductive atom in the loop. These kind of conditions are too strong, however, to encode up-to techniques for bisimulation. A commonly used up-to technique for bisimulation, say for CCS, is the up-to context technique, which uses the up-to function $\mathcal{F}(\mathcal{R}) = \{(C[P], C[Q]) \mid (P, Q) \in \mathcal{R}\}$, where C is a process context. So, for example, to establish $P+Q \sim R+Q$, one can simplify this first to the problem of checking $P \sim R$ via the up-to function $\mathcal{F}$. This kind of simplification via up-to context is exploited in [2], for example, to obtain a better bisimulation checking algorithm. An example of using "up-to context" is given in [9, Section A].

To capture bisimulation up-to, we need to encode up-to functions as logical theories, and use them to simplify a goal, before doing loop checking. This leads to inconsistency if done naively, even when the theories are valid. For example, since the processes $a.0$ and $b.0$ are not bisimilar, the formula $\mathrm{bisim}(a, b) \supset \bot$ should be provable. Now consider the following circular proof:

$$\frac{\dfrac{}{\mathrm{bisim}(a,b) \supset \bot \longrightarrow \mathrm{bisim}(a,b)} \; loop \qquad \dfrac{}{\mathrm{bisim}(a,b) \supset \bot, \bot \longrightarrow \mathrm{bisim}(a,b)} \; \bot\mathcal{L}}{\mathrm{bisim}(a,b) \supset \bot \longrightarrow \mathrm{bisim}(a,b)} \; \supset\mathcal{L}$$

where the leftmost leaf is the same as the root sequent. If this were admitted as a proof, then one can prove $\bot$. Indeed, this kind of loop is forbidden in sound circular proof systems [4,14] and is an example of a non-progressing loop. Unfortunately, as we mentioned above, forbidding circular proofs outright leads to a restricted system where bisimulation up-to algorithms cannot be encoded directly. An important part of the design of the tabled proof system is to rule out unsound loops while still being able to encode up-to techniques. This involves a careful management of tabled entries from which we deduce good loops.

In our tabled proof system, we capture the notion of a loop in a derivation by extending sequents with *history contexts*. We consider only tabling of atoms and universally quantified atoms. We distinguish three types of co-inductive atoms: *proved atoms*, *history atoms*, and *open atoms*. Only the first two types of atoms can appear in a table. Open atoms can only appear in the goal formula (i.e., the formula on the right-hand side of a sequent) or a theory, and are used to indicate atoms that are yet to be proved or disproved. When atoms occur in sequents, the history atoms will be annotated with ∘ while open atoms are annotated with ∗. History atoms and open atoms are syntactic devices used only in the tabled proof system; they have no meaning inside Linc. As the name suggests, history atoms are those encountered during proof search, for which a co-inductive rule has been applied. If a predicate symbol is co-inductively defined, then its history atoms are used to establish a post-fixed point. We consider only history atoms that are co-inductive.[1] A formula is ∗-free (resp., history free) if it has no occurrences of

---

[1] Inductive history atoms can be added, and their use would be to table *disproved* atomic goals. We leave the treatment of inductive history atoms to future work.

open atoms (resp., history atoms). Given a set $\mathcal{P}$ of formulas, we denote with $\mathcal{P}^\circ$ the set of history atoms in $\mathcal{P}$. Given a predicate p, we denote with $\mathcal{P} \setminus p$ the set $\mathcal{P}$ with all atoms of the form $p\vec{t}$ removed.

Sequents have for form $\mathcal{P}; \mathcal{T}; \Gamma \longrightarrow C; \mathcal{P}'$, where Γ is a set of level-0 $*$-free and history-free formulas; C is a level-1 history-free formula; $\mathcal{P}$ and $\mathcal{P}'$ are multisets of $*$-free atoms or universally quantified atoms; and $\mathcal{T}$ is a *theory*, i.e., a set of closed formulas. The set $\mathcal{P}$ and $\mathcal{P}'$ are bookkeeping devices essentially. Operationally, the sequent can be understood as follows: in the beginning of proof search for the sequent, $\mathcal{P}$ contains the current table entries, and when proof search concludes successfully, $\mathcal{P}'$ contains the new table entries generated by the proof search.

Depending on the tabling strategy, theories can be lemmas (provable in, say, Linc) or rewriting rules on open atoms (which correspond to up-to functions), or a mixture of both. When no history atoms are present, the informal reading of such a sequent is as follows: assuming $\mathcal{T}$ and $\mathcal{P}$ are provable in Linc, then $\Gamma \longrightarrow C$ is provable in Linc and its proof contains subproofs of atoms in $\mathcal{P}'$. When history atoms are present, the interpretation of the sequent is more complicated. Roughly, assuming we only have one co-inductively defined predicate symbol, say p, and the only history atoms are those of p, then $\mathcal{P}^\circ \cup (\mathcal{P}')^\circ$ forms a post fixed point of (the operator associated with) p. The precise interpretation will be given when we formally prove the soundness result for each strategy.

The inference rules involving these richer sequents are given in Figure 2. We consider only unification problems that have most general unifiers, e.g., first-order or higher-order pattern unification: in this way, eq$\mathcal{L}$ has at most one premise. In branching rules, the accumulated history or proved atoms on the right-hand side of a sequent in one branch are passed on to the other branch. When using this proof system for proof search, this set will be populated deterministically in a depth-first search strategy. The most interesting rule is ν_R. Here, reading the rule upwards, one replaces the co-inductive predicate p with p^*, and add $p^\circ \vec{t}$ to the history context on the left to allow it to be used to detect loops. When proof search is done, the history context on the right will be populated with history atoms. The intention is that these history atoms will form a post-fixed point (up-to) of p; hence when the proof search concludes, we replace each history atom p° on the right with p, signifying that every element in the post-fixed point is contained in the largest fixed point of p.

Notice that our sequent calculus does not have explicit structural rules (contraction and weakening) since these rules have been internalized in other rules. We have also omitted the cut rule. We currently do not know whether cut is admissible, but it is not important for this work as we only are interested in soundness. Notice also that if a sequent has a non-empty left-hand (Γ) context, then it can be the conclusion of only left-introduction rules: furthermore, since Γ can only contain level-0 formulas, there is no need for left introduction rules for implications and universal quantifiers.

Let $p_1, \ldots, p_n$ be the set of all co-inductive predicates that are defined in the logic. We denote by $\mathcal{L}$ the set $\{\forall \vec{x_1}(p_1^\circ \vec{x_1} \supset p_1^* \vec{x_1}), \ldots, \forall \vec{x_n}(p_n^\circ \vec{x_n} \supset p_n^* \vec{x_n})\}$. That

$$\frac{\mathbb{S}(\mathcal{P},\mathcal{T})\vdash_I A}{\mathcal{P};\mathcal{T};\cdot\longrightarrow A;\cdot}\ init \qquad \frac{}{\mathcal{P};\mathcal{T};\bot,\Gamma\longrightarrow B;\cdot}\ \bot\mathcal{L} \qquad \frac{}{\mathcal{P};\mathcal{T};\cdot\longrightarrow\top;\cdot}\ \top\mathcal{R}$$

$$\frac{\mathcal{P};\mathcal{T};B,C,\Gamma\longrightarrow D;\mathcal{P}'}{\mathcal{P};\mathcal{T};B\wedge C,\Gamma\longrightarrow D;\mathcal{P}'}\ \wedge\mathcal{L} \qquad \frac{\mathcal{P};\mathcal{T};\cdot\longrightarrow B;\mathcal{P}_1 \quad \mathcal{P},\mathcal{P}_1;\mathcal{T};\cdot\longrightarrow C;\mathcal{P}'}{\mathcal{P};\mathcal{T};\cdot\longrightarrow B\wedge C;\mathcal{P}',\mathcal{P}_1}\ \wedge\mathcal{R}$$

$$\frac{\mathcal{P};\mathcal{T};B,\Gamma\longrightarrow D;\mathcal{P}_1 \quad \mathcal{P},\mathcal{P}_1;\mathcal{T};C,\Gamma\longrightarrow D;\mathcal{P}'}{\mathcal{P};\mathcal{T};B\vee C,\Gamma\longrightarrow D;\mathcal{P}',\mathcal{P}_1}\ \vee\mathcal{L} \qquad \frac{\mathcal{P};\mathcal{T};\cdot\longrightarrow B_i;\mathcal{P}'}{\mathcal{P};\mathcal{T};\cdot\longrightarrow B_1\vee B_2;\mathcal{P}'}\ \vee\mathcal{R}$$

$$\frac{\mathcal{P};\mathcal{T};B\longrightarrow C;\mathcal{P}'}{\mathcal{P};\mathcal{T};\cdot\longrightarrow B\supset C;\mathcal{P}'}\ \supset\mathcal{R} \qquad \frac{\mathcal{P};\mathcal{T};\cdot\longrightarrow B[y/x];\mathcal{P}'}{\mathcal{P};\mathcal{T};\cdot\longrightarrow\forall x.B;\mathcal{P}'}\ \forall\mathcal{R}$$

$$\frac{\mathcal{P};\mathcal{T};B[y/x],\Gamma\longrightarrow C;\mathcal{P}'}{\mathcal{P};\mathcal{T};\exists x.B,\Gamma\longrightarrow C;\mathcal{P}'}\ \exists\mathcal{L} \qquad \frac{\mathcal{P};\mathcal{T};\cdot\longrightarrow B[t/x];\mathcal{P}'}{\mathcal{P};\mathcal{T};\cdot\longrightarrow\exists x.B;\mathcal{P}'}\ \exists\mathcal{R}$$

$$\frac{\mathcal{P};\mathcal{T};\Gamma[\rho]\longrightarrow C[\rho];\mathcal{P}'}{\mathcal{P};\mathcal{T};s=t,\Gamma\longrightarrow C;\mathcal{P}'}\ \mathrm{eq}\mathcal{L},\rho=\mathrm{mgu}(s,t) \qquad \frac{}{\mathcal{P};\mathcal{T};\cdot\longrightarrow t=t;\cdot}\ \mathrm{eq}\mathcal{R}$$

$$\frac{}{\mathcal{P};\mathcal{T};s=t,\Gamma\longrightarrow C;\cdot}\ \mathrm{eq}\mathcal{L},s\ \text{and}\ t\ \text{not unifiable.}$$

$$\frac{\mathcal{P};\mathcal{T};B\,p\vec{t},\Gamma\longrightarrow C;\mathcal{P}'}{\mathcal{P};\mathcal{T};p\vec{t},\Gamma\longrightarrow C;\mathcal{P}'}\ \mathrm{def}\mathcal{L} \qquad \frac{\mathbb{S}(\mathcal{P},\mathcal{T})\not\vdash_I p\vec{t} \quad \mathcal{P};\mathcal{T};\cdot\longrightarrow B\,p\vec{t};\mathcal{P}'}{\mathcal{P};\mathcal{T};\cdot\longrightarrow p\vec{t};\mathcal{P}',\forall\vec{x}.p\,\vec{t}}\ \mathrm{def}\mathcal{R}$$

$$\frac{\mathbb{S}(\mathcal{P},\mathcal{T})\not\vdash_I p^*\vec{t} \quad \mathcal{P},p^{\circ}\vec{t};\mathcal{T};\cdot\longrightarrow B\,p^*\vec{t};\mathcal{P}'}{\mathcal{P};\mathcal{T};\cdot\longrightarrow p^*\vec{t};\mathcal{P}',p^{\circ}\vec{t}}\ \nu_R^*$$

$$\frac{\mathbb{S}(\mathcal{P}\setminus p^{\circ},\mathcal{T})\not\vdash_I p\vec{t} \quad (\mathcal{P}\setminus p^{\circ}),p^{\circ}\vec{t};\mathcal{T};\cdot\longrightarrow B\,p^*\vec{t};\mathcal{P}'}{\mathcal{P};\mathcal{T};\cdot\longrightarrow p\vec{t};\mathcal{P}'[p/p^{\circ}],p\vec{t}}\ \nu_R$$

Fig. 2. Inference rules for the tabled proof system. In $\mathrm{def}\mathcal{L}$ and $\mathrm{def}\mathcal{R}$, $p\,\vec{x}\overset{\triangle}{=}B\,p\,\vec{x}$, and in ν_R^* and ν_R, $p\,\vec{x}\overset{\nu}{=}B\,p\,\vec{x}$ and $\vec{t}$ are ground terms.

is, $\mathcal{L}$ allows one to backchain from an open atom to a history atom. Adding $\mathcal{L}$ as theories to the tabled proof system allows one to loop on co-inductive atoms.

The function $\mathbb{S}$ used in Figure 2 is determined by the tabling strategies:

Strategy I: $\mathbb{S}(\mathcal{P},\mathcal{T})=\mathcal{P}\setminus\mathcal{P}^{\circ}$ Strategy III: $\mathbb{S}(\mathcal{P},\mathcal{T})=\mathcal{P}\cup\mathcal{L}$
Strategy II: $\mathbb{S}(\mathcal{P},\mathcal{T})=(\mathcal{P}\setminus\mathcal{P}^{\circ})\cup\mathcal{T}$ Strategy IV: $\mathbb{S}(\mathcal{P},\mathcal{T})=\mathcal{P}\cup\mathcal{T}$

We shall refer to these functions as, respectively, $\mathbb{S}_1$, $\mathbb{S}_2$, $\mathbb{S}_3$ and $\mathbb{S}_4$. The proof systems for these strategies are defined as follows: the proof systems $\mathcal{TD}_1$ and $\mathcal{TD}_2$ are proof systems obtained by using, respectively, $\mathbb{S}_1$ and $\mathbb{S}_2$, and whose rules include all the inference rules in Figure 2 except ν_R^* and ν_R. The proof systems $\mathcal{TD}_3$ and $\mathcal{TD}_4$ are proofs systems obtained using, respectively, $\mathbb{S}_3$ and $\mathbb{S}_4$, and whose rules include all the inference rules in Figure 2 except $\mathrm{def}\mathcal{R}$.

The relation $\vdash_I$ refers to the deducibility relation of intuitionistic logic (without fixed points). When $\mathcal{T}$ is restricted to formulas containing just $\supset$, $\wedge$, and $\forall$ the relation $\vdash_I$ is implemented by λProlog [7].

4   Constructing Post-Fixed Point from Tables

In the following, given two lists of terms $\vec{s} = s_1, \ldots, s_n$ and $\vec{t} = t_1, \ldots, t_n$, we write $\vec{s} = \vec{t}$ to denote the formula $(s_1 = t_1) \wedge (s_2 = t_2) \wedge \cdots \wedge (s_n = t_n)$. For simplicity, we shall assume that all co-inductive predicates have the same arity. We denote by $\mathcal{P}^\bullet$ the set $\mathcal{P} \setminus \mathcal{P}^\circ$.

Theorem 1. *Suppose $\mathcal{P}; \mathcal{T}; \Gamma \longrightarrow C; \mathcal{P}'$ is derivable in $\mathcal{T}\mathcal{D}_3$, where Γ is history-free and C contains no negative occurrences of history atoms. Let $\{p_1, \ldots, p_n\}$ be the set of co-inductive predicates occuring in $\mathcal{P}, C$ and $\mathcal{P}'$. Then there exist invariants $S_1, \ldots, S_n$ such that*

- *the sequent $(\mathcal{P}^\bullet, \Gamma \longrightarrow C[S_1/p_1^*, \ldots, S_n/p_n^*])$ is derivable in Linc,*
- *for each $B \in (\mathcal{P}')^\bullet$, the sequent $(\mathcal{P}^\bullet \longrightarrow B)$ is derivable in Linc, and*
- *for each $p_i^\circ \vec{t} \in (\mathcal{P}')^\circ$, where $p_i \, \vec{x} \stackrel{\nu}{=} D_i \, p_i \, \vec{x}$, the sequent $(\mathcal{P}^\bullet \longrightarrow D_i \, S_i \, \vec{t})$ is derivable in Linc.*

Proof. (Outline.) Given sequent $\mathcal{P}; \mathcal{T}; \Gamma \longrightarrow C; \mathcal{P}'$, the abstraction S_i

$$S_i = \lambda \vec{x}. \bigwedge (\mathcal{P})^\bullet \wedge \bigvee \{(\vec{x} = \vec{t}) \mid p_i^\circ \, \vec{t} \in \mathcal{P} \cup \mathcal{P}'\},$$

forms a post-fixed point of the definition of p_i, i.e., $D \, S_i \, \vec{x} \longrightarrow S_i \, \vec{x}$.   □

5   Co-inductive Tabling Modulo Theories

In a naive algorithm for bisimulation checking, one can construct a bisimulation set by progressively unfolding transitions from a given pair of processes, until one arrives at stuck processes or encounters a previously seen pair of processes. This is very similar to how proof search with strategy III works. The up-to techniques add to this the possibility of simplifying the continuations of a pair of processes, before doing the loop checking. For example, a typical simplification rule is the context closure, e.g., when one encounters a new pair to be checked $((P \mid R), (Q \mid R))$, instead of unfolding these, we simplify it to (P, Q) and proceed. This kind of simplification before loop checking is in general unsound; see [11] for an example. An important line of research in the up-to techniques is in characterizing sound simplification rules.

To capture up-to techniques in our tabling proof system, we need a mechanism to apply simplification to an open co-inductive goal before doing loop checking. This can be done simply by backchaining on the co-inductive goal. Since open co-inductive goals are marked with $*$, to be able to backchain on them, we need to allow $*$-atoms in the theory component of a sequent. However, when the theory $\mathcal{T}$ contains $*$-atoms, it is not possible in general to construct a post fixed point from tabled entries as they are no longer closed under fixed point unfolding. This is because the theory $\mathcal{T}$ may allow one to deduce $*$-atoms that have not been encountered during proof search (hence those particular atoms would not have been unfolded). Soundness in this case is conditional on an additional statement, which happens to coincide with the (logical interpretation) of the soundness condition for up-to techniques [10].

To simplify the presentation, we shall restrict to one co-inductive definition in the following. We shall refer to this definition simply as $p\vec{x} \overset{\nu}{=} D\,p\,\vec{x}$. So we have only one kind of history atoms and one kind of $*$ atoms, i.e., those of the form $p^\circ\vec{t}$ and $p^*\vec{t}$. The set $\mathcal{L}$ in this case contains exactly one formula, i.e., $\forall\vec{x}(p^\circ\vec{x} \supset p^*\vec{x})$.

To formalize the up-to techniques, we need to quantify over relations and functions. Thus we introduce HOLinc, the extension of Linc that contains higher-order quantifiers. In other words, the logic we have now is an intuitionistic higher-order logic (i.e., the intuitionistic version of Church's Simple Theory of Types) with fixed points and (free) equality. The latter two can be encoded in higher-order logic, so we essentially only work within higher-order logic.

Definition 1. *An up-to theory is a set $\mathcal{T}$ of higher-order hereditary Harrop (hH) formulas such that the head of each clause is of the form $p^*\vec{t}$. We assume that $\mathcal{L} \subseteq \mathcal{T}$, and the only place where history atoms occur in $\mathcal{T}$ is in this subset.*

Definition 2. *If $\mathcal{T}$ is an up-to theory, it induces the function*

$$\mathcal{F}_\mathcal{T} = \lambda\mathcal{R}\lambda\vec{x}.\forall q. \bigwedge \mathcal{T}[q/p^*, \mathcal{R}/p^\circ] \supset q\vec{x}.$$

In more informal set-theoretic notation, $\mathcal{F}_\mathcal{T}$ can be written as:

$$\mathcal{F}_\mathcal{T}(\mathcal{R}) = \{\vec{x} \mid \forall q(\bigwedge \mathcal{T}[q/p^*, \mathcal{R}/p^\circ] \supset q\vec{x}) \text{ is provable in HOLinc. }\}$$

The adequacy of this encoding of up-to functions is the result of the completeness of goal-directed proof for hH fragment of higher-order logic; see [7].

Definition 3. *An up-to theory $\mathcal{T}$ is sound if the following formula, named $Snd(\mathcal{T})$ holds: $\forall\mathcal{R}.(\forall\vec{x}.(\mathcal{R}\vec{x} \supset D\,(\mathcal{F}_\mathcal{T}\mathcal{R})\,\vec{x})) \supset (\forall\vec{x}.\mathcal{R}\vec{x} \supset p\,\vec{x}).$*

Theorem 2. *Suppose $\mathcal{P}; \mathcal{T}; \Gamma \longrightarrow C; \mathcal{P}'$ is derivable in $\mathcal{TD}_4$. Then there exists an invariant S such that*

- *the sequent $(Snd(\mathcal{T}), \mathcal{P}^\bullet, \Gamma \longrightarrow C[\mathcal{F}_\mathcal{T}S/p^*])$ is derivable in HOLinc,*
- *for each $B \in (\mathcal{P}')^\bullet$, the sequent $(\mathcal{P}^\bullet \longrightarrow B)$ is derivable in HOLinc, and*
- *for each $p\vec{t} \in (\mathcal{P}')^\circ$, the sequent $(\mathcal{P}^\bullet \longrightarrow D\,(\mathcal{F}_\mathcal{T}S)\,\vec{t})$ is derivable in HOLinc.*

Proof. (Outline.) Given $\mathcal{P}; \mathcal{T}; \Gamma \longrightarrow C; \mathcal{P}'$, the abstraction $\lambda\vec{x}.\bigvee\{(\vec{x} = \vec{u}) \mid p^\circ\vec{u} \in \mathcal{P} \cup \mathcal{P}'\}$ can be shown to be a post-fixed point "up-to" $\mathcal{F}_\mathcal{T}$. □

Corollary 1. *Let $\mathcal{T}$ be an up-to theory. If $\cdot; \mathcal{T}; \cdot \longrightarrow B; \mathcal{P}$ is derivable in $\mathcal{TD}_4$, for some $\mathcal{P}$, then $Snd(\mathcal{T}) \longrightarrow B$ is derivable in HOLinc.*

Thus strategy IV is sound for, say, bisimulation checking if one can discharge the assumption $Snd(\mathcal{T})$, a task that can often be tedious to do. We are currently developing some of these proofs in the (higher-order version of the) theorem prover Abella, which is an interactive prover based on a logic similar to Linc.

6 Compositions of Up-to Functions

One important line of research in the up-to techniques in bisimulation is that of compositions of up-to functions. More precisely, one is interested in characterizing when the composition of two sound up-to functions gives rise to a sound up-to function. Such results allow one to combine simple functions to form powerful sound composite functions. We show next that composition of up-to functions can be defined via a notion of composition of up-to theories.

Definition 4. *Let $\mathcal{T}_1$ and $\mathcal{T}_2$ be up-to theories. Their composition, written $\mathcal{T}_1 \circ \mathcal{T}_2$, is defined as $\mathcal{T}_1 \circ \mathcal{T}_2 = \mathcal{T}_1[F/p^\circ]$ where $F = \lambda \vec{x}.(\forall q. \bigwedge \mathcal{T}_2[q/p^*] \supset q\,\vec{x})$.*

The following lemma states that this definition of composition of theories is adequate, i.e., it respects the composition of logical up-to functions.

Lemma 1. *$\mathcal{F}_{\mathcal{T}_1} \circ \mathcal{F}_{\mathcal{T}_2}$ and $\mathcal{F}_{\mathcal{T}_1 \circ \mathcal{T}_2}$ define the same function.*

In practice, up-to techniques are often used by interleaving applications of several up-to functions. However, proving that such interleaving is sound is obviously more complicated than proving soundness of restricted compositions. In the logical encodings, interleaving of two theories $\mathcal{T}_1$ and $\mathcal{T}_2$ can be captured simply by joining the theories, i.e., $\mathcal{T}_1 \cup \mathcal{T}_2$. We show next that soundness of tabled proof search in the up-to theory $\mathcal{T}_1 \cup \mathcal{T}_2$ can be reduced to soundness of proof search under their composition $\mathcal{T}_1 \circ \mathcal{T}_2$, under certain conditions.

To prove the following results, it is convenient to view a theory as an inference rule. This is straightforward when the theories are Horn clauses. The Horn clause

$$\forall \vec{x}.(A_1 \wedge \cdots \wedge A_n) \supset p^*\vec{t} \quad \text{can be written as the rule} \quad \frac{A_1 \quad \cdots \quad A_n}{p^*\vec{t}} \,,$$

where $A_1, \ldots, A_n$ are atoms and where $\vec{x}$ become schematic variables of the inference rule. Let $\mathcal{D}(\mathcal{T})$ denote the set of inference rules for a given Horn theory $\mathcal{T}$. Then $\mathcal{P}, \mathcal{T} \vdash_I p^* \vec{t}$ holds iff there is a derivation of $p^*\vec{t}$ from $\mathcal{P}$ in the inference system $\mathcal{D}(\mathcal{T})$. We say that an inference rule r_1 *permutes over* another inference rule r_2 iff every derivation of $p^*\vec{t}$, for any $\vec{t}$, where r_2 appears immediately above r_1 can be transformed into another derivation of $p^*\vec{t}$ where r_1 appears above r_2. Given $\mathcal{D}(\mathcal{T}_1)$ and $\mathcal{D}(\mathcal{T}_2)$, we say that $\mathcal{D}(\mathcal{T}_1)$ permutes over $\mathcal{D}(\mathcal{T}_2)$ iff every rule of $\mathcal{D}(\mathcal{T}_1 \setminus \mathcal{T}_2)$ permutes over every rule of $\mathcal{D}(\mathcal{T}_2 \setminus \mathcal{T}_1)$.

Lemma 2. *Let $\mathcal{T}_1$ and $\mathcal{T}_2$ be two Horn up-to theories such that $\mathcal{D}(\mathcal{T}_2)$ permutes over $\mathcal{D}(\mathcal{T}_1)$. Then $(\mathcal{P}, \mathcal{T}_1, \mathcal{T}_2 \vdash_I p^*\vec{t})$ iff $(\mathcal{P}, \mathcal{T}_1 \circ \mathcal{T}_2 \vdash_I p^*\vec{t})$, for every set of $*$-free atoms $\mathcal{P}$ and every $\vec{t}$.*

Theorem 3. *Let $\mathcal{T}_1$ and $\mathcal{T}_2$ be two Horn up-to theories such that $\mathcal{T}_1 \circ \mathcal{T}_2$ is sound and that $\mathcal{D}(\mathcal{T}_2)$ permutes over $\mathcal{D}(\mathcal{T}_1)$. If $\cdot; \mathcal{T}_1, \mathcal{T}_2; \cdot \longrightarrow B; \mathcal{P}$ is derivable in $\mathcal{T}\mathcal{D}_4$, for some $\mathcal{P}$, then $Snd(\mathcal{T}_1 \circ \mathcal{T}_2) \longrightarrow B$ is derivable in HOLinc.*

Proof. This follows from Theorem 2 and Lemma 2.   □

Note that Theorem 3 does not imply that $\mathcal{F}_{\mathcal{T}_1 \cup \mathcal{T}_2}$ is sound given that $\mathcal{F}_{\mathcal{T}_1 \circ \mathcal{T}_2}$ is sound; it only implies that, for the purpose of proving a co-inductive goal in the tabled proof system, one can freely combine $\mathcal{T}_1$ and $\mathcal{T}_2$ without losing soundness. This is useful in practice where one could combine different up-to techniques freely but only need to prove soundness for a restricted form of composition.

Below, we shall use $\sim$ to denote the predicate *bisim*.

Example 1. Consider the CCS example again. Let $\mathcal{T}_1$ be the up-to theory formalizing context closure, and let $\mathcal{T}_2$ be the up-to theory formalizing reflexive and transitive closure. The inference rules of $\mathcal{D}(\mathcal{T}_1)$ are the rules $\{b, re, tr\}$ and the rules of $\mathcal{D}(\mathcal{T}_2)$ are $\{b, cng\}$, where b, re, tr, cng are as follows:

$$\frac{s \sim^{\circ} t}{s \sim^* t} \, b \qquad \frac{}{t \sim^* t} \, re \qquad \frac{s \sim^* u \quad u \sim^* t}{s \sim^* t} \, tr \qquad \frac{s \sim^* t}{C[s] \sim^* C[t]} \, cng$$

and where $C[\,]$ is a process context. It can be easily shown that cng permutes up over re and tr, for example:

$$\frac{\dfrac{s \sim^* u \quad u \sim^* t}{s \sim^* t} \, tr}{C[s] \sim^* C[t]} \, cng \quad \rightsquigarrow \quad \frac{\dfrac{s \sim^* u}{C[s] \sim^* C[u]} \, cng \quad \dfrac{u \sim^* t}{C[u] \sim^* C[t]} \, cng}{C[s] \sim^* C[t]} \, tr$$

So we can freely mix $\mathcal{T}_1$ and $\mathcal{T}_2$ in proving particular instances of bisimilarity, but we only need to prove soundness of the composition $\mathcal{T}_1 \circ \mathcal{T}_2$.

If the up-to theory $\mathcal{T}$ contains occurrences of the co-inductive predicate p, then we can consider using previously proved facts, say $\mathcal{T}'$, about p to prove subgoals of the form $p\vec{t}$. The use of lemmas is orthogonal to the soundness condition for up-to techniques, as stated in the following theorem.

Theorem 4. *Let $\mathcal{U}$ be a set of $*$-free and history-free formulas that are valid in HOLinc. Suppose $\mathcal{P}; \mathcal{U}, \mathcal{T}; \Gamma \longrightarrow C; \mathcal{P}'$ is derivable in $\mathcal{TD}_4$. Then there exists an invariant S such that*

- *the sequent $(Snd(\mathcal{T}), \mathcal{P}^{\bullet}, \Gamma \longrightarrow C[\mathcal{F}_\mathcal{T} S/p^*])$ is derivable in HOLinc,*
- *for each $B \in (\mathcal{P}')^{\bullet}$, the sequent $(\mathcal{P}^{\bullet} \longrightarrow B)$ is derivable in HOLinc, and*
- *for each $p\vec{t} \in (\mathcal{P}')^{\circ}$, the sequent $\mathcal{P}^{\bullet} \longrightarrow D\,(\mathcal{F}_\mathcal{T} S)\,\vec{t})$ is derivable in HOLinc.*

Proof. This proof follows the proof of Theorem 2, except we use the following invariant: given sequent $\mathcal{P}; \mathcal{U}, \mathcal{T}; \Gamma \longrightarrow C; \mathcal{P}'$, define $S = \lambda \vec{x}. \bigwedge \mathcal{U} \wedge \bigvee \{(\vec{x} = \vec{u}) \mid p^{\circ}\vec{u} \in \mathcal{P} \cup \mathcal{P}'\}$. $\qquad \square$

The composition result (Theorem 3) can be slightly modified to take into account uses of lemmas. As we shall see later, this leads to a rather pleasant result concerning compositions with up-to bisimilarity.

Lemma 3. *Let $\mathcal{U}$ be a set of Horn clauses which are also lemmas of HOLinc. Let $\mathcal{T}_1$ and $\mathcal{T}_2$ be two Horn up-to theories such that $D(\mathcal{U} \cup \mathcal{T}_2)$ permutes over $D(\mathcal{U} \cup \mathcal{T}_1)$. Then $(\mathcal{P}, \mathcal{U}, \mathcal{T}_1, \mathcal{T}_2 \vdash_I p^*\vec{t})$ iff $(\mathcal{U}, \mathcal{P}, \mathcal{T}_1 \circ \mathcal{T}_2 \vdash_I p^*\vec{t})$, for every set of $*$-free atoms $\mathcal{P}$ and every $\vec{t}$.*

Theorem 5. *Let $\mathcal{U}$ be a set of lemmas of* HOLinc. *Let $\mathcal{T}_1$ and $\mathcal{T}_2$ be two Horn up-to theories such that $\mathcal{T}_1 \circ \mathcal{T}_2$ is sound and that $\mathcal{D}(\mathcal{U} \cup \mathcal{T}_2)$ permutes over $\mathcal{D}(\mathcal{U} \cup \mathcal{T}_1)$. If $\cdot; \mathcal{U}, \mathcal{T}_1, \mathcal{T}_2; \cdot \longrightarrow B; \mathcal{P}$ is derivable in $\mathcal{T}D_4$, for some $\mathcal{P}$, then $Snd(\mathcal{T}_1 \circ \mathcal{T}_2) \longrightarrow B$ is derivable in* HOLinc.

Example 2. Let $\mathcal{T}_1$ be the theory encoding up-to bisimilarity and let $\mathcal{T}_2$ be the theory encoding up-to context-closure for CCS. The inference rules of $\mathcal{T}_1$ consist of the rule b (see Example 1) and the following rule:

$$\frac{s \sim u \quad u \sim^* v \quad v \sim t}{s \sim^* t} \; bs$$

The composition $\mathcal{T}_1 \circ \mathcal{T}_2$ is shown to be sound in, e.g., [10]. Since bisimilarity in CCS is closed under arbitrary contexts, we can prove the lemma below (left) in HOLinc: the inference rule corresponding to that lemma is on the right:

$$\forall C \forall x, y \; (x \sim y \supset C[x] \sim C[y]) \qquad \frac{s \sim t}{C[s] \sim C[t]} \; bcng$$

where C denotes a process context. Let $\mathcal{U}$ be a set of Horn lemmas that includes this lemma. We show that $\mathcal{D}(\mathcal{U} \cup \mathcal{T}_2)$ permutes over $\mathcal{D}(\mathcal{U} \cup \mathcal{T}_1)$. It is enough to show that the rule cng (see Example 1) permutes over bs:

$$\frac{\dfrac{s \sim u \quad u \sim^* v \quad v \sim t}{s \sim^* t} \; bs}{C[s] \sim^* C[t]} \; cng \qquad \rightsquigarrow$$

$$\frac{\dfrac{s \sim u}{C[s] \sim C[u]} \; bcng \quad \dfrac{s \sim^* u}{C[u] \sim^* C[v]} \; cng \quad \dfrac{v \sim t}{C[v] \sim C[t]} \; bcng}{C[s] \sim^* C[t]} \; bs$$

This shows that, rather surprisingly, we can apply the congruence rule first, before applying up-to bisimilarity, without losing soundness, even though the meta theory only allows one to apply congruence rules last. This can potentially lead to a shorter proof as the congruence rule allows simplification of processes.

7 Conclusion and Future Work

We have shown a range of strategies for incorporating tables into proof search, where the most advanced strategy allows us to capture the up-to techniques for bisimilarity. For all strategies, we show that tabled proofs can be soundly interpreted as a proper proof in the same logic and formal proof certificates can be constructed from each successful proof search. Our encoding of up-to techniques also enables us to derive a new result in the composition of up-to techniques, allowing one to freely compose up-to techniques while only needing to prove soundness of a limited form of composition.

Orthogonal to all these strategies is the question of whether one should allow quantified formulas (existentially or universally) in the table. Such a possibility can arise if for example one can prove a goal $(p\ a\ X)$ for any X, e.g., simply because X is not used in the definition of p. Then a natural interpretation of this is to say that we have actually proved $\forall x.p\ a\ x$. While this kind of quantified tabled entries is harmless in Strategy I and II, it is less clear whether it is sound for Strategy III and IV. We shall leave this as future work.

We have concentrated on strong bisimulation as an application in this paper, but the framework we established here should apply to weak bisimulation as well, at least as far as the cyclic structure of proofs is concerned. The theory of weak-bisimulation up-to is a lot of more complex than the strong bisimulation up-to and less uniform, e.g., some obvious up-to functions (e.g., up-to weak bisimilarity) is unsound [12]. In terms of formalization in our framework, however, this complexity is mostly isolated in the theory part, i.e., in establishing $Snd(\mathcal{T})$. We plan to investigate weak-bisimilarity in immediate future work.

References

1. Baelde, D., Gacek, A., Miller, D., Nadathur, G., Tiu, A.: The Bedwyr system for model checking over syntactic expressions. In: Pfenning, F. (ed.) CADE 2007. LNCS (LNAI), vol. 4603, pp. 391–397. Springer, Heidelberg (2007)
2. Bonchi, F., Pous, D.: Checking NFA equivalence with bisimulations up to congruence. In: Proceedings of the 40th Annual ACM SIGPLAN-SIGACT Symposium on Principles of Programming Languages, pp. 457–468. ACM (2013)
3. Brotherston, J., Gorogiannis, N., Petersen, R.L.: A generic cyclic theorem prover. In: Jhala, R., Igarashi, A. (eds.) APLAS 2012. LNCS, vol. 7705, pp. 350–367. Springer, Heidelberg (2012)
4. Brotherston, J., Simpson, A.: Complete sequent calculi for induction and infinite descent. In: 22nd Symp. on Logic in Computer Science, pp. 51–62 (2007)
5. Jaffar, J., Santosa, A.E., Voicu, R.: A CLP proof method for timed automata. In: RTSS, pp. 175–186. IEEE Computer Society (2004)
6. McDowell, R., Miller, D., Palamidessi, C.: Encoding transition systems in sequent calculus. Theoretical Computer Science 294(3), 411–437 (2003)
7. Miller, D., Nadathur, G.: Programming with Higher-Order Logic. Cambridge University Press (June 2012)
8. Miller, D., Nigam, V.: Incorporating tables into proofs. In: Duparc, J., Henzinger, T.A. (eds.) CSL 2007. LNCS, vol. 4646, pp. 466–480. Springer, Heidelberg (2007)
9. Miller, D., Tiu, A.: Extracting proofs from tabled proof search: Extended version. Technical report, HAL-INRIA (2013), http://hal.inria.fr/hal-00863561
10. Pous, D., Sangiorgi, D.: Enhancements of the bisimulation proof method. In: Sangiorgi, D., Rutten, J. (eds.) Advanced Topics in Bisimulation and Coinduction, pp. 233–289. Cambridge University Press (2011)
11. Sangiorgi, D.: On the bisimulation proof method. Mathematical Structures in Computer Science 8(5), 447–479 (1998)
12. Sangiorgi, D., Milner, R.: The problem of "weak bisimulation up to". In: Cleaveland, W.R. (ed.) CONCUR 1992. LNCS, vol. 630, pp. 32–46. Springer, Heidelberg (1992)
13. Simon, L., Mallya, A., Bansal, A., Gupta, G.: Coinductive logic programming. In: Etalle, S., Truszczyński, M. (eds.) ICLP 2006. LNCS, vol. 4079, pp. 330–345. Springer, Heidelberg (2006)

14. Sprenger, C., Dam, M.: On global induction mechanisms in a μ-calculus with explicit approximations. ITA 37(4), 365–391 (2003)
15. Tiu, A.: A Logical Framework for Reasoning about Logical Specifications. PhD thesis, Pennsylvania State University (May 2004)
16. Tiu, A., Miller, D.: Proof search specifications of bisimulation and modal logics for the π-calculus. ACM Trans. on Computational Logic 11(2) (2010)
17. Tiu, A., Momigliano, A.: Cut elimination for a logic with induction and co-induction. Journal of Applied Logic (2012)
18. Walukiewicz, I.: Completeness of Kozen's axiomatisation of the propositional μ-calculus. Inf. Comput. 157(1-2), 142–182 (2000)

Formalizing the SAFECode Type System

Daniel Huang and Greg Morrisett

Harvard University, Cambridge MA 02138, USA
dehuang@fas.harvard.edu,
greg@eecs.harvard.edu

Abstract. The Secure Virtual Architecture (SVA) provides an object-level integrity policy, similar to type-safety, for languages such as C and C++, and thus rules out a wide range of common vulnerabilities. SVA uses an enhanced version of the Low-Level Virtual Machine (LLVM) compiler called SAFECode to enforce the policy through a combination of static and dynamic type-checks. However, this results in a relatively large trusted computing base (TCB). SVA reduces the TCB with an unverified type-checker that relies upon a paper-and-pencil proof of type-soundness for a core-language. As a further step towards increasing the assurance of the compiler, we present a mechanized proof of soundness and a verified type-checker for a realistic subset of the SAFECode type system developed using the Coq Proof Assistant.

Keywords: verification, SAFECode, LLVM, memory safety.

1   Introduction

Most of our computing infrastructure is coded using low-level languages such as C/C++. Unsurprisingly, it is easy to make simple mistakes in these languages that lead to well-known vulnerabilities. In principle, recoding the infrastructure in a type-safe language would eliminate many of these vulnerabilities, but the costs of doing so seem to outweigh the benefits.

An attractive alternative is to bring the benefits of type-safety to legacy code by combining static analyses and run-time checks to automatically enforce a type-safety policy. There are many challenges in doing this effectively, as static analyses are generally too weak to reason effectively about real C/C++ programs, resulting in many false positives. The cost of inserting run-time checks and maintaining the meta-data needed to support those checks can also be prohibitively expensive. Recently, a number of systems have successfully combined the benefits of static analysis, dynamic checks, program optimization, and clever run-time representations to produce viable solutions [3,10,11,5,4].

One such system, SAFECode [5], uses sophisticated analyses and optimizations to eliminate run-time checks. However, this adds the SAFECode compiler to the trusted computing base (TCB). We could try to prove that the analyses and transformations (and subsequent optimizations) are correct as in the CompCert project [8]. Perhaps more easily, we can build a verified checker that

G. Gonthier and M. Norrish (Eds.): CPP 2013, LNCS 8307, pp. 211–226, 2013.
© Springer International Publishing Switzerland 2013

Type	$\tau ::=$	$\texttt{int} \mid \texttt{char} \mid \texttt{unknown} \mid \tau * \rho \mid \texttt{handle}(\rho, \tau)$
Statements	$S ::=$	$\epsilon \mid \texttt{S;S} \mid x = \texttt{E} \mid \texttt{store E, E} \mid \texttt{storeToU x, E, E}$
		$\mid \texttt{storec E, E} \mid \texttt{storecToU E, E} \mid \texttt{poolfree(E, E)}$
		$\mid \texttt{poolinit}(\rho, \tau)x\{\texttt{S}\} \mid \texttt{pool}\{\texttt{S}\}\texttt{pop}(\rho)$
Expressions	$\texttt{E} ::=$	$\texttt{var} \mid \texttt{V} \mid \texttt{E } op \texttt{ E} \mid \texttt{load E} \mid \texttt{loadFromU x, E} \mid \texttt{loadc E}$
		$\mid \texttt{loadcFromU E} \mid \texttt{cast E to } \tau \mid \texttt{poolalloc(x, E)}$
		$\mid \texttt{(x, \&E[E])} \mid \texttt{castint2pointer x, E to } \tau$
Value	$\texttt{V} ::=$	$uninit \mid \texttt{Int} \mid \texttt{region}(\rho)$

Fig. 1. Core-language presented in original SAFECode paper

attempts to prove that rewritten and optimized code respects the SAFECode security policy. The goal of this paper is to increase the assurance of the SAFE-Code compiler by formalizing a realistic subset of the language and its type system, presenting a mechanically-checked proof of soundness, and building a verified checker that can be used to check code emitted by SAFECode.

2   Overview of SAFECode

Our work builds upon a previous paper describing the SAFECode system [5], which enforces an object-level integrity policy similar to, but weaker than traditional notions of type-safety. Conceptually, SAFECode instruments all dangerous operations such as loads and stores with dynamic checks. To justify the elimination of unnecessary run-time checks, the paper formalized a core-language, type system and gave a paper-pencil proof of soundness for the typing rules.

We have reproduced their core-language in Figure 1. SAFECode uses regions, similar to the approach pioneered by Tofte and Talpin [12] and later refined in Cyclone [6]. Like these previous systems, a pointer type $\tau * \rho$ is indexed by a region variable ρ indicating the region of memory it references. However, SAFECode only places objects of the same type in a given region, allowing region metadata to support efficient run-time casts. Objects whose type cannot be statically determined are put in untyped regions. The type system tracks which regions are accessible, and hence, which pointers can be safely dereferenced.

More concretely, SAFECode provides a lexically scoped construct of the form $\texttt{poolinit}(\rho, \ \tau)\{\cdots\}$, which allocates a new region (or pool) to exclusively hold values of type τ and binds the region to ρ. Regions are typically represented as lists of pages that can be dynamically grown as new objects are allocated. Initially, the pages are zero-filled and zero is assumed to be a valid value for any type. In particular, dereferencing address zero will result in a trapped error (segmentation fault). Within the scope of the $\texttt{poolinit}$, programs can allocate objects of type τ in ρ using $\texttt{poolalloc}$, which returns a pointer of type $\tau * \rho$. Such pointers can be dereferenced (via $\texttt{load}$), updated (via $\texttt{store}$), or used to deallocate the object (via $\texttt{poolfree}$) while in the scope of ρ. Memory reclaimed in a region can be recycled for use at the same type. At the end of $\texttt{poolinit}$'s

$\Delta : \rho \rightharpoonup \tau$	$\Gamma : x \rightharpoonup \tau$		
r1: i32		1.	`poolinit(r1, i32) {`
r2: i32*r1		2.	`  poolinit(r2, i32*r1) {`
		3.	`    x = poolalloc(r1, 4);`
	x: i32*r1	4.	`    y = poolalloc(r2, 4);`
	y: i32*r1*r2	5.	`    store x, y;`
		6.	`    z = load y;`
	z: i32*r1	7.	`    poolfree(z);`
		8.	`    p = (i32*r1) 42;`
	p:i32*r1	9.	`    w1 = load z;`
	w1: i32	10.	`    w2 = load p;`
	w2: i32	11.	`  }`
		12.	`}`

Fig. 2. Example program and typing derivation

scope, the entire region is deallocated and its memory can be safely recycled for use in other regions to hold values of potentially different types. SAFECode also supports checked operations $\texttt{loadU}(\rho,q)$, $\texttt{storeU}(\rho,v,q)$, and checked casts. These operations do not require that q is ascribed the static type $\tau * \rho$. Rather, a run-time check is performed to see if q is a valid, τ-aligned pointer into ρ. If the check fails, then the program is terminated.

Figure 2 illustrates a SAFECode program in the core-language. In the example, r1 is a statically-typed region that holds integer values and region r2 is a statically-typed region holding i32*r1 values. The typing context tracks the set of region variables in scope and their types (Δ) as well as the set of variables and their types (Γ). The table to the left of the code summarizes the context used to check each line. For example, after line 1, region r1 is in scope and is assumed to hold values of type i32. By line 5, we assume variable x has type i32*r1 and y has type i32*r1*r2. The store instruction type-checks because y is a pointer value into region r2, x is of the appropriate type i32*r1, and r2 holds values of type i32*r1. The following load instruction type-checks similarly.

Lines 7 through 10 illustrate how type-homogenous regions enable SAFECode to relax traditional notions of type-safety. For instance, the dangling pointer dereference on line 9 type-checks because the region r1 is still live and type-homogeneity guarantees it will produce a value of type int. SAFECode also allows arbitrary integers to be casted to pointers. Doing so may necessitate a run-time check that can fail, but if the integer actually is an address of the appropriate type in the appropriate region, the cast will succeed. SAFECode accepts the above program as well-typed and guarantees for all possible executions that x and z always point into region r1, and that y always points into region r2. Furthermore, dereferencing x and z guarantees an integer, and dereferencing y guarantees a pointer into region r1 (or else zero values).

Currently, there is a huge gap between the original presentation and the actual SAFECode implementation, which is a large amount of C/C++ code implementing a LLVM bitcode transformation. At a first approximation, there is a mismatch between the C-like core-language presented in the paper and the implementation which is at the level of the LLVM IR. Moreover, the lack of any real language features (e.g. structs, control flow, procedure calls, etc.) in the core-language makes the argument of type soundness less compelling in the context of the implementation. For example, the lack of control flow and ability to express interesting data structures drastically reduces the complexity of reasoning about a region's lifetime. However, the actual system supports almost the entirety of LLVM, including extra features such as region-polymorphic functions that induce a LIFO-ordering of regions not expressible in the original model.

In the rest of this paper, we describe a verified checker that we have constructed for the SAFECode compiler using the Coq Proof Assistant [1]. In particular, we describe (a) our formalization of the syntax and semantics of the SAFECode variant of the LLVM [7] intermediate language, (b) a new declarative type system that formalizes the SAFECode policy, (c) a proof that the type system is sound with respect to our semantics, (d) an executable type-checker for SAFECode, (e) a proof that the type-checker is correct with respect to the declarative typing rules, and (f) our experiences in type-checking code generated by the SAFECode compiler.

In addition to scaling the language to the actual implementation, we also hope that our reformulation of SAFECode is cleaner than the one originally presented. For example, the original small-step semantics contains 40 operational transitions (even though it is not very expressive), exposes the details of region and memory meta-data in the operational semantics leading to a less modular proof, has memory leaks, and contains many constructs with duplicate functionality such as load, loadc, loadFromU and loadcFromU. While none of these prove fatal to the soundness of the type system, we believe a cleaner model reduces clutter and provides a better intuition for how the proof of soundness is related to the actual system implementation.

3   Language and Operational Semantics

Here, we describe the formal language and semantic model that we have constructed in Coq, but take the liberty of using conventional notation to describe the ideas instead of the details of the Coq code.[1]

3.1   Language

The language is derived from the LLVM IR to mirror the SAFECode implementation as closely as possible and is summarized in Figure 3. Types τ include arbitrary width integers (in), single/double precision floats ($d32$ and $d64$), typed-pointers ($\tau * \rho$), untyped-pointers ($U(b) * \rho$) to b bytes, arrays ($[\tau \times n]$), region-polymorphic named types, and region-polymorphic functions. Named types are used to support aggregate and (iso-)recursive data structures. For example, to define a linked-list node parameterized over a region ρ, we can write the type $node = \langle \rho \rangle \{i32;\ node * \rho\}$. The notation $name\langle \overline{\rho} \rangle$ indicates that a named type

---

[1] http://people.fas.harvard.edu/~dehuang/projects/sc-formalism.zip

x: local variable, ρ: region variable, f: function, ℓ: block label, i_n: n-bit integer constant, n: integer size, d: single/double precision float

type	$\tau ::=$	$\text{in} \mid \text{d32} \mid \text{d64} \mid \tau * \rho \mid U(b) * \rho \mid name\langle\overline{\rho}\rangle \mid [\tau \times n] \mid \forall\overline{\rho}.\overline{\tau}_{\overline{p}} \to \tau_{\overline{p}}$
name env	$\Upsilon ::=$	$name \rightharpoonup \langle\overline{\rho}\rangle\{\tau_{1\overline{p}}; \cdots ; \tau_{n\overline{p}}\}$
operand	$o ::=$	$x \mid i_n \mid f \mid d \mid \text{undef}(\tau) \mid \text{blockaddr}(f, \ell) \mid \text{null}$
insn	$\iota ::=$	$x = o_1 \text{ binop } o_2 \mid x = o_1 \text{ icmp } o_2 \mid x = o_1 \text{ fbinop } o_2 \mid$

$$x = o_1 \text{ fcmp } o_2 \mid x = \text{iconv } \tau_1 \text{ } o \text{ to } \tau_2 \mid x = \text{fconv } \tau_1 \text{ } o \text{ to } \tau_2$$
$$x = \text{ptrtoint } \tau_1 \text{ } o \text{ to } \tau_2 \mid x = \text{inttoptr } \tau_1 \text{ } o \text{ to } \tau_2 \mid$$
$$x = \text{extractvalue } \tau_1 \text{ } o, \tau_2 \text{ } i_n \mid x = \text{insertvalue } \tau_1 \text{ } o_1, \tau_2 \text{ } o_2 \text{ } i_n \mid$$
$$x = \text{bitcast } \tau_1 \text{ } o \text{ to } \tau_2 \mid x = \text{select } \tau_1 \text{ } o_1, \tau_2 \text{ } o_2, \tau_3 \text{ } o_3 \mid \text{exit} \mid$$
$$x = \text{getelementptr } \tau_1 \text{ } o_1, \tau_2 \text{ } o_2 \mid x = \text{getelementptrU } \tau \text{ } o_1 \text{ } o_2 \mid$$
$$x = \text{load } \tau \text{ } o \mid \text{store } \tau_1 \text{ } o_1, \tau_2 \text{ } o_2 \mid$$
$$x = \text{poolcheck } \rho \text{ } \tau \text{ } o \mid x = \text{poolcheckU } \rho \text{ } c \text{ } o \mid \text{poolfree } \tau \text{ } o$$

terminator	$tm ::=$	$\text{return } \tau \text{ } x \mid \text{br } \ell \mid \text{br } o \text{ } \ell_1 \text{ } \ell_2 \mid \text{switch } \tau \text{ } o, \ell \text{ } \overline{\tau * o * \ell} \mid$

$$\text{indirectbr } \tau \text{ } o, \overline{\ell} \mid x = \text{poolalloc } \tau, \rho, i, \ell \mid$$
$$x = \tau \text{ call } o \langle\overline{\rho}\rangle(\overline{o}), \ell \mid x = \tau \text{ unsafecall } o \langle\overline{\rho}\rangle(\overline{o}), \ell \mid$$

ϕ *node*	$\Phi ::=$	$x = \phi(\overline{o})_{\overline{\ell}} : \tau$
block	$blk ::=$	$[\Phi_1; \ldots \Phi_n; \iota_1; \ldots \iota_m; tm]$
function	$fn ::=$	$\tau \text{ } f\langle\overline{\rho}_p\rangle(\overline{x} : \overline{\tau}_{\overline{p}_p})\{$
		$\qquad \overline{\rho}_l = \text{poolinit } \overline{\tau}_{\overline{p}_p \cup \overline{p}_l};$
		$\qquad \text{body} : \ell \to blk$
		$\}$
fn table	$F :$	$f \rightharpoonup fn$

Fig. 3. Abstract syntax for our SAFECode language. We use $\overline{x}$ to denote a list of x's. The subscripts on types indicate which regions the types can mention.

is instantiated with the regions specified by $\overline{\rho}$. A global environment Υ is used to associate names with their definitions. Union types can be represented using LLVM's encoding as a byte struct whose size is the maximum size of all the types in the union.

The syntax supports nested aggregates (i.e., structs) through names, but internally, we only manipulate flattened *primitive* types. For example, given named types Foo = {i32; i32*} and Bar = {Foo; i32}, the flattened representation of Bar is {i32; i32*; i32}. Flattening aggregates in this fashion supports more (type-safe) projections, and avoids the need for complicated path expressions to calculate offsets. Given a name environment Υ, we define a partial function $\flat(\tau)$ that flattens τ into a vector of primitive (non-aggregate) types.

$$\flat(\text{in}) = [\text{in}] \qquad \flat(\text{d32}) = [\text{d32}] \qquad \flat(\text{d64}) = [\text{d64}] \qquad \flat(\tau * \rho) = [\tau * \rho]$$

$$\flat(U(b) * \rho) = [U(b) * \rho] \qquad \flat([\tau \times n]) = \flat(\tau) \mathbin{+\!\!+} \cdots \mathbin{+\!\!+} \flat(\tau)$$

$$\flat(name\langle\overline{\rho}\rangle) = (\flat(\tau_1) \mathbin{+\!\!+} \cdots \mathbin{+\!\!+} \flat(\tau_n))\{\overline{\rho}/\overline{\rho}'\} \text{ when } \Upsilon(name) = \langle\overline{\rho}'\rangle\{\tau_1; \cdots ; \tau_n\}$$

Note that $\flat(-)$ stops when it encounters a pointer, and is thus well-founded when recursive uses of names are limited to positions under a pointer (as in C). Here, we have omitted details of padding and alignment, which are covered in our Coq development.

Basic blocks consist of a sequence of ϕ-nodes followed by a sequence of deterministic instructions, and end with a terminator instruction. Instructions manipulate operands which are either variables or constants, and are mostly derived from the LLVM IR. The instructions are organized into two categories: The first contains instructions modeled deterministically (i.e., functionally). The second category contains LLVM terminator instructions (i.e., control-flow operators) and instructions modeled non-deterministically (i.e., axiomatically) such as `poolalloc`. Unlike LLVM, we consider a `call` to terminate a block, as a `call` will generally have non-deterministic behavior. Another difference with LLVM is that our `getelementptr` instruction does not perform multistep indexing because the type environment for aggregate data structures is already flattened.

Functions abstract over a set of caller-provided regions, and begin by defining a set of local regions which are scoped over the lifetime of the function invocation. The syntax prevents new regions from being allocated in the function body. While the actual SAFECode system allows regions to be allocated anywhere, we opt for this design because in practice, the compiler usually emits `poolinit` instructions at the beginning of a function. We encode `allocas` (stack allocations) as a `poolalloc` and `poolfree`.

One difference between our language and the SAFECode implementation is that we introduce a dedicated untyped pointer $U(b) * \rho$ to make it easier to express that we can safely dereference b bytes, but do not know what those bytes are. As a result, the syntax provides two versions of `poolcheck/U` and `getelementptr/U` to work with types in the typed-region case and bytes in the untyped-region case.

3.2   Representation of Run-Time Values

All run-time values are represented as a list of bit strings:

$$val \quad ::= \quad [v_1; v_2; \ldots; v_n]$$
$$v ::= \text{bit}_n(i) \quad (i \in [0..2^n))$$

For example, the 16-bit integer 0xF00D is represented as $[\text{bit}_{16}(\text{0xF00D})]$, whereas a struct containing a 16-bit integer and 32-bit pointer {0xBEEF; 0x0BADF00D} is represented as $[\text{bit}_{16}(\text{0xBEEF}), \text{bit}_{32}(\text{0x0BADF00D})]$. Note that pointers and integers (of the appropriate size) have the same representation.

There are other possible representations. For instance, CompCert [8,9] treats a pointer as a symbolic block number and an offset within that block, (b, o). This "Swiss cheese" model effectively enforces object isolation: we simply cannot get to an object at address a through pointer arithmetic on the base address of a different object b. Such a model makes sense when trying to define the formal semantics of C which makes actions, such as treating an integer as a pointer undefined.

In our case, SAFECode allows integers to be used as pointers provided that the compiler can statically prove that it is safe or it is guarded with a dynamic check. To support this behavior, we found the simplest approach for our model was to treat all run-time values as bit strings.

Local environment	$env : \text{var} \rightharpoonup \text{val}$	Primitive types	$\mathcal{K} := \{\texttt{in}, \texttt{d32}, \texttt{d64}, \tau * \rho,$
Region instantiation	$\sigma : \mathcal{P} \rightharpoonup \mathcal{R}$		$U(b) * \rho, \forall \rho.\overline{\tau} \to \tau\}$
Memory	$M : \texttt{mem\_t}$	Region names	$\mathcal{P} := \{\rho_1, \dots, \rho_n\}$
Live region set	$\mathcal{L} : \text{set } \mathcal{R}$	Run-time regions	$\mathcal{R} := \{r_1, \dots, r_n\}$
Heap typing	$\Sigma : \texttt{heap\_t}$	Heap type	$\texttt{heap\_t} := \mathbb{Z} \rightharpoonup (\mathcal{K}, \mathcal{R})$
	Execution context	$E ::= (\text{fn}, \text{blk}, env, \Sigma, \sigma, \mathcal{L})$	
Execution stack	$S : \text{list } E$	Machine state	$ms ::= (M, E, S)$

Fig. 4. Components of the abstract machine

3.3 Abstract Machine

The components of our abstract machine are presented in Figure 4. A machine state is represented by the tuple (M, E, S), where M is the current memory, E is the current execution context, and S is the control stack of execution contexts. An execution context contains information relevant to the computation in the current stack frame, including the function definition (f), a currently executing basic block (b), and an environment (env) mapping variables to run-time values. The last three components of E are used to support regions. The first of these is the heap typing Σ, which maps addresses to a pair of a primitive type and region. Conceptually, the heap typing holds the run-time meta-data, allowing the system to perform a run-time check. Next, σ contains a mapping of the region variables written in code to their actual run-time regions. The component $\mathcal{L}$ represents the set of live regions.

3.4 Memory Model and Memory Management

At the level of our abstract machine, we do not want to specify how regions are represented, nor the meta-data that is needed to manage memory. Rather, we parameterized our development over an abstract memory management interface specified axiomatically. To ensure that the axioms are consistent, we implemented a simple allocation strategy and proved that the strategy satisfied the axioms.

We treat memory as a partial map from integers to bytes, paired with some allocator-specific meta-data of abstract type:

$$\texttt{mem\_t} : (\mathbb{Z} \rightharpoonup \texttt{byte}) \times \texttt{metadata\_t}$$

Conceptually, this meta-data can encode information such as the size of memory or a specific allocation strategy. In our instance of the memory model, we use the meta-data to ensure that all addresses in use fall within the range of 64-bit machine integers to model a finite memory with 2^{64} addresses.

We summarize the memory operations and sketch the pre- and post-conditions in Figures 5, 6 and 7. The $\texttt{mload}$ function reads $\texttt{sizeof}(\tau)$ bytes from the specified address and returns an optional value. The $\texttt{mstore}$ function writes $\texttt{sizeof}(\tau)$ bytes coming from the specified value to the specified address and

$$
\begin{array}{ll}
\texttt{mload}: & \texttt{mem\_t} \times \mathbb{Z} \times \mathcal{K} \to \texttt{val}_\bot \\
\texttt{mstore}: & \texttt{mem\_t} \times \mathbb{Z} \times \mathcal{K} \times \texttt{val} \to \texttt{mem\_t}_\bot \\
\texttt{mpoolalloc}: & \texttt{mem\_t} \times \texttt{set}\, \mathcal{R} \times \texttt{heap\_t} \times \overline{\mathcal{K}} \times \mathbb{N} \times \mathcal{R} \to (\mathbb{Z} * \texttt{mem\_t})_\bot \\
\texttt{mpoolfree}: & \texttt{mem\_t} \times \mathbb{Z} \to \texttt{mem\_t}_\bot \\
\texttt{mpoolinit}: & \texttt{mem\_t} \times \sigma \times \texttt{set}\, \mathcal{R} \times \texttt{heap\_t} \times (\mathcal{R} * \mathcal{K}) \to (\texttt{mem\_t} * \texttt{heap\_t} * \mathcal{R})_\bot \\
\texttt{mpooldel}: & \texttt{mem\_t} \times \texttt{set}\, \mathcal{R} \times \texttt{heap\_t} \times \texttt{set}\, \mathcal{R} \to (\texttt{mem\_t} * \texttt{heap\_t} * \texttt{set}\, \mathcal{R})_\bot \\
\texttt{mcheck}: & \texttt{heap\_t} \times \mathbb{Z} \times \mathcal{R} \times \overline{\mathcal{K}} \to \texttt{bool}
\end{array}
$$

Fig. 5. Memory signature. $\overline{X}$ indicates a list of values drawn from the domain X.

$$
\begin{array}{ll}
\texttt{mload}\ M\ 0\ \tau = \bot & \texttt{mstore}\ M\ 0\ \tau\ v = \bot \\
\end{array}
$$

$$\texttt{mload}\ (\texttt{mstore}\ M\ a\ \tau\ v)\ a\ \tau = v$$

$$\texttt{mload}\ (\texttt{mstore}\ M\ a_1\ \tau_1\ v)\ a_2\ \tau_2 = \texttt{mload}\ M\ a_2\ \tau_2\ \text{when}$$

$$[a_1, a_1 + \texttt{sizeof}(\tau_1))\ \text{disjoint}\ [a_2, a_2 + \texttt{sizeof}(\tau_2))$$

$$\texttt{mload}\ (\#2\ \texttt{mpoolalloc}\ M\ \mathcal{L}\ \overline{\Sigma\ \tau'})\ a\ \tau = v \implies \texttt{mload}\ M\ a\ \tau = v$$

$$\texttt{mload}\ (\texttt{mpoolfree}\ M\ a)\ a'\ \tau = \texttt{mload}\ M\ a'\ \tau$$

Fig. 6. Selected memory operation equations. We write $\#2$ for second projection.

optionally returns the updated memory. Loading or storing to the null pointer 0 results in failure. We note the types are used strictly for size information.

The operations `mpoolalloc` and `mpoolfree` are used to allocate and free memory within a specified region. Unlike conventional allocation, `mpoolalloc` does not generate fresh locations, i.e., the heap typing remains invariant. Instead, it checks to see if there is a block of memory mapped in the heap typing at the specified type and region. If successful, `mpoolalloc` returns a pointer to that memory and can update its internal meta-data. The specification allows repeated calls to `mpoolalloc` for the same type to return the same pointer. In practice, one would use the internal meta-data for a freshness guarantee. Similarly, `mpoolfree` does not change the heap typing, but reflects that its meta-data might have changed to reclaim a set of addresses. This allows reclaimed addresses to be recycled for allocations of the same type.

The operation `mpoolinit` allocates fresh locations for a specified region. It first generates a region name fresh from $\mathcal{L}$. It then allocates fresh locations for that region, zeroes out the allocated memory, and updates the heap typing to map the appropriate addresses to the appropriate types and region.[2] The operation `mpooldel` is the inverse of `mpoolinit`. It runs through the input heap typing and frees up all addresses mentioning the specified regions. Lastly, the operation `mcheck` models a run-time check. Given a heap typing, it verifies whether an address belongs to a region at the correct type. Note that `mcheck` works at the level of primitive types, so if the check returns true for a specified address,

---

[2] In our Coq development, we break this operation into two parts, one for fresh region creation, and one for region allocation to support mutually recursive regions.

$$\frac{\{ M(a) = b_1 \wedge \ldots \wedge M(a + \texttt{sizeof}(\tau)) = b_n \}}{\texttt{mload}\ M\ a\ \tau = v}$$
$$\{ v = \texttt{endian}\ [b_1, \ldots, b_n] \}$$

$$\frac{\{ \}}{\texttt{poolalloc}\ M\ \mathcal{L}\ \varSigma\ \overline{\tau}\ n\ r = (a, M')}$$
$$\{ \texttt{mcheck}\ \varSigma\ a\ r\ \overline{\tau} = \text{true} \}$$

$$\frac{\{ \exists v_2, \texttt{mload}\ M\ a\ \tau = v_2 \}}{\texttt{mstore}\ M\ a\ \tau\ v_1 = M'}$$
$$\{ M(a) = b_1 \wedge \ldots \wedge M(a + \texttt{sizeof}(\tau)) = b_n$$
$$\text{where}\ v_1 = \texttt{endian}(b_1, \ldots, b_n) \}$$

$$\frac{\{ \}}{\texttt{mcheck}\ \varSigma\ a\ r\ [\tau_1, \ldots, \tau_n] = \text{true}}$$
$$\{ \varSigma(a) = (\tau_1, r),$$
$$\varSigma(a + \texttt{sizeof}(\tau_1)) = (\tau_2, r), \ldots \}$$

$$\frac{\{ \}}{\texttt{poolinit}\ M\ \sigma\ \mathcal{L}\ \varSigma\ (\rho, \tau) = (M', \varSigma', r)}$$
$$\{ a_1, \ldots, a_n\ \text{fresh} \wedge r \notin \mathcal{L} \wedge b(\tau) = [\tau_1, \ldots, \tau_m] \wedge$$
$$\varSigma' = \{ a_i \mapsto (\tau_1[\sigma], r) \} \uplus \{ a_i + \texttt{sizeof}(\tau_1) \mapsto (\tau_2[\sigma], r) \} \uplus \ldots \uplus \varSigma \wedge$$
$$\texttt{mload}\ M\ a_i\ \tau = \bot\ \text{for}\ i = 1, \ldots, n \wedge \texttt{mload}\ M'\ a_i\ \tau = 0\ \text{for}\ i = 1, \ldots, n \}$$

$$\frac{\{ r \in \mathcal{L} \}}{\texttt{mpooldel}\ M\ \mathcal{L}\ \varSigma\ r = (M', \varSigma', \mathcal{L}')}$$
$$\{ \mathcal{L} = r \cup \mathcal{L}' \wedge \varSigma = (a_1 \mapsto (\tau, r)) \uplus \ldots \uplus (a_n \mapsto (\tau, r)) \uplus \varSigma' \wedge$$
$$\texttt{mload}\ M'\ a_1\ \tau = \bot \wedge \ldots \wedge \texttt{mload}\ M'\ a_n\ \tau = \bot \}$$

Fig. 7. Axiomatic specification of memory operations. We use $M(a)$ to abbreviate the first projection of M at a, and $\tau[\sigma]$ to instantiate τ with regions specified in σ.

region and list of primitive types, it will also return true for any truncation of the sequence, holding the other parameters constant.

3.5 Operational Semantics

We structure the semantics as an evaluation function (Coq function) and a small-step relation (inductive predicate) on abstract machine states.

Figure 8 shows selected definitions from our `evalblock` function, which computes the resulting memory and local environment after evaluating all deterministic instructions in a basic block. The parts of the abstract machine not mentioned such as the control stack remain invariant when executing a basic block. There are two distinguished failure states `Err` and `Halt`. The `Err` state denotes a stuck state that our soundness theorem will rule out. A transition to the `Err` state can occur if we look up a variable that is not bound in the environment. A transition to the `Halt` state indicates that the run-time has prevented an unsafe memory operation. For example, a call to `poolcheck` may fail, halting the system. The functions describing instruction evaluation are straightforward, and make use of the definitions from our memory model in the case of memory instructions, or machine arithmetic in the case of binary operations.

Figure 9 shows almost all of our operational rules, omitting just the failure transitions for `call` and `poolalloc`, and the `unsafecall` and branch cases. A function call may fail if we run out of memory to hold local regions, while a poolalloc may fail if a specific region runs out of memory. The small-step semantics is compact because it pushes most of the work into the evaluation function.

evalblock nil M env =
  Ok(M, env)
evalblock ($\iota :: \bar\iota$) M env =
  match eval ι M env with
    | Ok(M', env') =>
      evalblock $\bar\iota$ M' env'
    | ans => ans

eval [x = poolcheck ρ τ o] M env =
  match eval_op o env with
    | Ok(bit(a)) =>
      if mcheck Σ a $\sigma(\rho)$ $\tau[\sigma]$
      then Ok($M, env[x \mapsto \text{ptr}(a)]$)
      else Halt
    | _ => Err

Fig. 8. Selected evaluation rules. We use the notation $\tau[\sigma]$ to indicate applying a region instantiation map σ to τ.

This is beneficial for model validation since the evaluation function is already executable. We only need to write an interpreter for the terminator instructions such as branch and return (which we can't do in Coq since it might diverge), and prove that the driver for terminator instructions respects the relation. Furthermore, if we grow a language by adding additional instructions, these instructions are unlikely to be terminator instructions, so validation should continue to scale.

4   Type System

Our goal to support real SAFECode programs means that the type system is quite complex. For the sake of brevity, we present a subset of the rules here and elide many details that are present in our Coq-development, such as sub-typing relations, primitive type manipulations, and contexts related to ϕ-node typing.

4.1   Typing Rules

At its core, the type system tracks region lifetimes and ensures that type-homogeneity is preserved for typed-regions. The key idea is that pointers into live regions can always be safely dereferenced at the appropriate type. The rules used to define the typing judgments are sketched in Figure 10. Throughout, we assume a context including the function table F mapping function names to their definitions, and a type environment Υ mapping named types to their definitions. In addition, the judgments use region contexts Δ to determine the region variables in scope, variable contexts Γ mapping variables to their types, and a label environment Ψ mapping block labels to their preconditions.

The well-formedness rules for types and instruction operands are straightforward. A type is well-formed ($\Delta \vdash \tau$) if all the regions that appear free in the type are in the region context Δ. Function types are required to be locally closed. That is, the argument and return types may only depend upon the region variables bound by the function. The second judgement ($\Delta \vdash o : \tau$) determines when operands are well-formed at a type.

EVAL
$$\frac{\text{evalblock } E.b.\bar{\iota} \; M \; E.env = (M', env')}{(M, E, S) \rightarrow (M', E[env := env'], S)}$$

EVAL-HALT
$$\frac{\text{evalblock } E.b.\bar{\iota} \; M \; E.env = \text{Halt}}{(M, E, S) \rightarrow \text{Halt}}$$

RETURN
$$\frac{\text{mpooldel}(M, \mathcal{L}, \Sigma) = (M', \Sigma'', \mathcal{L}'') \qquad env(x_1) = v}{\begin{array}{c}(M, (f, [b.\iota = \text{return } \tau \; x_1], env, \Sigma, \sigma\mathcal{L}), \\ (f', [b'.nd = x_2 = \text{call } \tau' \; o'\langle \bar{\rho}\rangle(\bar{x}), l')], env', \Sigma', \sigma', \mathcal{L}') :: S) \rightarrow \\ (M', (f', [b'.tm = \text{br } l'], (x_2 \mapsto v) \cup env', \Sigma'', \sigma', \mathcal{L}''), S)\end{array}}$$

CALL
$$\frac{\begin{array}{c}F(o) = \tau \; f'\langle\bar{\rho}_p\rangle(\bar{y} : \bar{\tau}_{\bar{\rho}_p})\{\bar{\rho}_l = \text{poolinit } \bar{\tau}_{\bar{\rho}_p \cup \bar{\rho}_l}; \text{body} : \ell \rightarrow \text{blk}\} \\ b' = f'.entry \qquad env' = env[\{\bar{y} \mapsto env(\bar{x})\} \qquad \{\bar{\rho}_p \mapsto \sigma(\bar{\rho})\} = \sigma'_e \\ \{\bar{\rho}_l \mapsto rgns'\} \cup \sigma'_e = \sigma' \qquad \text{mpoolinit}(M, \sigma', \mathcal{L}, \Sigma, \bar{\rho}_l * \bar{\tau}_{\bar{\rho}_p \cup \bar{\rho}_l}) = (M', \Sigma', rgns')\end{array}}{\begin{array}{c}(M, (f, [b.\iota = (x_1 = \text{call } \tau' \; o'\langle\bar{\rho}\rangle(\bar{x}), l')].env, \Sigma, \sigma, \mathcal{L}), S) \rightarrow \\ (M, (f', b', env', \Sigma', \sigma'_b, rgns' \cup \mathcal{L}), (f, [b.\iota = (x_1 = \text{call } \tau' \; o'\langle\bar{\rho}\rangle(\bar{x}), l')], env, \Sigma, \sigma', \mathcal{L}) :: S)\end{array}}$$

POOLALLOC
$$\frac{\flat(\tau) = \bar{\tau} \qquad \text{mpoolalloc}(M, lo, \mathcal{L}, \Sigma, \bar{\tau}[\sigma], n, \sigma(r)) = (a, M')}{\begin{array}{c}(M, (f, [b.nd = (x = \text{poolalloc } \tau, \rho, n, \ell)], env, \Sigma, \sigma, \mathcal{L}), S) \rightarrow \\ (M, (f, [b.tm = \text{br } \ell], env[(x \mapsto \text{ptr}(a))], \Sigma, \sigma, \mathcal{L}), S)\end{array}}$$

Fig. 9. Selected operational rules. The function table F remains constant throughout operation and is not shown in the rules. The notation $a.b$ projects b from a.

The judgment for typing instructions has the form $\Delta; \Gamma \vdash \iota : \Gamma'$. For example, the load rule checks that the operand has pointer type $\tau * \rho$ and that the pointer type is well-formed. The postcondition guarantees that a τ value has been loaded and passes that context forward to the next instruction. The postcondition for **poolcheck** says that no matter what the operand is, we can add the checked type into the context. There is a run-time check to ensure that this rule is sound.

The **poolalloc** rule is conceptually similar to a malloc rule where the result is a pointer of the correct type. The rule additionally checks that the type requested corresponds to the region's type. Thus, we cannot allocate an i32 in a region that holds {i32; i32}, although we can load an i32 out. Note that a request for n objects of type τ only reveals that the pointer to the front of that object is valid. This captures the essence of SAFECode that type-safety is guaranteed at the region-level, not for individual pointers. The call rule fully instantiates the function's polymorphic regions and then checks that the arguments have the appropriate types.

The typing rule for a function declaration (Figure 10) imposes a LIFO ordering on region lifetimes. The rule uses two region contexts Δ_p and Δ_l to accomplish this. Δ_p mentions only the regions the function is polymorphic in, while Δ_l extends Δ_p with locally allocated regions. We type the function signature and return type under Δ_p and the function body under context Δ_l. This ensures that regions never escape from callees to callers and that the only regions in scope of a function body are live. Typing for a function body (ommited) is done in a straightforward manner by typing all basic blocks in the body using the rules for

$$\boxed{\Delta \vdash \tau} \hspace{6cm} \text{Well-formed type}$$

$$\frac{}{\Delta \vdash \text{in}} \qquad \frac{}{\Delta \vdash \text{d32}} \qquad \frac{}{\Delta \vdash \text{d64}} \qquad \frac{\Delta \vdash \tau \quad \rho \in \text{dom}(\Delta)}{\Delta \vdash \tau * \rho} \qquad \frac{\rho \in \text{dom}(\Delta)}{\Delta \vdash U(b) * \rho}$$

$$\frac{\varUpsilon(name) = \overline{\tau} \quad \forall \rho \in \overline{\rho}, \rho \in dom(\Delta)}{\Delta \vdash name\langle \overline{\rho} \rangle} \qquad \frac{\overline{\rho} \vdash \overline{\tau} \rightarrow \tau}{\Delta \vdash \forall \overline{\rho}.\tau \rightarrow \tau} \qquad \frac{\Delta \vdash \tau \quad n \neq 0}{\Delta \vdash [\tau \times n]}$$

$$\boxed{\Gamma \vdash o : \tau} \hspace{4cm} \text{Well-formed operand (selected rules)}$$

$$\frac{\{\tau' * \rho, U(b) * \rho\} \notin \tau}{\Gamma \vdash \text{undef}(\tau) : \tau} \qquad \frac{F(f) = f\langle \overline{\rho} \rangle (\overline{x} : \overline{\tau})\{body\} \rightarrow \tau}{\Gamma \vdash f : \forall \overline{\rho}.\overline{\tau} \rightarrow \tau} \qquad \frac{\Gamma(x) = \tau}{\Gamma \vdash \text{reg } x : \tau}$$

$$\boxed{\Delta; \Gamma \vdash \iota : \Gamma} \hspace{4cm} \text{Well-formed instruction (selected rules)}$$

$$\frac{\Delta; \Gamma \vdash o : \tau * \rho \quad \Delta \vdash \tau * \rho}{\Delta; \Gamma \vdash x = \text{load } (\tau * \rho)\, o : \Gamma[x \mapsto \tau]} \qquad \frac{\Delta; \Gamma \vdash o : \tau' \quad \Delta \vdash \tau * \rho}{\Delta; \Gamma \vdash x = \text{poolcheck } \rho\, \tau\, o : \Gamma[x \mapsto \tau * \rho]}$$

$$\boxed{\Delta; \Psi; \Gamma \vdash tm} \hspace{3cm} \text{Well-formed terminator instruction (selected rules)}$$

$$\frac{\Gamma \vdash o : \tau \quad \Delta \vdash \tau}{\Delta; \Psi; \Gamma \vdash \text{return } \tau\, o} \qquad \frac{\Delta; \Psi; \Gamma[x \mapsto \tau * \rho] \vdash \text{br } \ell \quad \Delta(\rho) = \tau}{\Delta; \Psi; \Gamma \vdash x = \text{poolalloc } \tau, \rho, n, \ell}$$

$$\frac{\Gamma \vdash o : \forall \rho_p.\, \overline{\tau}_{\rho_p} \rightarrow \tau'_{\rho_p} \quad \Gamma \vdash \overline{o} : \overline{\tau}[\overline{\rho}/\overline{\rho_p}]}{\forall \rho \in \overline{\rho}.\, \rho \in dom(\Delta) \quad \tau = \tau'[\overline{\rho}/\overline{\rho_p}] \quad \flat(\tau) = \overline{\tau} \quad \Delta; \Psi; \Gamma[x \mapsto \tau] \vdash \text{br } \ell}{\Delta; \Psi; \Gamma \vdash x = \tau \text{ call } o\langle \overline{\rho} \rangle(\overline{o}), \ell}$$

$$\boxed{\Psi; \Delta_p; \Delta_l \vdash \text{func}} \hspace{4cm} \text{Well-formed function}$$

$$\frac{\overline{\rho}_p \cap \overline{\rho}_l = \emptyset \quad \Delta_p \vdash \overline{\rho}_p \quad \Delta_l \vdash \overline{\rho}_l \quad \Delta_p \subseteq \Delta_l}{\forall \tau \in \overline{\tau}_l.\, \Delta_l \vdash \tau \quad \Delta_p \vdash \tau \quad \forall \tau \in \overline{\tau}_1.\, \Delta_1 \vdash \tau \quad \forall \ell \in dom(\text{body}).\, \Psi; \Delta_l \vdash \text{body}(\ell)}{\Psi; \Delta_p; \Delta_l \vdash \tau\, f\langle \overline{\rho}_p \rangle (\overline{x} : \overline{\tau}_1)\{\overline{\rho}_l = \text{poolinit } \overline{\tau}_2; \text{body} : \ell \rightarrow \text{blk}\}}$$

$$\boxed{F_\Psi; F_\Delta \vdash \text{prog}} \hspace{4cm} \text{Well-formed program}$$

$$\frac{F_\Psi(f_i) = \Psi_f \quad F_\Delta(f_i) = (\Delta_p, \Delta_l) \quad \Psi_f; \Delta_p; \Delta_l \vdash f_i, \text{ for } i = 1, \ldots, n}{F_\Psi; F_\Delta \vdash \{f_1, \ldots, f_n\}}$$

Fig. 10. Selected typing rules

deterministic and terminator instructions. The top-level typing rule ensures that all functions are well-formed. It introduces two new contexts $F_\Psi : f \rightharpoonup \Psi$ and $F_\Delta : f \rightharpoonup (\Delta_p, \Delta_l)$. The former maps a function to its basic block preconditions. The latter maps a function to its appropriate region contexts. The top-level mapping ensures that mutually recursive functions use consistent contexts.

4.2   Type Soundness

The most difficult part of the proof reduces to arguing about the LIFO structure of region lifetimes to ensure that pointers only point into live regions. As functions can only be declared at the top-level in LLVM, we do not need to worry about regions escaping through closures. The key invariant to highlight is that the heap typing stack is well-formed.

Definition (Well-formed stack: heap typing). For any two adjacent execution contexts E_1 and E_2, where E_1 is the callee frame and E_2 is the caller frame,

$$
\begin{array}{cl}
\text{(a)} & E_2.\Sigma \subseteq E_1.\Sigma \\
\text{(b)} & E_2.\mathcal{L} \cup \{r_1 \ldots r_n\} = E_1.\mathcal{L} \wedge E_2.\mathcal{L} \cap \{r_1 \ldots r_n\} = \emptyset \\
\text{(c)} & E_2.\mathcal{L} \vdash E_2.\Sigma \\
\text{(d)} & E_1.\mathcal{L} \vdash E_1.\Sigma
\end{array}
$$

In words, the heap typing increases monotonically (in terms of addresses mapped) as we move from caller to callee frames in the execution stack. Similarly, the live region set grows monotonically. Lastly, the regions mentioned in an execution context's heap typing are closed under its respective live region set.

We now state the main lemmas that are used in the proof of type soundness.

Lemma 1. (Progress and Preservation of basic block evaluation) If $F_\Psi; F_\Delta \vdash \{f_1, \ldots, f_n\} \wedge \vdash (M, E, S)$, then either `evalblk` $E.b.\bar{\imath}$ M $E.env = \mathrm{Ok}(M', env') \wedge \vdash (M', E[env := env'], S)$ or `evalblk` $E.b.\bar{\imath}$ M $E.env = \mathrm{Halt}$.

The proof is mechanized in Coq, but the interesting bit is that our mixed semantics allows us to prove progress and preservation simultaneously as proving that our evaluation function `evalblk` preserves the invariants also implies that we must be able to take a step. One nice property of this structure is if we extend the language by adding additional instructions to `evalblk`, we only need to modify our proof of type soundness in one place. This was particularly useful when we were scaling our language out to handle real programs, as many instructions that we added later (e.g. `insertvalue`) required minimal proof changes.

Lemma 2. (Preservation) If $F_\Psi; F_\Delta \vdash \{f_1, \ldots, f_n\} \wedge \vdash (M, E, S) \wedge (M, E, S) \rightarrow (M', E', S')$, then $\vdash (M', E', S')$.

Recall that the core of the proof reduces to arguing about LIFO region lifetimes. As our abstract memory interface specifies that `mpoolinit` (used on a function call) and `mpooldel` (used on a return) are inverses of each other with respect to the heap typing, the proof reduces to invoking this fact to argue that a callee returns the heap-typing to the state expected by the caller.

Lemma 3. (Progress) If $F_\Psi; F_\Delta \vdash \{f_1, \ldots, f_n\} \wedge \vdash (M, E, S)$, then either $(M, E, S) \rightarrow (M', E', S') \wedge \vdash (M', E', S')$ or $(M, E, S) \rightarrow \text{Halt}$.

This lemma has few cases because of the small operational semantics and is straightforward to prove. Our soundness result implies that a pointer with type $\tau * \rho$ in a well-typed program always points into region ρ and references a τ.

5   Evaluation

The previous section described a declarative type system and argued that it is sound. We have also built an algorithmic type system tc (i.e., a type-checker as a function) and proved that it respects the declarative typing rules:

Lemma 4. (Type-checker sound) If $\text{tc}(F_\Psi, F_\Delta, \{f_1, \ldots, f_n\}) = \text{true}$, then $F_\Psi; F_\Delta \vdash \{f_1, \ldots, f_n\}$.

The type-checker is straightforward to write and prove sound. It can be extracted as an executable OCaml program to serve as a verified type checker for SAFECode. Unfortunately, the SAFECode compiler emits code that does not adhere strictly to the SAFECode type system. For ease of code generation, the compiler erases almost all region and type information from the LLVM bitcode. This significantly increases the difficulty of applying our type-checker as we now need to perform type-inference.

We had to write two pieces of code to close this gap. First, we wrote an LLVM pass in C++ that crawls SAFECode's internal structures that annotates the resulting code with region information. Second, we wrote an OCaml pass that performs type-inference and translates the input LLVM bitcode into our representation. The bulk of the OCaml pass is dedicated to reconstructing the types that SAFECode erases. In principle, we should not need to write this pass because conceptually, the SAFECode compiler produces this typing derivation when instrumenting code. Our type-checker checks the output of these two pieces of code. The Coq formalization is about 12000 lines of code, while our OCaml translation and inference pass is about 4800.

5.1   Experimental Results

In addition to the two pieces of code, we had to make a few simplifications that possibly introduce unsoundness into the system to type-check real code. First, we translate calls to library functions where we do not have the code or are not instrumented by SAFECode into unsafecalls. We also cannot type variable argument functions and certain LLVM intrinsic functions and translate those to unsafecalls as well.[3] Lastly, to keep our type system from becoming too unwieldy, we choose not to add a typing context to handle aliases. In practice, SAFECode sometimes calls poolcheck on a variable x, and later uses an alias

---

[3] Consequently, preservation holds only on code that does not contain unsafecalls.

of x without a check. As a workaround, every time a `poolcheck` is encountered during translation, we emit an extra `poolcheck` for x's aliases.

We ran our type-checker on micro-benchmarks included with the SAFECode distribution and on the Olden benchmarks [2], a pointer-intensive test suite on which the original SAFECode system was evaluated. We discovered some bugs with the current SAFECode[4] instrumentation of a few programs in the Olden benchmarks, mostly with region instantiations. In the program *bh*, we found that a call to a region-polymorphic function was not instantiated with a region. In the program *em3d*, we found that a pointer to a function-local pool was allocated and returned to the caller, violating the LIFO region invariant. We also discovered some false-positives. For example, in the program *perimeter*, SAFECode performs an interval-analysis over multiple program paths to determine that an array index variable is statically in-bounds. This is a limitation of our type-checker, as it cannot reason about the above analysis. Although our type-checker is still a prototype, it is already effective at finding bugs.

6   Related and Future Work

Zhao et al. [13] presented a semantics for the LLVM IR formalized in Coq and used it to prove the correctness of a closely related, but alternative technique for enforcing spacial memory safety on C code called Softbound [10]. On the one hand, their proof is more impressive because it shows the correctness of the transformation. On the other hand, their model for LLVM's IR cannot handle idioms that arise in real C/C++ programs, (e.g., casting a pointer to an integer and then back) because their treatment of memory is too high-level. Furthermore, our type-checker can be used not only to validate the initial transformation of the code, but also the code that comes out of subsequent optimizations.

In addition to SAFECode, there is a rich history of region-based type systems, first pioneered by Tofte and Talpin [12] and later refined in Cyclone [6] that our type system draws inspiration from. In many regards, our type system is much simpler because regions can only be passed "downwards" to functions and never returned. Furthermore, SAFECode does not support lexically scoped closures or existential types, so there is no need for type-and-effects systems. Languages such as Cyclone had many more cases in which region names could escape function scope, and thus required a much more complicated type system. In contrast, the regions in SAFECode and our type system are type-homogenous and allow explicit deallocation of memory in these regions, operations which were not permitted except in restricted cases in those languages.

In our future work, we hope to thoroughly test our semantics to make sure that it is compatible with the actual semantics implemented by the LLVM compiler and SAFECode run-time system. However, we structured our semantics to lighten the burden of validation as explained in Section 3.

---

[4] The system we evaluated our type-checker on is the most current implementation and not the one presented in the original paper.

Acknowledgements. We thank Gregory Malecha, John Criswell, Joseph Tassarotti, Stephen Chong and our reviewers for their helpful discussions.

References

1. Coq Proof Assistant, http://coq.inria.fr/
2. Carlisle, M.C.: Olden: Parallelizing Programs with Dynamic Data Structures on Distributed-Memory Machines. PhD thesis (1996)
3. Castro, M., Costa, M., Martin, J.-P., Peinado, M., Akritidis, P., Donnelly, A., Barham, P., Black, R.: Fast Byte-Granularity Software Fault Isolation. In: Proc., SOSP 2009 (2009)
4. Criswell, J., Lenharth, A., Dhurjati, D., Adve, V.: Secure Virtual Architecture: A Safe Execution Environment for Commodity Operating Systems. In: Proc., SOSP 2007 (2007)
5. Dhurjati, D., Kowshik, S., Adve, V.: SAFECode: Enforcing Alias Analysis for Weakly Typed Languages. In: Proc., PLDI 2006 (2006)
6. Grossman, D., Morrisett, G., Jim, T., Hicks, M., Wang, Y., Cheney, J.: Region-Based Memory Management in Cyclone. In: Proc., PLDI 2002 (2002)
7. Lattner, C., Adve, V.: LLVM: A Compilation Framework for Lifelong Program Analysis & Transformation. In: Proc. International Symposium on Code Generation and Optimization, CGO 2004 (March 2004)
8. Leroy, X.: Formal verification of a realistic compiler. Commun. ACM 52(7) (July 2009)
9. Leroy, X., Blazy, S.: Formal Verification of a C-like Memory Model and Its Uses for Verifying Program Transformations. J. Autom. Reason. 41(1) (July 2008)
10. Nagarakatte, S., Zhao, J., Martin, M.M.K., Zdancewic, S.: SoftBound: Highly Compatible and Complete Spatial Memory Safety for C. In: Proc. PLDI 2009 (2009)
11. Necula, G.C., McPeak, S., Weimer, W.: CCured: Type-Safe Retrofitting of Legacy Code. In: Proc. POPL (2002)
12. Tofte, M., Talpin, J.-P.: Region-based memory management. Inf. Comput. 132(2) (February 1997)
13. Zhao, J., Nagarakatte, S., Martin, M.M.K., Zdancewic, S.: Formalizing the LLVM Intermediate Representation for Verified Program Transformations. In: Proc. POPL 2012 (2012)

Certifiably Sound Parallelizing Transformations

Christian J. Bell*

Princeton University
cbell@cs.princeton.edu
http://www.cs.princeton.edu/

Abstract. Sustaining scalable performance trends in the multicore era
has led many compiler researchers to develop a host of optimizations
to parallelize sequential programs. At the same time, formal methods
researchers have pushed compiler verification technology forward to the
point that real compilers may be checked for correctness by proving that
the compiler preserves a simulation relation between the source and target languages. We join these two lines of research by proving a general
parallelizing transformation schema sound for an extension of the Calculus of Communicating Systems (CCS) with semaphores and sequential
composition. When source programs contain internal nondeterminism,
we have found that the simulation relations that underlie the most prominent verified compilers, like CompCert, are too strong to admit a large
class of parallelizing transformations. Thus we prove soundness with respect to a new simulation relation, called *eventual simulation*, that resolves this issue and is equivalent to weak bisimulation when no internal
nondeterminism is present. All formal details presented are proven and
mechanically checked using the Coq Proof Assistant.

1   Introduction

Parallelizing optimizations allow programmers to take advantage of the increasing number of cores found in modern CPUs with little additional effort. These
optimizations free the programmer from dealing with the inherent complexities
of writing multithreaded code directly and bring new vigor to a large base of
existing sequential source code by the simple act of recompilation.

Compiler verification guarantees that a compiler does not contain bugs, or
at least does not introduce bugs into compiled programs. Well-known examples include CompCert (and variations such as the Verified Software Toolchain,
XCERT, and CompCertTSO) and Vellvm [5][1][11][10][13]. CompCert translates
a C source program into successive stages of intermediate languages until it finally generates PowerPC, ARM, or x86 assembly code. Each translation is proven
correct with respect to a behavioral equivalence called *weak bisimulation* (henceforth referred to as just "bisimulation"); by transitivity, the source and target
assembly programs will bisimulate each other.

---

* This work was supported in part by the NSF under grants 1047879 and 1016937.
  Any opinions, findings, and recommendations are those of the author and do not
  necessarily reflect the views of the NSF.

G. Gonthier and M. Norrish (Eds.): CPP 2013, LNCS 8307, pp. 227–242, 2013.
© Springer International Publishing Switzerland 2013

Bisimulation is a member of a large class of simulation relations that have a co-inductive proof method and preserve many strong properties of program behavior, such as the interactions with an environment (e.g. the operating system, users, and libraries) and deadlocking behavior. It may also be augmented with termination and divergence sensitivity without much difficulty. Two programs are bisimilar if they can mimic each other indefinitely.

Parallelizing optimizations have been studied extensively, but research has been primarily focused on performance considerations and on developing the supporting static analyses. Our goal is to prove the correctness of optimizations like DOALL, DOACROSS, and Decoupled Software Pipelining (DSWP) [7][9], which transform a loop into multiple parallel loops, and may introduce synchronization to communicate data and control flow dependencies. Toward this end, we have proven the soundness of a highly general parallelizing transformation.

Admitting parallelization presumes that the threading primitives – fork, join, and synchronization – and the scheduler are not directly observable. In contrast, the Verified Software Toolchain [1] conservatively treats these primitives as observable system calls; in this setting, parallelization is clearly not admissible.

Combining parallelization and internal nondeterminism – the choices a source program makes that are not directly visible to the observer – raises an interesting challenge because parallelization may cause the nondeterminism to interleave with observable actions. Even when benign, a weak simulation (henceforth referred to as just "simulation") is not preserved by this behavior. In many cases, the source of the internal nondeterminism is "unspecified behavior" that the compiler may refine. A potential solution is to first refine all internal nondeterminism, after which parallelization will preserve a bisimulation.

Internal nondeterminism can be intentional,[1] however, and refining it may either be incorrect (depending on the language specification) or cause the program to run slower (e.g., by removing concurrency). Or it may be desirable to perform the refinement in a latter phase of compilation; after parallelization. Thus we use a new type of simulation relation, which we call *eventual simulation*, that allows the compiler to preserve internal nondeterminism throughout parallelization.

Our main contributions are:

- We prove soundness of a general-purpose parallelizing transformation schema for an extension of CCS, called CCS-Seq, with respect to a new simulation relation called eventual simulation. Additionally, we prove that the termination properties are correctly preserved.
- Because eventual simulation is not symmetric, we propose using *contrasimulation* [3] in verified compilers. It is implied by eventual simulation and is generally a congruence for imperative languages. Without internal nondeterminism, contrasimulation reduces to bisimulation and thus our proof of soundness for parallelization is still directly applicable.

---

[1] A specification could permit the program to invoke an unobservable third party to make a choice, such as a random number generator or (indirectly) through the interleavings chosen by the scheduler.

– We mechanically formalize and prove this work – CCS-Seq, eventual simulation, contrasimulation, the proof of parallelization, and all other details – in the Coq Proof Assistant.[2] All definitions and lemmas are in a "mathematical" notation, yet are otherwise identical to our Coq development.

In Section 2, we begin with an illustrative example of parallelizing a simple program with internal nondeterminism. Then we give the formal tools necessary to compare the behaviors of programs and introduce eventual simulation to state the correctness of this transformation. CCS-Seq is defined in Section 3, for which we present the parallelizing transformation schema. In Sections 4 and 5 we introduce contrasimulation and then show how using "delayed observations" allows it to be used in more situations. Soundness proofs and formal definitions are given in Section 6. The remaining sections discuss related work and conclude.

2   A Simple Parallelizing Transformation

We begin by showing how a simple program might be parallelized. This example introduces a few of the basic concepts that we will be using throughout this paper – transition diagrams, labeled transition systems, observable versus unobservable actions, internal nondeterminism, and comparing program behaviors using simulations and bisimulations. Finally, we define eventual simulation and prove that it holds for this example.

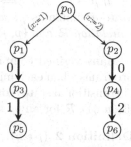

(a) sequential (2.1)

The following sequential program randomly assigns either 1 or 2 to x, outputs 0, and then outputs x.

$$x := \text{either 1 or 2; print 0; print x} \qquad (2.1)$$

A potential result of parallelizing the program is:

$$(x := \text{either 1 or 2} \parallel \text{print 0}); \text{print x}, \qquad (2.2)$$

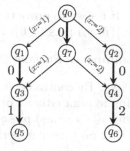

(b) parallel (2.2)

which may also output 0 before choosing a value for x. The either-or statement is an instance of internal nondeterminism. Although the parallelized program is intuitively "correct", the original program does not simulate it because choosing a random value can be reordered with console output (an observable action).

Figure 1 presents the semantics of each program as transition diagrams where nodes represent program states and directed edges represent program transitions. The initial states are represented by the root nodes; p_0

Fig. 1. Semantics of the programs as transition diagrams

and q_0 correspond to (2.1) and (2.2), respectively. Unobservable (silent) steps that correspond with a line of source code are labeled in parentheses. For example, choosing to assign either 1 or 2 to x is labeled ($x := 1$) or ($x := 2$), respectively. In later examples, silent steps may not be labeled at all. Observable actions (console output) have bold edges.

---
[2] The full Coq development is at http://www.cs.princeton.edu/~cbell/par/

Labeled transition systems. A labeled transition system (LTS) is defined by a triple, (S, L, δ), where S is the set of states, L is the set of observable actions, and $\delta \subseteq S \times (L \cup \{\tau\}) \times S$ is the transition relation. A step from state p to p' that performs action $\alpha \in L \cup \{\tau\}$ is defined if $(p, \alpha, p') \in \delta$ and is denoted by $p \xrightarrow{\alpha} p'$. τ is reserved for internal (silent) transitions; we write $p \to p'$ instead of $p \xrightarrow{\tau} p'$.

The transitive reflexive closure of $\to$ is denoted by $\Rightarrow$. We write $p \xRightarrow{\alpha} p'$ for $\alpha \in L \cup \{\tau\}$ when p performs action α by taking zero or more steps to p'; i.e. $p \xRightarrow{a} p'$ for $a \in L$ iff $p \Rightarrow \cdot \xrightarrow{a} \cdot \Rightarrow p'$ and $\xRightarrow{\tau}$ is equivalent to $\Rightarrow$. A weak transition from p to p' that performs actions $\vec{a} = [a_0, \ldots, a_n] \in L^*$ is defined as $p \Rightarrow \cdot \xRightarrow{a_0} \cdots \xRightarrow{a_n} p'$ and is denoted by $p \xRightarrow{\vec{a}} p'$. Weak transitions over an empty list of actions may take multiple silent steps rather than just zero steps.

Two programs can be compared by showing that one mimics the other indefinitely, which is defined with respect to a LTS.

Definition 1 $(p \leq q)$. $\mathcal{R} \subseteq S \times S$ *is a* simulation *when for any* $(p, q) \in \mathcal{R}$,
- *if* $p \xrightarrow{\alpha} p'$, *then* $q \xRightarrow{\alpha} q'$ *and* $(p', q') \in \mathcal{R}$ *for some* q'.

State q simulates p, *written* $p \leq q$ *(or equivalently* $q \geq p$*), iff there exists a* simulation $\mathcal{R}$ *and* $(p, q) \in \mathcal{R}$.

Many verified compilers are founded on bisimulation. It holds between two programs when each mimics (simulates) the other indefinitely, such that all pairs of transitional states continue to be bisimilar. $\mathcal{R}^{-1}$ is the inverse of $\mathcal{R}$: $(q, p) \in \mathcal{R}^{-1}$ iff $(p, q) \in \mathcal{R}$ for any p and q.

Definition 2 $(p \approx q)$. $\mathcal{R}$ *is a* bisimulation *when* $\mathcal{R}$ *and* $\mathcal{R}^{-1}$ *are simulations.* p *and* q *are* bisimilar, *written* $p \approx q$, *iff* $(p, q) \in \mathcal{R}$ *for some bisimulation* $\mathcal{R}$.

It is easy to prove that state q_0 simulates p_0 in Fig. 1. The simulation relation is $\{(p_i, q_i) \mid 0 \leq i \leq 6\}$. However, simulation does not hold in the other direction:

Lemma 1. *Program* (2.1) *does not simulate* (2.2): $p_0 \not\geq q_0$.

Proof. By contradiction: assume $p_0 \geq q_0$. We take step $q_0 \xrightarrow{0} q_7$ without committing to print either 1 or 2. By assumption, p_0 must be able to mimic this action, thus $p_0 \xRightarrow{0} p'$ and $p' \geq q_7$ for some p'. Fig. 1a shows that p' is either p_3 or p_4. In either case, q_7 may perform an action that p' is not capable of, thus $p' \not\geq q_7$. □

Because simulation fails in this direction, we choose a weaker relation for parallelization. In particular, one that preserves as many of the strong properties of bisimulation as possible, such as a co-inductive proof method and the fact that related programs continue to mimic each other during execution. Note that when the simulation fails, it is possible for the parallel program to eventually step to a state where simulation can be reestablished: from state q_7 to either q_3 or q_4. It turns out that this "eventuality" holds for parallelization in general, and thus we formalize this idea as *eventual simulation*.

Definition 3 $(p \lesssim q)$. $\mathcal{R}$ *is an* eventual simulation *when for any* $(p, q) \in \mathcal{R}$,
- *if* $p \xRightarrow{\vec{a}} p'$, *then* $p' \Rightarrow p''$, $q \xRightarrow{\vec{a}} q''$, *and* $(p'', q'') \in \mathcal{R}$ *for some* p'' *and* q''.

State q eventually simulates p, written $p \precsim q$ (or $q \succsim p$), if there exists a simulation $\mathcal{R}$ whose inverse is an eventual simulation and $(p, q) \in \mathcal{R}$.

Eventual similarity is like bisimilarity – and unlike plain similarity – in that both programs mimic each other indefinitely, but we still call it a "similarity" because it is asymmetric. It differs from them by considering multiple actions at once. Bisimilarity implies eventual similarity, eventual similarity is reflexive and transitive, and crucially, it holds for a large class of parallelizing transformations in addition to our simple example. Further properties are explored in Section 4.

Lemma 2. *If $p \approx q$, then $p \succsim q$ (and $p \precsim q$ because $\approx$ is symmetric).*

Lemma 3. *$\succsim$ is reflexive and transitive.*

Lemma 4. *The programs in Fig. 1 are eventually similar: $p_0 \precsim q_0$.*

Proof. We select relation $\mathcal{R} = \{(p_i, q_i) \mid 0 \le i \le 6\}$. Trivially, $(p_0, q_0) \in \mathcal{R}$ and $\mathcal{R}$ is a simulation. Finally, we prove that $\mathcal{R}^{-1}$ is an eventual simulation. The interesting case is for $(p_0, q_0) \in \mathcal{R}$, when $q_0 \overset{0}{\Rightarrow} q_7$. In response, we have p_0 follow by $p_0 \overset{0}{\Rightarrow} p_3$ and $q_7 \Rightarrow q_3$. (We may have instead chosen to step to p_4 and q_4.)   □

3   CCS-Seq

We now investigate parallelization for an extension of the Calculus of Communicating Systems (CCS) [6]. CCS is widely used as a model for analyzing bisimulation relations and the behavior of programs and systems with multiple concurrent agents acting in concert via message passing and synchronization. However, we must extend CCS with a sequential composition operator in order to model parallelizing transformations. Furthermore, implementing asynchronous communication on top of CCS-style synchronous channels is tedious and not modular (requiring auxiliary threads for buffering), so we replace its channels with semaphores. We refer to this language as CCS-Seq.

$$\alpha ::= \tau \mid a \mid \bar{a}$$
$$P ::= \mathbf{0} \mid P + P \mid P \mid P \mid \alpha.P \mid !P \mid P; P \mid va{:}n.P$$

Metavariables P, Q, M, N, and R refer to processes; a is the name of a semaphore; α is an action (τ is internal), and n is a natural number. Figure 2 lists the operational semantics. Action prefixing, $\alpha.P$, emits action α and resolves to P. $\mathbf{0}$ is a terminated process (we abbreviate $\alpha.\mathbf{0}$ as α), $P \mid Q$ is parallel composition, $P; Q$ is sequential composition, and $P + Q$ represents a choice between executing either P or Q. A process may create infinite, parallel copies of itself by replication: $!P$. Although we do not use replication directly in this paper, its presence gives the language "teeth" – so that proving termination properties is not trivial (for this purpose, a termination rule for replication is unnecessary).

Restriction, $va{:}n.P$, declares that a is a semaphore, local to P, with state n. It is a way of introducing a fresh semaphore name that is hidden from any

$$\frac{P \xrightarrow{\alpha} P'}{P;Q \xrightarrow{\alpha} P';Q} \qquad \frac{P \xrightarrow{\alpha} P'}{P\,|\,Q \xrightarrow{\alpha} P'\,|\,Q} \qquad \frac{Q \xrightarrow{\alpha} Q'}{P\,|\,Q \xrightarrow{\alpha} P\,|\,Q'} \qquad \frac{P \xrightarrow{\alpha} P'}{P+Q \xrightarrow{\alpha} P'} \qquad \frac{Q \xrightarrow{\alpha} Q'}{P+Q \xrightarrow{\alpha} Q'}$$

$$\frac{P \xrightarrow{\alpha} P' \quad \alpha \notin \{a,\bar{a}\}}{va\!:\!n.P \xrightarrow{\alpha} va\!:\!n.P'} \qquad \frac{P \xrightarrow{\bar{a}} P'}{va\!:\!n.P \to va\!:\!(n+1).P'} \qquad \frac{P \xrightarrow{a} P'}{va\!:\!(n+1).P \to va\!:\!n.P'}$$

$$\frac{}{\alpha.P \xrightarrow{\alpha} P} \qquad \frac{P\,\|\,!P \xrightarrow{\alpha} P'}{!P \xrightarrow{\alpha} P'} \qquad \frac{}{0;P \to P} \qquad \frac{}{va\!:\!n.0 \to 0} \qquad \frac{}{0\,|\,0 \to 0} \qquad \frac{}{0+0 \to 0}$$

Fig. 2. Operational semantics for CCS-Seq

observer or process outside of P. When P emits action $\bar{a}$ or a, the semaphore is incremented or decremented, respectively, and the observed action is τ. If the semaphore count is zero, then P cannot emit a to decrement the semaphore until the count becomes nonzero. We will reason about programs with an arbitrary number of semaphores, so we define a vectorized form of restriction.

Definition 4. $va_1\!:\!n_1.\dots.va_k\!:\!n_k.P$ *is abbreviated as* $\varUpsilon\vec{a}\!:\!\vec{n}.P$.

We define a LTS for CCS-Seq in the usual way. Processes are synonymous with states, an action is either a or $\bar{a}$ for any semaphore a, and the set of single steps defined in Fig. 2 is the transition relation.

Since we have added sequential semantics, bisimulation alone is not a congruence for sequential composition. For example, even though $0 \approx !\tau$, it is the case that $0; a \not\approx !\tau; a$. This is easy to fix by adding *termination sensitivity*.

Definition 5. *A relation* $\mathcal{R}$ *is* one-way termination sensitive *when for any* $(p,q) \in \mathcal{R}$, *if* p *is* halted *(for CCS-Seq, if* $p = 0$*), then* $q \Rightarrow p$. $\mathcal{R}$ *is termination sensitive if* $\mathcal{R}$ *and* $\mathcal{R}^{-1}$ *are one-way termination sensitive.*

Definition 6 $(p \approx_\downarrow q)$. *States* p *and* q *are termination sensitive bisimilar, written* $p \approx_\downarrow q$, *if there exists a termination sensitive bisimulation* $\mathcal{R}$ *such that* $(p,q) \in \mathcal{R}$.

Definition 7 $(p \precsim_\downarrow q)$. *State* q *termination sensitive eventually simulates* p, *written* $p \precsim_\downarrow q$ *(or* $q \succsim_\downarrow p$*), if there exists a termination sensitive simulation* $\mathcal{R}$, *whose inverse is an eventual simulation, such that* $(p,q) \in \mathcal{R}$.

Lemma 5 (Compositional properties of $\approx$, $\approx_\downarrow$, $\precsim$, **and** $\precsim_\downarrow$**).** *Where* $\equiv$ *ranges over* $\{\approx, \approx_\downarrow, \precsim, \precsim_\downarrow\}$; *if* $P \equiv Q$ *then:* $P\,|\,R \equiv Q\,|\,R$, $\alpha.P \equiv \alpha.Q$, $!P \equiv !Q$, $va\!:\!n.P \equiv va\!:\!n.Q$, $R;P \equiv R;Q$, *and* $\tau.P + R \equiv \tau.Q + R$. *If* $P \approx_\downarrow Q$, *then* $P;R \approx_\downarrow Q;R$. *If* $P \precsim_\downarrow Q$, *then* $P;R \precsim_\downarrow Q;R$.

Before presenting a general parallelization transformation for sequential composition, we warm up with a simpler form of parallelization in the following lemma. By targeting the sequentialism found in action prefixing, the lemma suggests that eventual simulation may have some uses in plain CCS as well.

Lemma 6. $\tau.(P\,|\,Q) + \tau.(P\,|\,R) \precsim_\downarrow P\,|\,(\tau.Q + \tau.R)$.

Proof. We choose $\mathcal{R} = \bigcup_P(\tau.(P\,|\,Q) + \tau.(P\,|\,R), P\,|\,(\tau.Q + \tau.R)) \cup \mathcal{I}$, where $\mathcal{I}$ is the identity relation. Showing that $\mathcal{R}$ is a simulation and that $\mathcal{I}$ is an eventual simulation is trivial. We show that the first part of $\mathcal{R}$ is an eventual simulation. The right program may either 1) choose between Q or R, or 2) avoid choosing and only run P: $P\,|\,(\tau.Q + \tau.R) \overset{\alpha}{\Rightarrow} P'\,|\,(\tau.Q + \tau.R)$. *Case 1:* the left program can converge to the same state. *Case 2:* we arbitrarily pick Q such that $P'\,|\,(\tau.Q + \tau.R) \Rightarrow P'\,|\,Q$; the left program can then converge to the same state. □

If we choose $P = 0.0$, $Q = 1.0$, and $R = 2.0$, then sequential program (2.1) roughly corresponds to $\tau.(0\,|\,1) + \tau.(0\,|\,2)$ and parallel program (2.2) roughly corresponds to $0\,|\,(\tau.1 + \tau.2)$. (The "rough" difference is that action 0 is allowed to interleave with actions 1 and 2 in more ways than in Fig. 1.)

3.1 The Parallelization Transformation

The key idea of Lemma 6 is that the more-parallel program may be able to perform some action (by executing P) without making an internal choice (between Q and R). However, the more-sequential program will not be able to simulate this (by running P) before first committing itself to one of these choices. Eventual simulation holds because the more-parallel program can take extra steps to resolve the same choices so that both programs converge to the same state.

This same idea applies to our key result: a general parallelizing transformation between sequential and parallel programs. An obvious schema (despite the subtle premises) for a parallelizing transformation converts two programs in sequence into two programs in parallel, which we describe here. (Section 6 goes into further detail and provides proof sketches.)

Proposition 1. *If P may always silently terminate (modulo $\vec{a}$), P and Q do not both decrement any of the same semaphores, and either P or Q never performs an observable action (modulo $\vec{a}$), then $\Upsilon\vec{a}{:}\vec{n}.(P;Q) \lesssim_\downarrow \Upsilon\vec{a}{:}\vec{n}.(P\,|\,Q)$.*

We specify a list of actions, $\vec{a}$, to facilitate unobservable communication between P and Q; when we state that an execution is "silent", we mean that only hidden actions (i.e. those named by $\vec{a}$) may be performed. If P "may always silently terminate", then no matter how P executes (even performing observable actions), we can always ask it to then silently transition to a terminated state. This allows the sequential program, in response to the parallel program executing Q before P terminates, to "catch up" by forcing both to terminate P (possibly making some arbitrary internal choices in doing so) and converging to the same state. (Although this premise is complex, it is more general than simply not allowing P to diverge at all and requiring P to be completely silent).

However, it is not enough for P to terminate in isolation because Q will interleave with P. The second premise ensures that Q cannot block P by stealing a semaphore and causing it to deadlock.[3] The last premise, where either P or

---

[3] If the language were extended with shared queues of values, this condition would also prevent Q from interfering by stealing values intended for P.

Q must be silent, prevents parallelization from resulting in new interleavings of observable actions because such a difference would be trivial to detect.

We also prove a transformation that combines two parallel programs. Two processes, P_1 and P_2, *coterminate* (modulo $\vec{a}$) when (1) $P_1 \mid P_2 \overset{\alpha}{\Rightarrow} \mathbf{0} \mid P_2'$ implies P_2' may always silently terminate (modulo $\vec{a}$); and when (2) $P_1 \mid P_2 \overset{\alpha}{\Rightarrow} P_1' \mid \mathbf{0}$ implies P_1' may always silently terminate (modulo $\vec{a}$).

Proposition 2. *If P_1 and P_2 coterminate (modulo $\vec{a}$), P_1 and P_3 do not decrement any of the same semaphores as P_2 and P_4, and P_2 and P_4 never perform an observable action (modulo $\vec{a}$), then $\Upsilon \vec{a} : \vec{n}.((P_1 \mid P_2); (P_3 \mid P_4)) \precsim_\downarrow \Upsilon \vec{a} : \vec{n}.((P_1; P_3) \mid (P_2; P_4))$.*

Proposition 2 is strictly more general than Prop. 1 because it allows P_1 and P_2 to coordinate termination (or not terminate at all). It results in the parallelization of P_3 with P_2 and P_4 with P_1.

Now we show that Prop. 1 is sufficient to parallelize a CCS-Seq implementation of program (2.1) into (2.2).

Lemma 7. *Given*

$$M = \Upsilon[e, f] : [0, 0]. \left(\Upsilon[c, d] : [0, 0]. \left(\tau.\bar{c} + \tau.\bar{d}; \bar{0}; c.\bar{e} + d.\bar{f} \right); e.\bar{1} + f.\bar{2} \right)$$

$$N = \Upsilon[e, f] : [0, 0]. \left(\Upsilon[c, d] : [0, 0]. \left((\tau.\bar{c} + \tau.\bar{d}) \mid (\bar{0}; c.\bar{e} + d.\bar{f}) \right); e.\bar{1} + f.\bar{2} \right),$$

where M and N correspond to (2.1) and (2.2), respectively: $M \precsim_\downarrow N$.

Proof. By Prop. 1 and congruence. $\tau.\bar{c} + \tau.\bar{d}$ silently terminates; its actions, $\bar{c}$ and $\bar{d}$, are hidden by the semaphore restriction. Finally, $\tau.\bar{c} + \tau.\bar{d}$ does not decrement any semaphores and thus the process does not interfere with $\bar{0}; c.\bar{e} + d.\bar{f}$.   □

4   Pursuing Symmetry: Contrasimulation is a Congruence

$\succsim$ is reflexive, transitive, and compositional, but it is not a congruence because it lacks symmetry. Adding symmetry would enable some useful optimization strategies, like commuting two blocks of instructions by first parallelizing them, swapping the threads, and then applying parallelization in reverse to sequentialize them. Although symmetry is not always desirable (e.g. when refining unspecified behavior), there is no clear benefit to the asymmetry in eventual similarity. In fact, it is even asymmetric in the wrong direction – it allows more interleavings to be *added*, not refined. To obtain symmetry, we might attempt to define a relation where eventual simulation holds in both directions.

Definition 8. *Programs p and q are eventually bisimilar, written $p \overset{\approx}{\sim} q$, if there exists an $\mathcal{R}$ such that $\mathcal{R}$ and $\mathcal{R}^{-1}$ are eventual simulations and $(p, q) \in \mathcal{R}$.*

Lemma 8. *If $p \precsim_\downarrow q$ or $p \succsim_\downarrow q$, then $p \overset{\approx}{\sim} q$.*

However, $\overset{\approx}{\sim}$ is not transitive for divergent LTSs, as demonstrated in Fig. 3, limiting its use to languages without infinite loops or recursion. (It is transitive

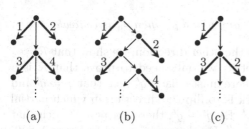

(a)   (b)   (c)

Fig. 3. Counterexample of transitivity for $\overset{\cdots}{\approx}$. Each transition diagram depicts an infinite series of states where the actions continue to grow. Although a $\overset{\cdots}{\approx}$ b and b $\overset{\cdots}{\approx}$ c, it is not the case that a $\overset{\cdots}{\approx}$ c.

for LTSs that do not diverge.) Parrow and Sjödin worked on a similar problem that also needed a coarser view of internal nondeterminism than bisimulation afforded [8]. They developed *coupled simulation* to relate the behavior of multiway distributed internal choice to a reference implementation that resolves all choices in one synchronous step. Coupled simulation is finer than eventual simulation and is also not transitive for divergent LTSs. Should they need transitivity, they suggested use of contrasimulation [3], which contains coupled simulation.

Definition 9 $(p \approx_c q)$. $\mathcal{R}$ *is a* contrasimulation *when for any* $(p, q) \in \mathcal{R}$,
- *if* $p \overset{\vec{a}}{\Rightarrow} p'$, *then* $q \overset{\vec{a}}{\Rightarrow} q'$ *and* $(p', q') \in \mathcal{R}^{-1}$ *for some* q'.
Note the reversal of $\mathcal{R}$. *State* q *partially contrasimulates* p *iff there exists a contrasimulation* $\mathcal{R}$ *such that* $(p, q) \in \mathcal{R}$. *States* p *and* q *are* contrasimilar, *written* $p \approx_c q$, *iff there exists a contrasimulation* $\mathcal{R}$ *such that* $(p, q) \in \mathcal{R} \cap \mathcal{R}^{-1}$.

Lemma 9. *If* $p \overset{\cdots}{\approx} q$, *then* $p \approx_c q$.

When two programs are contrasimilar, they *take turns* simulating each other indefinitely, starting with either program. Unlike bisimilarity, the relation between two programs only needs to be symmetric for the initial states. Crucially, contrasimulation is an equivalence.

Lemma 10. *Contrasimulation is reflexive, symmetric, and transitive.*

As a sanity check, contrasimulation is stronger than *trace equivalence*:

Definition 10 $(p \approx_{tr} q)$. p *and* q *are* trace equivalent, *written* $p \approx_{tr} q$, *when*
- *if* $p \overset{\vec{a}}{\Rightarrow} p'$, *then there exists a* q' *such that* $q \overset{\vec{a}}{\Rightarrow} q'$
- *if* $q \overset{\vec{a}}{\Rightarrow} q'$, *then there exists a* p' *such that* $p \overset{\vec{a}}{\Rightarrow} p'$.

Lemma 11. *If* $p \approx_c q$, *then* $p \approx_{tr} q$.

Trace equivalence is a sufficient soundness criterion in some situations, but it is useful to prove a simulation relation because trace equivalence is not a congruence (e.g. for parallel composition) and the co-inductive proof method can be easier to work with. Of course, when the LTS is deterministic, these relationships are all equivalent to bisimulation. But we can prove that contrasimulation is equivalent to bisimulation when only internal transitions are deterministic, which suggests that it is not significantly weaker than necessary in order to deal with the interleaving of internal nondeterminism with observable actions.

Theorem 1. *If $p \to p'$ implies $p \approx p'$ for any p and p', then $\approx_c$ is equivalent to $\approx$.*

Proof. Lemmas 8 and 9 prove $\approx \subseteq \approx_c$. In the other direction, we show that $\mathcal{R} = \approx_c$ is a bisimulation. Because it is symmetric, we only need to prove that $\mathcal{R}$ is a simulation. Assume $p \approx_c q$ and $p \overset{\alpha}{\to} p'$; there must exist a q' such that $q \overset{\alpha}{\Rightarrow} q'$ and p' partially contrasimulates q'. The trick is to flip the direction in which partial contrasimulation holds, implying $p' \approx_c q'$. By $q' \Rightarrow q'$, there exists a p'' such that $p' \Rightarrow p''$ and q' partially contrasimulates p''. By the premise and Lemmas 2, 8, 9, and 10, $p'' \approx p'$, q' partially contrasimulates p', and thus $p' \approx_c q'$.   □

As a corollary, $\gtrsim$ collapses to $\approx$ when there is no internal nondeterminism. We define termination sensitive contrasimulation and then show that contrasimilarity has the same compositional properties as bisimilarity and eventual similarity.

Definition 11 $(p \approx_{\downarrow c} q)$. *States p and q are termination sensitive contrasimilar, written $p \approx_{\downarrow c} q$, iff there exists a one-way termination sensitive contrasimulation $\mathcal{R}$ such that $(p, q) \in \mathcal{R} \cap \mathcal{R}^{-1}$.*

Lemma 12 (Compositional properties of $\approx_c$ and $\approx_{\downarrow c}$). *Where $\equiv$ ranges over $\{\approx_c, \approx_{\downarrow c}\}$; if $P \equiv Q$ then: $P \mid R \equiv Q \mid R$, $\alpha.P \equiv \alpha.Q$, $!P \equiv !Q$, $va{:}n.P \equiv va{:}n.Q$, $R; P \equiv R; Q$, and $\tau.P + R \equiv \tau.Q + R$. If $P \approx_{\downarrow c} Q$, then $P; R \approx_{\downarrow c} Q; R$.*

Like coupled similarity, contrasimilarity is congruent for + when the processes are equally *stable*.

Definition 12 (stable p). *p is stable if there does not exist a p' such that $p \to p'$.*

Lemma 13. *If $P \approx_c Q$ and (stable P iff stable Q), then $P + R \approx_c Q + R$. (And likewise for $\approx_{\downarrow c}$).*

Compiled languages do not usually allow "mixed choice", where one option is stable and the other is not, so both bisimulation and contrasimulation are often full congruences in practice. Thus we can build correct, *modular* optimizations based on contrasimulation (or bisimulation) using the above congruence results. Because bisimulation is finer than contrasimulation, all of its algebraic properties for CCS (and CCS-Seq) hold for contrasimulation.

Voorhoeve and Mauw investigate further properties of contrasimulation and describe an axiomatization for CCS [12]. Their axiomatization relates stable internal choice for an observable action into the action followed by internal choice.

Lemma 14. $a.P + a.Q \approx_c a.(\tau.P + \tau.Q)$.
Interestingly, this holds for $\gtrsim$ as well.

Combined with a few algebraic properties of bisimulation, like $\tau.P + P \approx \tau.P$, Lemma 14 proves equivalence between programs (2.1) and (2.2).

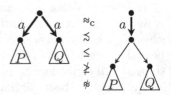

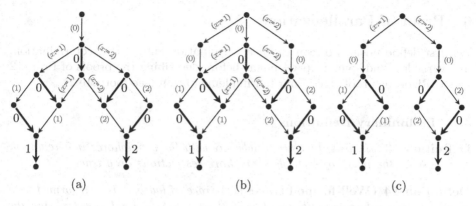

Fig. 4. Delayed observation semantics for programs (5.1), (2.2), and (2.1)

5 Delayed Observations

Contrasimulation effectively allows a program to delay an internal choice until after an observable action. But this does not allow the observation to be fully commuted. Consider the sequential program that prints 0 *before* choosing a value for x,

Fig. 5

$$\text{print 0; } x := \text{either 1 or 2; print x,} \qquad (5.1)$$

whose semantics are given by Fig. 5. Intuitively, (5.1) and (2.2) should be equivalent, but contrasimilarity does not hold between their corresponding transition diagrams, Figs. 1b and 5. If Fig. 1b chooses x := 1, then Fig. 5 will be unable to commit to the same choice without first observing 0. We have yet to find a satisfactory equivalence that holds for Fig. 5.

If (5.1), (2.2), and (2.1) were C programs, however, their semantics would be subtly different. The specification of many IO operations in C, such as `printf`, allows output to be buffered before being printed to the console. In other words, observable actions may be delayed.

We say that a LTS has *delayed observations* when output is queued before appearing on the screen at a nondeterministic point in the future. Figure 4 gives the semantics of programs (5.1), (2.2), and (2.1) using delayed observations. Figures 4b and 4c are contrasimilar. Moreover, Figs. 4a and 4b are now *bisimilar*.

Although contrasimulation cannot directly allow observations to be delayed until after internal choice, we can side-step the issue by choosing a semantics with delayed observations. In such a setting, Props. 1 and 2 can be used to parallelize programs such as (5.1). They also become somewhat easier to use: by delaying all observations until after termination, proving termination is enough to prove *silent* termination. However, a limitation remains: not all observations may be delayed. For example, C's `fflush` forces immediate observation, thus commuting it with internal choice would not maintain a contrasimulation.

6   Proof of Parallelization

We first define notions of convergence, termination entailment, cotermination, free variables, and some helper lemmas before describing the proofs of Props. 2 and 1. Rigorous proofs are in our full implementation (see footnote 1).

6.1   Preliminary Definitions

Definition 13. $\vec{\alpha} - \vec{a}$ is the trace of labels $\vec{\alpha}$ with the semaphores in $\vec{a}$ removed. It represents the result of multiple semaphore restrictions on a trace.

Definition 14 (Well-formed traces). A trace of labels $\vec{\alpha}$ is well-formed with respect to semaphore a with count n if there exists a final count n' for the semaphore such that the trace does not decrement the semaphore below a count of 0. We denote this as $n \overset{\vec{\alpha}}{\leadsto}_a n'$ and define it recursively on the structure of $\vec{\alpha}$.

$$
\begin{array}{ll}
n \overset{[]}{\leadsto}_a n' & \text{if } n' = n \\
n \overset{\alpha'::\vec{\alpha}}{\leadsto}_a n' & \text{if } \alpha' \notin \{a, \bar{a}\} \text{ and } n \overset{\vec{\alpha}}{\leadsto}_a n' \\
n + 1 \overset{a::\vec{\alpha}}{\leadsto}_a n' & \text{if } n \overset{\vec{\alpha}}{\leadsto}_a n' & \text{(decrements } a\text{)} \\
n \overset{\bar{a}::\vec{\alpha}}{\leadsto}_a n' & \text{if } n + 1 \overset{\vec{\alpha}}{\leadsto}_a n' & \text{(increments } a\text{)}.
\end{array}
$$

We then define a well-formed trace with respect to a list of semaphores, $\vec{n} \overset{\vec{\alpha}}{\leadsto}_{\vec{a}} \vec{n}'$, recursively on the structure of $\vec{a}$.

$$
\begin{array}{ll}
[] \overset{\vec{\alpha}}{\leadsto}_{[]} [] & \text{always} \\
(n :: \vec{m}) \overset{\vec{\alpha}}{\leadsto}_{a::\vec{b}} (n' :: \vec{m}') & \text{if } n \overset{\vec{\alpha} - \vec{b}}{\leadsto}_a n' \text{ and } \vec{m} \overset{\vec{\alpha}}{\leadsto}_{\vec{b}} \vec{m}'.
\end{array}
$$

Definition 14 appears only in the next definition. However, it is used extensively by helper lemmas in our Coq proof development to separate the details of how a particular process runs from how its semaphores are used. For example, to state that a sequential and parallelized program use their semaphores in the same way despite their syntactic difference.

Silent termination and cotermination were introduced in Section 3.1 for use in Props. 1 and 2. We now give concrete definitions; recall the notation for vectorized semaphore restriction from Definition 4.

Definition 15 (Silent termination). P silently terminates, written $P \Downarrow^{\vec{a}:\vec{n}}$, if for any P' and $\vec{\alpha}$, $P \overset{\vec{\alpha}}{\Rightarrow} P'$ implies $\Upsilon \vec{a}:\vec{n}'.P' \Rightarrow \mathbf{0}$ and $\vec{n} \overset{\vec{\alpha}}{\leadsto}_{\vec{a}} \vec{n}'$ for some $\vec{n}'$.

Definition 16 (Termination entailment & cotermination). P_1 entails the termination of P_2, written $P_1 \Downarrow\Downarrow^{\vec{a}:\vec{n}} P_2$, if $\Upsilon \vec{a}:\vec{n}.(P_1 \mid P_2) \overset{\vec{\alpha}}{\Rightarrow} \Upsilon \vec{a}':\vec{n}'.(\mathbf{0} \mid P_2')$ implies $P_2' \Downarrow^{\vec{a}':\vec{n}'}$. P_1 and P_2 coterminate, written $P_1 \Updownarrow^{\vec{a}:\vec{n}} P_2$, iff $P_1 \Downarrow\Downarrow^{\vec{a}:\vec{n}} P_2$ and $P_2 \Downarrow\Downarrow^{\vec{a}:\vec{n}} P_1$.

In order to state noninterference properties between processes, we define functions to find the sets of free variables used to increment semaphores, decrement semaphores, and the union of each within a process.

Definition 17 (Free observable actions).

$$\mathrm{fa}(\bar{a}.P) = \{\bar{a}\} \cup \mathrm{fa}(P) \qquad \mathrm{fa}(\mathbf{0}) = \{\} \qquad \mathrm{fa}(P_1 + P_2) = \mathrm{fa}(P_1) \cup \mathrm{fa}(P_2)$$
$$\mathrm{fa}(a.P) = \{a\} \cup \mathrm{fa}(P) \qquad \mathrm{fa}(!P) = \mathrm{fa}(P) \qquad \mathrm{fa}(P_1 \,|\, P_2) = \mathrm{fa}(P_1) \cup \mathrm{fa}(P_2)$$
$$\mathrm{fa}(\tau.P) = \mathrm{fa}(P) \qquad \mathrm{fa}(va{:}n.P) = \mathrm{fa}(P) \setminus a \setminus \bar{a} \qquad \mathrm{fa}(P_1; P_2) = \mathrm{fa}(P_1) \cup \mathrm{fa}(P_2)$$

Definition 18 (Free variables: increment, decrement, and both).

$$\mathrm{fv_V}(P) = \{a \,|\, \bar{a} \in \mathrm{fa}(P)\} \qquad \mathrm{fv_P}(P) = \{a \,|\, a \in \mathrm{fa}(P)\} \qquad \mathrm{fv}(P) = \mathrm{fv_V}(P) \cup \mathrm{fv_P}(P)$$

The semaphores that a process, P, can increment and decrement are respectively limited by $\mathrm{fv_V}(P)$ and $\mathrm{fv_P}(P)$.

6.2 Proof

This first lemma performs case analysis on a single step of a "sequential" program in order to show that the parallelized program can perform the same action.

Lemma 15. *If* $\Upsilon \vec{a}{:}\vec{n}.((P_1 \,|\, P_2); (P_3 \,|\, P_4)) \xrightarrow{\alpha} p'$, *then either*
- *there exists* $\vec{n}'$, P_1', *and* P_2' *such that*
  - $p' = \Upsilon \vec{a}{:}\vec{n}.((P_1' \,|\, P_2'); (P_3 \,|\, P_4))$ *and*
  - $\Upsilon \vec{a}{:}\vec{n}.(P_1 \,|\, P_2) \xrightarrow{\alpha} \Upsilon \vec{a}{:}\vec{n}'.(P_1' \,|\, P_2')$
  - *(and thus* $\Upsilon \vec{a}{:}\vec{n}.((P_1; P_3) \,|\, (P_2; P_4)) \xrightarrow{\alpha} \Upsilon \vec{a}{:}\vec{n}'.((P_1'; P_3) \,|\, (P_2'; P_4)))$,
- *or* $P_1 = P_2 = \mathbf{0}$ *and* $p' = \Upsilon \vec{a}{:}\vec{n}.\mathbf{0}; (P_3 \,|\, P_4)$.

In the following lemma, we look at an execution of the parallelized program over multiple steps and show that the sequential program can either simulate it directly, or that there exists a future state where they can converge.

Lemma 16. *If* $\Upsilon \vec{a}{:}\vec{n}.((P_1; P_3) \,|\, (P_2; P_4)) \xRightarrow{\vec{\alpha}} p'$, $P_1 \updownarrow^{\vec{a}{:}\vec{n}} P_2$, $\mathrm{fv}(P_2; P_4) \subseteq \vec{a}$, *and* $\mathrm{fv_P}(P_1; P_3) \cap \mathrm{fv_P}(P_2; P_4) = \varnothing$, *then either*
- *there exists* $\vec{n}'$, P_1', *and* P_2' *such that*
  - $p' = \Upsilon \vec{a}{:}\vec{n}'.((P_1'; P_3) \,|\, (P_2'; P_4))$,
  - $\Upsilon \vec{a}{:}\vec{n}.P_1 \,|\, P_2 \xRightarrow{\vec{\alpha}} \Upsilon \vec{a}{:}\vec{n}.(P_1' \,|\, P_2')$
  - *(and thus* $\Upsilon \vec{a}{:}\vec{n}.((P_1 \,|\, P_2); (P_3 \,|\, P_4)) \xRightarrow{\vec{\alpha}} \Upsilon \vec{a}{:}\vec{n}.((P_1' \,|\, P_2'); (P_3 \,|\, P_4)))$;
- *or there exists* p'' *such that*
  - $p' \Rightarrow p''$ *and*
  - $\Upsilon \vec{a}{:}\vec{n}.((P_1 \,|\, P_2); (P_3 \,|\, P_4)) \xRightarrow{\vec{\alpha}} p''$.

Proof. We consider three outcomes of running $\Upsilon \vec{a} : \vec{n}.((P_1; P_3) \,|\, (P_2; P_4)) \xRightarrow{\vec{\alpha}} p'$, where $*_n$ represents the final state that process P_n reaches (if it runs but does not terminate). We focus on the second case as the other two are relatively easy.

$$P_3 \,|\, (*_2; P_4) \cdots\cdots\!\!\twoheadrightarrow *_3 \,|\, (\mathbf{0}; P_4)$$
$$(P_1; P_3) \,|\, (P_2; P_4) \cdots\!\!\blacktriangleright (*_1; P_3) \,|\, (*_2; P_4) \qquad\qquad\qquad *_3 \,|\, *_4 \cdots\!\!\blacktriangleright \mathbf{0}$$
$$(*_1; P_3) \,|\, P_4 \cdots\cdots\!\!\blacktriangleright (\mathbf{0}; P_3) \,|\, *_4$$

$$\underbrace{\qquad\qquad\qquad\qquad\qquad}_{1} \qquad \underbrace{\qquad}_{2} \underbrace{\qquad\qquad}_{3}$$

Case 1. $P_1 | P_2 \xrightarrow{\vec{\alpha}_1} p_1 | p_2$ and $\vec{\alpha} = \vec{\alpha}_1 - \vec{a}$ for some p_1 and p_2. The sequential program runs $P_1 | P_2$ to match the actions without converging to the same state.

Case 2. The parallel program has the form of either $p_3 | (p_2; P_4)$ or $(p_1; P_3) | p_4$. We consider only the first (top) form; the second is similar. Because P_1 terminated, p_2 can silently terminate. This yields $\Upsilon\vec{a} : \vec{n}.(p_3 | (p_2; P_4)) \Rightarrow \Upsilon\vec{a} : \vec{n}'.(p_3 | P_4)$ for some $\vec{n}'$. However, we need P_2 to terminate *before* P_3 even runs in order for the sequential program to mimic the behavior. This is possible if P_2 did not emit observable actions (a premise of this lemma) and P_3 did not influence P_2 as they interleaved. The last could only have happened if P_3 incremented a semaphore on which P_2 would otherwise deadlock. Because P_2 was capable of terminating by the time P_3 ran, such deadlocking was impossible. Thus we know we can run $\Upsilon\vec{a} : \vec{n}.(P_1 | P_2) \xrightarrow{\vec{\alpha}_{12}} \Upsilon\vec{a} : \vec{n}'^0.(0 | 0)$, followed by $\Upsilon\vec{a} : \vec{n}'^0.(P_3 | P_4) \xrightarrow{\vec{\alpha}_3} \Upsilon\vec{a} : \vec{n}'.(p_3 | P_4)$, for some $\vec{n}'^0$ and such that $\vec{\alpha}$ is equal to some $\vec{\alpha}_{12}$ appended with $\vec{\alpha}_3$. Both programs can converge to state $p'' = \Upsilon\vec{a} : \vec{n}'.(p_3 | P_4)$.

Case 3. $P_1 | P_2 \xrightarrow{\vec{\alpha}_1} 0 | 0$, $P_3 | P_4 \xrightarrow{\vec{\alpha}_2} p_3 | p_4$, and $\vec{\alpha} = (\vec{\alpha}_1 \cdot \vec{\alpha}_2) - \vec{a}$ for some p_3, p_4, $\vec{\alpha}_1$, and $\vec{\alpha}_2$. The sequential program runs $P_1 | P_2$ to termination and then runs $P_3 | P_4$ to converge to the same state as the parallel program.   □

Theorem 2 (Proof of Prop. 2). *If*
- *P_1 and P_2 coterminate: $P_1 \updownarrow\!\!\updownarrow^{\vec{a}:\vec{n}} P_2$;*
- *the processes do not interfere: $\mathrm{fv_P}(P_1; P_3) \cap \mathrm{fv_P}(P_2; P_4) = \varnothing$; and*
- *P_2 and P_4 cannot be observed: $\mathrm{fv}(P_2) \cup \mathrm{fv}(P_4) \subseteq \vec{a}$,*

then $\Upsilon\vec{a} : \vec{n}.\left(\begin{pmatrix} (P_1 | P_2); \\ (P_3 | P_4) \end{pmatrix} \right) \lesssim_{\downarrow} \Upsilon\vec{a} : \vec{n}.\left(\begin{pmatrix} P_1; \\ P_3 \end{pmatrix} \middle| \begin{pmatrix} P_2; \\ P_4 \end{pmatrix} \right).$

Proof. We choose $\mathcal{R} = \{(p,q) \mid \exists \vec{a}, \vec{n}, P_1, P_2. \ p = \Upsilon\vec{a} : \vec{n}.((P_1; P_3) | (P_2; P_4)) \ \wedge$ $q = \Upsilon\vec{a} : \vec{n}.((P_1 | P_2); (P_3 | P_4)) \ \wedge \ P_1 \updownarrow\!\!\updownarrow^{\vec{a}:\vec{n}} P_2 \ \wedge \ \mathrm{fv}(P_2) \cup \mathrm{fv}(P_4) \subseteq \vec{a} \ \wedge \ \mathrm{fv_P}(P_1;$ $P_3) \cap \mathrm{fv_P}(P_2; P_4) = \varnothing\} \cup \{(p,q) \mid p \approx_{\downarrow} q\}$, and show that $\mathcal{R}$ is a termination sensitive simulation, $\mathcal{R}^{-1}$ is an eventual simulation, and that $(\Upsilon\vec{a} : \vec{n}.((P_1 | P_2);$ $(P_3 | P_4)), \Upsilon\vec{a} : \vec{n}.((P_1; P_3) | (P_2; P_4))) \in \mathcal{R}$. The last condition is trivial. The remaining step is to consider all pairs $(p,q) \in \mathcal{R}$ and show that they behave accordingly for termination sensitivity, simulation and eventual simulation.

Case 1. There exists $\vec{a}$, $\vec{n}$, P_1, and P_2 such that $p = \Upsilon\vec{a} : \vec{n}.((P_1; P_3) | (P_2; P_4))$, $q = \Upsilon\vec{a} : \vec{n}.((P_1 | P_2); (P_3 | P_4))$, etc. Termination sensitivity holds because neither p nor q are halted. To satisfy simulation, we assume that $p \xrightarrow{\alpha} p'$ and must show that there exists a matching q' such that $q \xrightarrow{\alpha} q'$ and $(p', q') \in \mathcal{R}$. This follows directly from Lemma 15. Finally, we must satisfy eventual simulation. Assuming $q \xrightarrow{\vec{\alpha}} q'$, we show that there exists a p'' and q'' such that $p \xrightarrow{\alpha} p''$, $q' \Rightarrow q''$, and $(p'', q'') \in \mathcal{R}$. (Notice that this flips the direction of eventual simulation because we started with it holding for $\mathcal{R}^{-1}$.) This follows from Lemma 16.

Case 2: $p \approx_{\downarrow} q$. Termination sensitivity holds because $\approx_{\downarrow}$ is termination sensitive. Simulation and (inverse) eventual simulation hold because $\approx_{\downarrow}$ implies both.   □

Corollary 1 (Proof of Prop. 1). *If*
- *P_1 silently converges: $P_1 \Downarrow^{\vec{a}:\vec{n}}$;*
- *the processes do not interfere:* $\mathrm{fv_P}(P_1) \cap \mathrm{fv_P}(P_2) = \varnothing$; *and*
- *either P_1 or P_2 cannot be observed:* $\mathrm{fv}(P_1) \subseteq \vec{a}$ *or* $\mathrm{fv}(P_2) \subseteq \vec{a}$,

then $\Upsilon\vec{a}:\vec{n}.(P_1; P_2) \precsim_\downarrow \Upsilon\vec{a}:\vec{n}.(P_1 \mid P_2)$.

Proof. This reduces to proving either $\Upsilon\vec{a}:\vec{n}.((P_1 \mid 0);(0 \mid P_2)) \precsim_\downarrow \Upsilon\vec{a}:\vec{n}.((P_1;0) \mid (0;P_2))$ or $\Upsilon\vec{a}:\vec{n}.((0 \mid P_1);(P_2 \mid 0)) \precsim_\downarrow \Upsilon\vec{a}:\vec{n}.((0;P_1) \mid (P_2;0))$ by Theorem 2, depending on whether P_2 or P_1 is unobservable.   □

7   Related Work

Sound Parallelization. C. Hurlin proved partial correctness for an automated implementation of DOALL, where separation logic assertions provide both the specification to be preserved and the shape analysis [4]. M. Botinčan, M. Dodds et al. extended this proof-directed approach to support automated DOACROSS optimizations by injecting synchronization barriers; they prove a termination sensitive trace equivalence [2]. Our work supports more diverse dependency and synchronization patterns and proves a stronger correctness criterion. We view our proof theory and their automation techniques as being complementary.

Simulations. We are unaware of any prior mention of eventual simulation in the literature. After developing our proofs with respect to eventual simulation, we independently derived contrasimulation and its characteristic logic in order to regain symmetry and transitivity. However, van Glabbeek was the first to define contrasimulation [3]. Voorhoeve and Mauw investigated many properties of contrasimulation, describing its characteristic logic and axiomatization for CCS [12]. They established a notion of "good" and "bad" protocols and proved that contrasimulation can distinguish between them. Neither work discussed the possible applications of contrasimulation toward parallelizing transformations.

8   Conclusion

We have proven the soundness of a very general parallelizing transformation for CCS-Seq with respect to a new type of simulation relation, called eventual similarity, that allows internal nondeterminism to be preserved. Additionally, we identify contrasimilarity as a congruence that contains eventual similarity when symmetry is needed. In the absence of internal nondeterminism, both eventual similarity and contrasimilarity are equivalent to bisimulation. Because of these properties, we believe [termination sensitive] contrasimilarity is a good definition of correctness to build a verified compiler upon.

An underlying goal of this study was to develop a clear theory from the patchwork correctness criteria that resulted from our first attempt to prove parallelization for an imperative language. We were surprised to find that buffered IO (i.e., delayed observations), which is used to increase performance and is often overlooked by concurrency researchers, also contributes to expanding the kinds of parallelization that we can achieve using contrasimulation. All proofs were done in the Coq Proof Assistant, which we found instrumental to managing the complexity of proving parallelization.

References

1. Appel, A.W.: Verified software toolchain - (invited talk). In: Barthe, G. (ed.) ESOP 2011. LNCS, vol. 6602, pp. 1–17. Springer, Heidelberg (2011)
2. Botinčan, M., Dodds, M., Jagannathan, S.: Proof-directed parallelization synthesis by separation logic. ACM Trans. Program. Lang. Syst. 35(2), 8:1–8:60 (2013)
3. van Glabbeek, R.J.: The linear time - branching time spectrum II. In: Best, E. (ed.) CONCUR 1993. LNCS, vol. 715, pp. 66–81. Springer, Heidelberg (1993)
4. Hurlin, C.: Automatic parallelization and optimization of programs by proof rewriting. In: Palsberg, J., Su, Z. (eds.) SAS 2009. LNCS, vol. 5673, pp. 52–68. Springer, Heidelberg (2009)
5. Leroy, X.: Formal verification of a realistic compiler. Communications of the ACM 52(7), 107–115 (2009)
6. Milner, R.: A Calculus of Communicating Systems. Springer-Verlag New York, Inc., Secaucus (1982)
7. Padua, D.A., Wolfe, M.J.: Advanced compiler optimizations for supercomputers. Communications of the ACM 29(12), 1184–1201 (1986)
8. Parrow, J., Sjödin, P.: The complete axiomatization of cs-congruence. In: Enjalbert, P., Mayr, E.W., Wagner, K.W. (eds.) STACS 1994. LNCS, vol. 775, pp. 555–568. Springer, Heidelberg (1994)
9. Rangan, R., Vachharajani, N., Vachharajani, M., August, D.I.: Decoupled software pipelining with the synchronization array. In: IEEE PACT, pp. 177–188 (2004)
10. Ševčik, J., Vafeiadis, V., Nardelli, F.Z., Jagannathan, S., Sewell, P.: Relaxed-memory concurrency and verified compilation. SIGPLAN Not. 46(1), 43–54 (2011)
11. Tatlock, Z., Lerner, S.: Bringing extensibility to verified compilers. SIGPLAN Not. 45(6), 111–121 (2010)
12. Voorhoeve, M., Mauw, S.: Impossible futures and determinism. Inf. Process. Lett. 80(1), 51–58 (2001)
13. Zhao, J., Nagarakatte, S., Martin, M.M.K., Zdancewic, S.: Formalizing the LLVM intermediate representation for verified program transformations. SIGPLAN Not. 47(1), 427–440 (2012)

Programming Type-Safe Transformations Using Higher-Order Abstract Syntax

Olivier Savary-Belanger[1], Stefan Monnier[2], and Brigitte Pientka[1]

[1] McGill University
[2] Université de Montréal

Abstract. Compiling syntax to native code requires complex code transformations which rearrange the abstract syntax tree. This can be particularly challenging for languages containing binding constructs, and often leads to subtle, hard to find errors. In this paper, we exploit higher-order abstract syntax (HOAS) to implement a type-preserving compiler for the simply-typed lambda-calculus, including transformations such as closure conversion and hoisting, in the dependently-typed language Beluga. Unlike previous implementations, which have to abandon HOAS locally in favor of a first-order binder representation, we are able to take advantage of HOAS throughout the compiler pipeline, so that we do not have to include any lemmas about binder manipulation. Scope and type safety of the code transformations are statically guaranteed, and our implementation directly mirrors the proofs of type preservation. Our work demonstrates that HOAS encodings offer substantial benefits to certified programming.

1 Introduction

Type-based verification methods support building correct-by-construction software, and hold the promise of dramatically reducing the costs of quality assurance. Instead of verifying properties post-hoc about software, we rely on rich type abstractions which can be checked statically during the development.

Compiler implementers have long recognized the power of types to establish key properties about complex code transformations. However, the standard approach is to type-check the intermediate representations produced by compilation. This amounts to *testing* the result of compilation via type-checking. In this paper, we explore the use of sophisticated type systems to implement a correct-by-construction compiler for the simply typed lambda-calculus, including translation to continuation-passing style (CPS), closure conversion and hoisting. We concentrate here on the last two phases which are particularly challenging since they rearrange the structure of the abstract syntax tree.

A central question when implementing code transformations is the representation of the source and target languages. Shall we represent binders via first-order abstract syntax using de Bruijn indices or names or higher-order abstract syntax (HOAS) where we map binders in our source and target language to binders in our meta-language? - Arguably HOAS is the more sophisticated representation technique, eliminating the need to deal with common and

G. Gonthier and M. Norrish (Eds.): CPP 2013, LNCS 8307, pp. 243–258, 2013.
© Springer International Publishing Switzerland 2013

notoriously tricky aspects such as renaming, fresh name generation and capture-avoiding substitution. However, while the power and elegance of HOAS encodings have been demonstrated in representing proofs, for example in the Twelf system [Pfenning and Schürmann, 1999], it has been challenging to exploit its power in program transformations which rearrange abstract syntax trees and move possibly open code fragments. Previous implementations (for example Chlipala [2008]; Guillemette and Monnier [2008]) have been unable to take advantage of HOAS throughout the full compiler pipeline and have to abandon HOAS in closure conversion and hoisting. In this work, we rely on the rich type system and abstraction mechanisms of the dependently-typed language BELUGA [Pientka and Dunfield, 2010; Cave and Pientka, 2012] to implement a type and scope preserving compiler for the simply-typed lambda-calculus using HOAS for all the stages. There are two key ingredients crucial to the success: First, we encode our source and target languages using HOAS within the logical framework LF [Harper et al., 1993] reusing the LF function space to model object-level binders. As a consequence, we inherit support for α-renaming, capture-avoiding substitution, and fresh name generation from LF. Second, we represent and embed open code fragments using the notions of contextual objects and first-class contexts. A contextual object, written as $[\Psi.M]$, characterizes an open LF object M which may refer to the bound variables listed in the context Ψ [Nanevski et al., 2008]. We internalize this notion on the level of types using the contextual type $[\Psi.A]$ which classifies the contextual objects $[\Psi.M]$ where M has type A in the context Ψ. By embedding contextual objects into computations, users can not only characterize abstract syntax trees with free variables, but also manipulate and rearrange open code fragments using pattern matching.

Our implementation of a type-preserving compiler is very compact avoiding tedious infrastructure for manipulating binders. Our code directly manipulates intrinsically typed source terms and is an *executable* version of the proof that the compiler is type-preserving.

We believe our work shows that programming with contextual objects offers significant benefits to certified programming. For the full development see: http://complogic.cs.mcgill.ca/beluga/cc-code.

2   Source Language: Simply Typed Lambda-Calculus

We describe first the source language of our compiler, the simply typed lambda-calculus (STLC) extended with n-ary tuples, selectors, let-expressions and unit.

(Type) T, S $::= S \to T \mid \text{code } S\, T \mid S \times T \mid \text{unit}$
(Source) $M, N ::= x \mid \lambda x.\, M \mid M\ N \mid \text{fst } M \mid \text{rst } M \mid (M_1, M_2)$
           $\mid \text{let } x = N \text{ in } M \mid ()$
(Context) Γ $::= \cdot \mid \Gamma, x : T$

Each of our type-preserving algorithms transforms the source language to a separate target language, but uses the same language for types. N-ary products are constructed using the binary product $S \times T$ and unit. In closure conversion, we will use n-ary tuples to describe the environment. Foreshadowing the subsequent explanation of closure conversion, we also add a special type code S T; this type only arises as a result in closure conversion where it describes closed functions.

We omit the typing rules for our source language, since they are standard.

The encoding of the source language into the logical framework LF is straightforward. In this paper, we are using the dependently typed language BELUGA, which supports writing LF specifications and programs about them. By indexing source terms by their types, we only represent well-typed terms.

```
datatype source : tp → type =
| lam   : (source S → source T) → source (arr S T)
| app   : source (arr S T) → source S → source T
| fst   : source (cross S T) → source S
| rst   : source (cross S T) → source T
| cons  : source S → source T → source (cross S T)
| nil   : source unit
| letv  : source S → (source S → source T) → source T;
```

In BELUGA's concrete syntax, the *kind* type declares an LF type family, as opposed to a computational data type. Binders in our object languages are represented via the LF function space. For example, the lam constructor takes as argument a function source S → source T and constructs an object of type source (arr S T). As a consequence, we inherit α-renaming from LF and substitution is modelled via function application. N-ary tuples are represented using the constructor cons and () is represented as nil, emphasizing that n-ary tuples are encoded as lists.

3   Closure Conversion

Closure conversion is a code transformation that makes the manipulation of closure objects explicit and results in a program whose functions are closed so that they can be hoisted to the top-level.

3.1   Target Language

Our target language for closure conversion contains, in addition to functions $(\lambda y. P)$, function application P Q, tuples (P, Q), selectors (fst and rst), and let-expressions (let $y = P$ in Q), two new constructs: 1) we can form a closure $\langle P, Q \rangle$ of an expression P with its environment Q, represented as an n-ary tuple. 2) we can break apart a closure P using let $\langle y_f, y_{env} \rangle = P$ in Q.

$$\text{(Target)} \quad P, Q ::= x \mid \lambda x. P \mid P \ Q \mid \text{fst } P \mid \text{rst } P \mid \text{let } x = P \text{ in } Q$$
$$\mid (P, Q) \mid () \mid \langle P, Q \rangle \mid \text{let } \langle y_f, y_{env} \rangle = P \text{ in } Q$$
$$\text{(Context)} \quad \Delta \quad ::= \cdot \mid \Delta, x : T$$

The essential idea of closure conversion is to make the evaluation context of functions explicit; variables bound outside of a function are replaced by projections from an environment variable. Given a source-level function of type $T \to S$, we return a closure $\langle \lambda y_c. P, Q \rangle$ consisting of a closed function $\lambda y_c. P$, where y_c

pairs the input argument y and the environment variable y_{env}, and its environment Q, containing all its free variables. Such packages are traditionally given an existential type such as $\exists l.(\text{code } (T \times l) \; S) \times l$ where l is the type of the environment. We instead reuse the source type $T \to S$ which also hides l and saves us from having to handle existential types in their full generality. The rules for t_pack and t_letpack are modelling implicitly the introduction and elimination rules for existential types. Moreover, we enforce that $\lambda x.P$ is closed. The remaining typing rules for the target language are mostly straightforward and summarized next.

$$\boxed{\Delta \vdash P : T} \quad \text{Target } P \text{ has type } T$$

$$\frac{\Delta, x : T \vdash P : S}{\Delta \vdash \lambda x. P : \text{code } T \; S} \; \text{t\_lam} \qquad \frac{\Delta \vdash P : \text{code } T \; S \quad \Delta \vdash Q : T}{\Delta \vdash P \, Q : S} \; \text{t\_app}$$

$$\frac{x : T \in \Delta}{\Delta \vdash x : T} \; \text{t\_var} \qquad \frac{\cdot \vdash P : \text{code } (T \times T_{env}) \; S \quad \Delta \vdash Q : T_{env}}{\Delta \vdash \langle P, Q \rangle : T \to S} \; \text{t\_pack}$$

$$\frac{\Delta \vdash P : T \to S \quad \Delta, y_f : \text{code } (T \times l) \; S, y_{env} : l \vdash Q : S}{\Delta \vdash \text{let } \langle y_f, y_{env} \rangle = P \text{ in } Q : S} \; \text{t\_letpack}^l$$

$$\frac{\Delta \vdash P : T \quad \Delta \vdash Q : S}{\Delta \vdash (P, Q) : T \times S} \; \text{t\_cons} \qquad \frac{}{\Delta \vdash () : \text{unit}} \; \text{t\_unit}$$

3.2   Closure Conversion Algorithm

Before describing the algorithm in detail, let us illustrate briefly the idea of closure conversion using an example. Our algorithm translates the program $(\lambda x.\lambda y.x + y) \; 5 \; 2$ to

```
let ⟨f₁, c₁⟩ =
    let ⟨f₂, c₂⟩ =
    ⟨ λe₂. let x = fst e₂ in let xₑₙᵥ = rst e₂ in
    ⟨λe₁. let y = fst e₁ in let yₑₙᵥ = rst e₂ in fst yₑₙᵥ + y, (x, ())⟩
    , () ⟩
    in f₂ (5, c₂)
in f₁ (2, c₁)
```

Closure conversion introduces an explicit representation of the environment, closing over the free variables of the body of an abstraction. We represent the environment as a tuple of terms, corresponding to the free variables in the body of the abstraction.

We define the algorithm for closure conversion using $[\![M]\!]_\rho$, where M is a source term which is well-typed in the context Γ and ρ a mapping of source variables in Γ to target terms in the context Δ. Intuitively, ρ maps source variables to the corresponding projection of the environment. It is defined as follows:

$$\boxed{\Delta \vdash \rho : \Gamma} \quad \rho \text{ maps variables from source context } \Gamma \text{ to target context } \Delta$$

$$\frac{}{\Delta \vdash id : \cdot} \text{ m\_id} \qquad \frac{\Delta \vdash \rho : \Gamma \quad \Delta \vdash P : T}{\Delta \vdash \rho, x \mapsto P : \Gamma, x : T} \text{ m\_dot}$$

For convenience, we write π_i for the i-th projection instead of using the selectors fst and rst. We give here only the cases for variables, functions and function applications.

$$[\![x]\!]_\rho \quad = \quad \rho(x)$$
$$[\![\lambda x.M]\!]_\rho \quad = \quad \langle\, \lambda x_c. \text{let } x = \text{fst } x_c \text{ in let } x_{env} = \text{rst } x_c \text{ in } P \,,\, P_{env} \,\rangle$$
$$\text{where } \{x_1, \ldots, x_n\} = \mathsf{FV}(\lambda x.M)$$
$$\text{and } \rho' = x_1 \mapsto \pi_1 \, x_{env}, \ldots, x_n \mapsto \pi_n \, x_{env}, x \mapsto y$$
$$\text{and } P_{env} = (\rho(x_1), \ldots, \rho(x_n)) \text{ and } P = [\![M]\!]_{\rho'}$$
$$[\![M \, N]\!]_\rho \quad = \quad \text{let } \langle x_f, x_{env} \rangle = P \text{ in } x_f \, (Q \,,\, x_{env}) \quad \text{where } P = [\![M]\!]_\rho \text{ and } Q = [\![N]\!]_\rho$$

To translate a source variable, we look up its binding in the map ρ. When translating a lambda-abstraction $\lambda x.M$, we first compute the set $\{x_1, \ldots, x_n\}$ of free variables occurring in $\lambda x.M$. We then form a closure consisting of two parts: 1) a term P which is obtained by converting M with the new map ρ' which maps variables $x_1, \ldots, x_n$ to their corresponding projection of the environment variable and x to itself, thereby eliminating all free variables in M. 2) an environment tuple P_{env}, obtained by applying ρ to each variable in $(x_1, \ldots, x_n)$.

When translating an application $M \, N$, we first translate M and N to target terms P and Q. Since the source term M denotes a function, the target term P will denote a closure. We unpack the closure obtaining x_f, the part denoting the function, and x_{env}, the part denoting the environment. We then apply x_f to the extended environment (Q, x_{env}).

Our goal is to implement the described algorithm as a recursive program which manipulates intrinsically well-typed source terms. This is non-trivial. To understand the general idea behind our program, we discuss how to prove that given a well-typed source term M we can produce a well-typed target term which is the result of converting M. The proof relies on several straightforward lemmas which correspond exactly to auxiliary functions needed in our implementation.

Auxiliary lemmas:

- **Strengthening:** If $\Gamma \vdash M : T$ and $\Gamma' = FV(M)$, then $\Gamma' \vdash M : T$ and $\Gamma' \subseteq \Gamma$
- **Weakening:** If $\Gamma' \vdash M : T$ and $\Gamma' \subseteq \Gamma$ then $\Gamma \vdash M : T$.
- **Context reification:** Given a context $\Gamma = x_1 : T_1, \ldots, x_n : T_n$, there exists a type $T_\Gamma = (T_1 \times \ldots \times T_n)$ and there is a $\rho = x_1 \mapsto \pi_1 \, x_{env}, \ldots, x_n \mapsto \pi_n \, x_{env}$ s.t. $x_{env} : T_\Gamma \vdash \rho : \Gamma$ and $\Gamma \vdash (x_1, \ldots x_n) : T_\Gamma$.
- **Map extension:** If $\Delta \vdash \rho : \Gamma$, then $\Delta, x : T \vdash \rho, x \mapsto x : \Gamma, x : T$.
- **Map lookup:** If $x : T \in \Gamma$ and $\Delta \vdash \rho : \Gamma$, then $\Delta \vdash \rho(x) : T$.
- **Map lookup (tuple):**
  If $\Gamma \vdash (x_1, \ldots, x_n) : T$ and $\Delta \vdash \rho : \Gamma$ then $\Delta \vdash (\rho(x_1), \ldots, \rho(x_n)) : T$.

We show here the key cases of the proof concentrating on lambda abstractions and variables.

Theorem 1. *If $\Gamma \vdash M : T$ and $\Delta \vdash \rho : \Gamma$ then $\Delta \vdash [\![M]\!]_\rho : T$*

Proof. By induction on the structure of the term M.

Case : $M = x$.

$\Gamma \vdash x : T$ and $\Delta \vdash \rho : \Gamma$	by assumption
$\Delta \vdash \rho(x) : T$	by Map lookup
$\Delta \vdash [\![x]\!]_\rho : T$	by definition

Case $M = \lambda x.M$

$\Gamma \vdash \lambda x.M : T \to S$ and $\Delta \vdash \rho : \Gamma$	by assumption
$\Gamma' \vdash \lambda x.M : T \to S$ and $\Gamma' \subseteq \Gamma$	
where $\Gamma' = FV(\lambda x.M)$	by Term strengthening
$\Gamma', x : T \vdash M : S$	by inversion on t_lam
$\Gamma' \vdash (x_1, \ldots, x_n) : T_{\Gamma'}$ and $x_{env} : T_{\Gamma'} \vdash \rho' : \Gamma'$	by Context reification
$\Gamma \vdash (x_1, \ldots, x_n) : T_{\Gamma'}$	by Term Weakening
$\Delta \vdash \rho : \Gamma$	by assumption
$(\rho(x_1), \ldots, \rho(x_n)) = P_{env}$	by assumption
$\Delta \vdash P_{env} : T_{\Gamma'}$	by Map lookup (tuple)
$x_{env} : T_{\Gamma'}, x : T \vdash \rho', x \mapsto x : \Gamma', x : T$	By Map extension
$x_{env} : T_{\Gamma'}, x : T \vdash P : S$	
where $P = [\![M]\!]_{\rho', x \mapsto x}$	by i.h. on M
$c : T \times T_{\Gamma'}, x : T, x_{env} : T_{\Gamma'} \vdash P : S$	by Term weakening
$c : T \times T_{\Gamma'}, x : T \vdash \text{let } x_{env} = \text{rst } c \text{ in } P : S$	by rule t_let
$c : T \times T_{\Gamma'} \vdash \text{let } x = \text{fst } x \text{ in let } x_{env} = \text{rst } c \text{ in } P : S$	by rule t_let
$\cdot \vdash \lambda c. \text{let } x = \text{fst } c \text{ in let } x_{env} = \text{rst } c \text{ in } P : \text{code } (T \times T_{\Gamma'})\, S$	by rule t_lam
$\Delta \vdash \langle \lambda c. \text{let } x = \text{fst } c \text{ in let } x_{env} = \text{rst } c \text{ in } P \ , \ P_{env} \rangle : T \to S$	by rule t_pack
$\Delta \vdash [\![\lambda x.M]\!]_\rho : T \to S$	by definition

□

3.3   Representation of Target Language in LF

We now describe the implementation of the closure conversion algorithm in BELUGA. We begin by defining the target language, showing the constructs for lambda-abstraction, application, creating a closure and taking a closure apart.

```
datatype target: tp → type =
| clam    : (target T → target S) → target (code T S)
| capp    : target (code T S) → target T → target S
| cpack   : target (code (cross T L) S) → target L → target (arr T S)
| cletpack: target (arr T S)
            → ({l:tp} target (code (cross T l)) S)→target l → target S)
            → target S;
```

The data-type definition directly reflects the typing rules with one exception: our typing rule t_pack enforced that P was closed. This cannot be achieved in the LF encoding, since the context of assumptions is ambient. As a consequence, hoisting, which relies on the fact that the closure converted functions are closed, cannot be implemented as a separate phase after closure conversion. We will come back to this issue in Section 4.

3.4 Type-Preserving Closure Conversion in Beluga: An Overview

The top-level closure conversion function cc translates a closed source term of type T to a closed target term of type T, which is encoded in BELUGA using the computation-level type [.source T] → [.target T]. We embed closed contextual LF object of type source T and target T into computation-level types via the modality [ ]. The . separates the context of assumptions from the conclusion. Since we are describing closed objects, the context is left empty.

However, when closure converting and traversing source terms, our source terms do not remain closed. We generalize the closure conversion function to translate well-typed source terms in a source context Γ to well-typed target terms in the target context Δ given a map of the source context Γ to the target context Δ. Δ will consists of an environment variable x_{env} and the variable x bound by the last abstraction, along with variables introduced by let bindings.

    cc': Map [Δ] [Γ] → [Γ. source T] → [Δ. target T]

Just as types classify terms, schemas classify contexts in BELUGA, similarly to world declarations in Twelf [Pfenning and Schürmann, 1999]. The schema tctx defines a context where the type of each declaration is an instance of target T; similarly the schema sctx defines a context where the type of each declaration is an instance of source T. While type variables appear in the typing rule t_letpack, they only occur locally and are always bound before the term is returned by our functions, such that they do not appear in the context variables indexing them.

        schema tctx = target T;
        schema sctx = source T;

We use the indexed recursive type Map to relate the target context Δ and source context Γ [Cave and Pientka, 2012]. In BELUGA's concrete syntax, the *kind* ctype indicates that we are not defining an LF datatype, but a recursive type on the level of computations. → is overloaded to mean computation-level strong functions rather than the LF function space. Map is defined recursively on the source context Γ directly encoding our definition $\Delta \vdash \rho : \Gamma$ given earlier.

        datatype Map:{Δ:tctx}{Γ:sctx} ctype =
        | Id :{Δ:tctx} Map [Δ] []
        | Dot: Map [Δ] [Γ] → [Δ. target S] → Map [Δ] [Γ,x:source S];

BELUGA reconstructs the type of free variables Δ, Γ, and S and implicitly abstracts over them. In the constructor Id, we choose to make Δ an explicit argument to Id, since we often need to refer to Δ explicitly in the recursive programs we are writing about Map. The next section presents the implementation of the necessary auxiliary functions, followed by cc'.

3.5   Implementation of Auxiliary Lemmas

Term strengthening and weakening Both operations rely on an inclusion relation
$\Gamma' \subseteq \Gamma$ where we preserve the order, which is defined using the indexed recursive
computation-level data-type SubCtx.

```
datatype SubCtx: {Γ':sctx} {Γ:sctx} ctype =
| WInit: SubCtx [] []
| WDrop: SubCtx [Γ'] [Γ] → SubCtx [Γ'] [Γ,x:source T]
| WKeep: SubCtx [Γ'] [Γ] → SubCtx [Γ',x:source T] [Γ,x:source T];
```

Given a source term M in Γ the function strengthen computes the strengthened
version of M which is well-typed in Γ' characterizing the free variables in M together with the proof SubCtx [Γ'] [Γ]. We represent the result using the indexed
recursive type StrTerm encoding the existential in the specification as a universal
quantifier using the constructor STm. The fact that Γ' describes exactly the free
variables of M is not captured by the type definition.

```
datatype StrTerm: {Γ:sctx} [.tp] → ctype =
| STm: [Γ'. source T] → SubCtx [Γ'] [Γ] → StrTerm [Γ] [.T];

rec strengthen: [Γ.source T] → StrTerm [Γ] [.T]
```

Just as in the proof of the term strengthening lemma, we cannot implement
this function directly. This is because, while we would like to perform induction
on the size of Γ, we cannot appeal to the induction hypothesis while maintaining
a well-scoped source term in the case of an occurring variable in front of Γ.
Instead, we implement str, which, intuitively, implements the lemma

$$\text{If } \Gamma_1, \Gamma_2 \vdash M : T \text{ and } \Gamma'_1, \Gamma_2 = \mathsf{FV}(M), \text{ then } \Gamma'_1, \Gamma_2 \vdash M : T \text{ and } \Gamma'_1 \subseteq \Gamma_1.$$

In BELUGA, contextual objects can only refer to one context variable - we
cannot simply write [Γ_1, Γ_2. source T]. To express this, we use a data-type wrap
which abstracts over all the variables in Γ_2. wrap is indexed by the type T of the
source term and the size of Γ_2. str then recursively analyses Γ_1, adding variables
occurring in the input term to Γ_2. The type of str asserts, through its index N,
the size of Γ_2.

```
datatype wrap: tp → nat → type =
| ainit: (source T) → wrap T z
| add: (source S → wrap T N) → wrap (arr S T) (suc N);
datatype StrTerm': {Γ:sctx} [.tp] → [.nat] → ctype =
| STm': [Γ'. wrap T N] → SubCtx [Γ'] [Γ] → StrTerm' [Γ] [.T] [.N];
rec str: [Γ. wrap T N] →  StrTerm' [Γ] [.T] [.N]
```

The function str is implemented recursively on the structure of Γ exploits
higher-order pattern matching to test whether a given variable x occurs in a
term M. As a consequence, we can avoid the implementation of a function which
recursively analyzes M and test whether x occurs in it. While one can implement
term weakening following similar ideas, we incorporate it into the variable lookup
function defined next.

Map extension and lookup The lookup function takes a source variable of type
T in the source context Γ and Map [Δ] [Γ] and returns the corresponding target
expression of the same type.

```
rec lookup: {#p:[Γ.source T]} Map [Δ] [Γ] → [Δ. target T] =
λ□#p ⇒ fn ρ ⇒ let (ρ: Map [Δ] [Γ]) = ρ in case [Γ. #p... ] of
| [Γ',x:source T. x]     ⇒ let Dot ρ' [Δ.M... ] = ρ in [Δ.M... ]
| [Γ',x:source S. #q... ] ⇒ let Dot ρ' [Δ.M... ] = ρ in lookup [Γ'.#q... ] ρ';
```

We quantify over all variables in a given context by {#p:[Γ.source T]} where #p
denotes a variable of type source T in the context Γ. In the function body, $\lambda^\square$-
abstraction introduces an explicitly quantified contextual object and fn-abstraction
introduces a computation-level function. The function lookup is implemented by
pattern matching on the context Γ and the parameter variable #p.

To guarantee coverage and termination, it is pertinent that we know that
an n-ary tuple is composed solely of source variables from the context Γ, in
the same order. We therefore define VarTup as a computational datatype for
such variable tuples. Nex v of type VarTup [Γ] [.L$_\Gamma$], where $\Gamma = x_1:T_1,\ldots,x_n:T_n$,
is taken to represent the source language tuple $(x_1,\ldots,x_n)$ of type $T_1 \times \ldots \times T_n$
in the context Γ.

```
datatype VarTup: {Γ:sctx} [.tp] → ctype =
| Emp: VarTup [] [.unit]
| Nex: VarTup [Γ] [.L] → VarTup [Γ,x:source T] [.cross T L];
```

The function lookupVars applies a map ρ to every variable in a variable tuple.

```
rec lookupVars: VarTup [Γ'] [.L_Γ'] → SubCtx [Γ'] [Γ] → Map [Δ] [Γ]
              → [Δ. target L_Γ']
```

lookupVars allows the application of a Map defined on a more general context
Γ provided that $\Gamma' \subseteq \Gamma$. This corresponds, in the theoretical presentation, to
weakening a variable tuple before applying a mapping on it.

extendMap, which implements the Map extension lemma, weakens a mapping
with the identity on a new variable x. It is used to extend the Map with local
variables, for example when we encounter a let binding construct.

```
rec extendMap: Map [Δ] [Γ] → Map [Δ,x:target S] [Γ,x:source S]
```

A Reification of the Context as a Term Tuple

The context reification lemma
is proven by induction on Γ; to enable pattern matching on the context Γ, we
wrap it in the indexed data-type Ctx.

```
datatype Ctx: {Γ:sctx} ctype =
| Ctx: {Γ:sctx} Ctx [Γ];
datatype CtxAsTup: {Γ:sctx} ctype =
| CTup: VarTup [Γ] [.L_Γ] → Map [x:target L_Γ] [Γ] → CtxAsTup [Γ];
rec reify: Ctx [Γ] → CtxAsTup [Γ]
```

The function reify translates the context Γ to a source term. It produces a
tuple containing variables of Γ in order, along with Map [x:target T$_\Gamma$] [Γ] describing the mapping between those variables and their corresponding projections. The type of reify enforces that the returned Map contains, for each of the

```
rec cc': Map [Δ] [Γ] → [Γ. source T] → [Δ. target T] =
fn ρ ⇒ fn m ⇒ case m of
| [Γ. #p... ] ⇒ lookup ρ [Γ. #p... ]
| [Γ. lam λx.M... x] ⇒
    let STm' [Γ'. add (λx. ainit (M' ... x))] rel = str [Γ. add λx.ainit (M ... x)] in
    let CTup [Γ'. E... ] (ρ'':Map [x_env:target T_Γ'] [Γ']) = reify (Ctx [Γ']) in
    let ρ' = extendMap ρ'' in
    let [x_env:target T_Γ',x:target T. (P x_env x)] = cc' ρ' [Γ',x:source _. M... x] in
    let [Δ. P_env... ] = lookupVars [Γ'. E... ] rel ρ in
      [ Δ. cpack (clam (λc. (clet (cfst c)
                                 (λx.(clet (crst c)
                                          (λx_env. P x_env x))))))
           (P_env... ) ]
```

Fig. 1. Implementation of closure conversion in BELUGA

variables in Γ, a target term of the same type referring solely to a variable x of type T_Γ. This means the tuple of variables of type T_Γ also returned by reify contain enough *information* to replace occurrences of variables in any term in context Γ perserving types - it contains either the variables themselves or terms of the same type.

3.6   Closure Conversion: Top-Level Function

The function cc' (see Fig. 1) implements our closure conversion algorithm recursively by pattern matching on objects of type [Γ. source T] . It follows closely the earlier proof (Thm. 1). We describe here on the cases for variables and lambda-abstractions omitting the case for applications. When we encounter a variable, we simply lookup its corresponding binding in ρ.

Given a lambda abstraction in context Γ and ρ which represents the map from Γ to Δ, we begin by strengthening the term to some context Γ'. We then reify the context Γ' to obtain a tuple E together with the new map ρ'' of type Map [x_{env}:target $T_{\Gamma'}$] [Γ']. Next, we extend ρ'' with the identity on the lambda-abstraction's local variable to obtain ρ', and recursively translate M using ρ', obtaining a target term in context x_{env},x. Abstracting over x_{env} and x gives us the desired closure-converted lambda-abstraction. To obtain the environment P_{env}, we apply ρ on each variables in E using lookupVars. Finally, we pack the converted lambda-abstraction and the environment P_{env} as a closure, using the constructor cpack.

Our implementation of closure conversion, including all definitions and auxiliary functions, consists of approximately 250 lines of code.

4   Hoisting

Hoisting is a code transformation that lifts the lambda-abstractions, closed by closure conversion, to the top level of the program. Function declarations in the program's body are replaced by references to a global function environment.

As we alluded to earlier, our encoding of the target language of closure conversion does not guarantee that functions in a closure converted term are indeed closed. While this information is available during closure conversion, it cannot easily be captured in our meta-language. We therefore extend our closure conversion algorithm to perform hoisting at the same time. Hoisting can however be understood by itself; we highlight here its main ideas.

Performing hoisting on the closure-converted program presented in Sec. 3

$$
\begin{aligned}
&\text{let } \langle f_1, c_1 \rangle = \\
&\quad \text{let } \langle f_2, c_2 \rangle = \\
&\quad\quad \langle \; \lambda e_2. \text{ let } x = \text{fst } e_2 \text{ in let } x_{\text{env}} = \text{rst } e_2 \text{ in} \\
&\quad\quad\quad \langle \lambda e_1. \text{ let } y = \text{fst } e_1 \text{ in let } y_{\text{env}} = \text{rst } e_2 \text{ in fst } y_{\text{env}} + y, (x, ()) \rangle \\
&\quad\quad\quad , \, () \; \rangle \\
&\quad\quad \text{in } f_2 \, (5 \, , c_2) \\
&\quad \text{in } f_1 \, (2, c_1)
\end{aligned}
$$

will return

$$
\begin{aligned}
&\text{let } l = (\lambda l_2.\lambda e_2.\text{let } x = \text{fst } e_2 \text{ in let } x_{\text{env}} = \text{rst } e_2 \text{ in } \langle \, (\text{fst } l_2) \, (\text{rst } l_2) \, , \, (x, ()) \, \rangle, \\
&\quad\quad\quad \lambda l_1.\lambda e_1.\text{let } y = \text{fst } e_1 \text{ in let } y_{\text{env}} = \text{rst } e_2 \text{ in fst } y_{\text{env}} + y, ()) \\
&\text{in let } \langle f_1, c_1 \rangle = \\
&\quad \text{let } \langle f_2, c_2 \rangle = \langle (\text{fst } l) \, (\text{rst } l), (\cdot) \rangle \\
&\quad \text{in } f_2 \, (5, c_2) \\
&\text{in } f_1 \, (2, c_1)
\end{aligned}
$$

4.1   Source and Target Languages - Revisited

We define hoisting on the target language of closure conversion and keep the same typing rules (see Fig. 3.1) with one exception: the typing rule for t_pack is replaced by the one below.

$$
\frac{l : T_f \vdash P : \text{code } (T \times T_x) \; S \quad \Delta, l : T_f \vdash Q : T_x}{\Delta, l : T_f \vdash \langle P, Q \rangle : T \to S} \; \text{t\_pack'}
$$

When hoisting is performed at the same time as closure conversion, P is not completely closed anymore, as it refers to the function environment 1. Only at top-level, where we bind the collected tuple as 1, will we recover a closed term. The distinction between t_pack and t_pack' is irrelevant in our implementation, as in our representation of the typing rules in LF the context is ambient.

We now define the hoisting algorithm as $[\![P]\!]_l = Q \bowtie E$. Hoisting takes as input a target term P and returns a hoisted target term Q together with its function environment E, represented as a n-ary of product type L. We write $E_1 \circ E_2$ for appending tuple E_2 to E_1 and $L_1 \circ L_2$ for appending the product type L_2 to L_1. We concentrate here on the cases for variables and closures.

$$
\begin{aligned}
&[\![x]\!]_l \quad\quad = x \bowtie () \\
&[\![\langle P_1, P_2 \rangle]\!]_l = \langle (\text{fst } l) \, (\text{rst } l), Q_2 \rangle \bowtie E \text{ where } Q_1 \bowtie E_1 = [\![P_1]\!]_l \\
&\quad\quad\quad\quad\quad\quad\quad\quad\quad\quad\quad\quad\quad\quad\quad \text{and } Q_2 \bowtie E_2 = [\![P_2]\!]_l \\
&\quad\quad\quad\quad\quad\quad\quad\quad\quad\quad\quad\quad\quad\quad\quad \text{and } E = (\lambda l.Q_1, \, E_1 \circ E_2)
\end{aligned}
$$

While the presented hoisting algorithm is simple to implement in an untyped setting, its extension to a typed language demands more care with respect to the form and type of the functions environment. As $\circ$ is only defined on n-ary tuples and product types and not on general terms and types, we enforce that the returned E and its type L are of the right form. We take $\Delta \vdash_l E : L$ to mean $\Delta \vdash E : L$ for a n-ary tuple E of product type L.

Auxiliary lemmas:
- **Append function environments**
  If $\Delta \vdash_l E_1 : L_1$ and $\Delta \vdash_l E_2 : L_2$, then $\Delta \vdash_l E_1 \circ E_2 : L_1 \circ L_2$.
- **Function environment weakening (1)**
  If $\Delta, l : L_{f_1} \vdash P : T$ and $L_{f_1} \circ L_{f_2} = L_f$, then $\Delta, l : L_f \vdash P : T$.
- **Function environment weakening (2)**
  If $\Delta, l : L_{f_2} \vdash P : T$ and $L_{f_1} \circ L_{f_2} = L_f$, then $\Delta, l : L_f \vdash P : T$.

Theorem 2. *If $\Delta \vdash P : T$ then $\cdot \vdash_l E : L_f$ and $\Delta, l : L_f \vdash Q : T$ for some L_f where $[\![P]\!]_l = Q \bowtie E$.*

Proof. By induction on the term P.

4.2   Auxiliary Functions

Defining environments Our hoisting algorithm uses operations such as $\circ$, which are only defined on n-ary tuples and on product types. To guarantee coverage, we define an indexed datatype encoding the judgement $\Delta \vdash_l E : L_f$, which asserts that environment E and its type L_f are of the right form.

```
datatype Env: {Lf:[.tp]} [.target Lf] → ctype =
| EnvNil: Env [.unit] [.cnil]
| EnvCons: {P:[.target T]}
           Env [.L] [.E]   → Env [.cross T L] [.ccons P E];
```

Appending function environments. When hoisting terms with more than one subterm, each recursive call on those subterms results in a different function environment; they need to be merged before combining the subterms again. This is accomplished by the function append which takes in Env [.L1] [.E1] and Env [.L2] [.E2], and constructs the function environment Env [.L1 ∘ L2] [.E1 ∘ E2]. As BELUGA does not support functions in types, we return some function environment E of type L, and a proof that E and L are the results of concatenating respectively E1 and E2, and L1 and L2.

```
datatype App: {T:[.tp]}{S:[.tp]}{TS:[.tp]} [.target T] → [.target S]
          → [.target TS] → ctype =
| AStart: Env [.S] [.Q] → App [.unit] [.S] [.S] [.cnil] [.Q] [.Q]
| ACons: App [.T] [.S] [.TS] [.P] [.Q] [.PQ]
         → App [.(cross T' T)] [.S] [.(cross T' TS)]
           [.(ccons P' P)] [.Q] [.(ccons P' PQ)];

datatype ExApp: {T:[.tp]}{S:[.tp]} [.target T] → [.target S] → ctype =
| AP: App [.L1] [.L2] [.L] [.E1] [.E2] [.E] → Env [.L] [.E]
    → ExApp [.L1] [.L2] [.E1] [.E2];

rec append: Env [.L1] [.E1]→ Env [.L2] [.E2]→ ExApp [.L1] [.L2] [.E1] [.E2]
```

App [.L$_1$] [.L$_2$] [.L] [.E$_1$] [.E$_2$] [.E] can be read as E$_1$ and E$_2$ being tuples of type L$_1$ and L$_2$, and concatening them yields the tuple E of type L.

Next, we show the type of the two lemmas about function environment weakening. They are a direct encoding of their specifications.

```
rec weakenEnv1: (Δ:tctx) App [.L₁] [.L₂] [.L] [.E₁] [.E₂] [.E]
                → [Δ, 1:target L₁. target T] → [Δ, 1:target L. target T]
rec weakenEnv2: (Δ:tctx) App [.L₁] [.L₂] [.L] [.E₁] [.E₂] [.E]
                → [Δ, 1:target L₂. target T] → [Δ, 1:target L. target T]
```

4.3   The Main Function

The top-level function hcc generalizes cc such that it performs hoisting at the same time as closure conversion. Again we only concentrate on the case for variables and lambda-abstraction to illustrate that only small changes are required. We generalize hcc and implement hcc' to closure convert and hoist open terms when given a map between the source and target context.

```
datatype HCCRet:{Δ:tctx} [.tp] → ctype =
| HRet: [Δ,1:target L_f. target T] → Env [.L_f] [.E] → HCCRet [Δ] [.T];
rec hcc': Map [Δ] [Γ] → [Γ. source T] → HCCRet [Δ] [.T] =
fn ρ ⇒ fn m ⇒ case m of
| [Γ. #p... ] ⇒
  let [Δ. Q... ] = lookup [Γ] [Γ. #p... ] ρ in
    HRet [Δ,1:target (prod unit). Q... ] EnvNil
| [Γ. lam λx.M... x] ⇒
  let STm' [Γ'.add λx.ainit (M'... x)] rel = str [Γ.add λx. ainit (M ... x)] in
  let CTup vt (ρ'':Map [x_env:target T_Γ'] [Γ']) = reify (Ctx [Γ']) in
  let [Δ. P_env... ] = lookupTup vt rel ρ in
let HRet r e = hcc'' (extendMap ρ'') [Γ',x:source _.M'..x] in (rho → ρ'' , g' →
    Γ')
  let [x_env:target T_Γ', x:target T, 1:target T_f. (Q x_env x 1)] = r in
  let e' = EnvCons [.clam λl. clam λc.
                        clet (cfst c) (λx.clet (crst c) (λx_env. Q x_env x 1))]
                e in
  let [.T'] = [.cross (code T_f (code (cross T T_Γ') S)) T_f]
  in HRet [Δ,1:target T'. cpack (capp (cfst 1) (crst 1))  (P_env... )] e'
;

rec hcc: [.source T] → [.target T] =
fn m ⇒ let HRet r (e: Env [._] [.E]) = hcc' (IdMap []) m  in
         let [1:target S. Q 1] = r in
            [.clet E (λl. Q 1)];
```

hcc calls hcc' with the initial map and the source term m of type T. It then binds, with clet, the function environment as 1 in the hoisted term, resulting in a closed target term of the same type.

hcc' converts a source term in the context Γ given a map between the source context Γ and the target context Δ following the algorithm described in Sec. 4. It returns a target term of type T which depends on a function environment 1 of some product type L$_f$ together with a concrete function environment of type L$_f$. The result of hcc' is described by the datatype HCCRet which is indexed by the target context Δ and the type T of the target term.

hcc' follows closely the structure of cc'. When we encounter a variable, we look it up in ρ and return the corresponding target term with an empty well-formed function environment EnvNil. When reaching a lambda-abstraction of type arr s T, we again strengthen the body lam λx.M ... x to some context Γ'. We then reify Γ' to obtain a variable tuple $(x_1, \ldots, x_n)$ and convert the strengthened M recursively using the map ρ extended with the identity. As a result, we obtain a closed target term Q together with a well-formed function environment e containing the functions collected so far. We then build the variable environment $(\rho(x_1), \ldots, \rho(x_n))$, extend the function environment with the converted result of M which is known to be closed, and return capp (cfst l) (crst l) where l abstracts over the current function environment.

Our implementation of hoisting adds in the order of 100 lines to the development of closure conversion and retains its main structure.

An alternative to the presented algorithm would be to thread through the function environment as an additional argument to hcc. This avoids the need to append function environments and obviates the need for weakenEvn1. Other properties around concat would however still have to be proven, some of which require multiple nested inductions; therefore, the complexity and length of the resulting implementation is similar or even larger.

5   Related Work

While HOAS holds the promise of dramatically reducing the overhead related to manipulating abstract syntax trees with binders, the implementation of a certified compiler, in particular the phases of closure conversion and hoisting, using HOAS has been elusive.

One of the earliest studies of using HOAS in implementing compilers was presented in Hannan [1995], where the author describes the implementation of a type-directed closure conversion in *Elf* [Pfenning, 1989], leaving open several implementation details, such as how to reason about variables equality.

Abella [Gacek, 2008] is an interactive theorem prover which supports HOAS, but not dependent types at the specification level. The standard approach would be to specify source terms, typing judgments, and the closure conversion algorithm, and then prove that it is type-preserving. However, one cannot obtain an executable program from the proof. Moreover, it is not obvious how to specify closure conversion algorithm since one of its arguments is the mapping ρ which itself inductively defined

Closely related to our work is Guillemette and Monnier [2007]'s implementation of a type-preserving closure conversion algorithm over STLC in Haskell. While HOAS is used in the CPS translation, the languages from closure conversion onwards use de Bruijn indices. Since the language targeted by their closure conversion syntactically enforces that functions are closed, it is possible for them to perform hoisting in a separate phase. In Guillemette and Monnier [2008], the authors extend the closure conversion implementation to System F.

Chlipala [2008] presents a certified compiler for STLC in Coq using parametric higher-order abstract syntax (PHOAS), a variant of weak HOAS. He however annotates his binders with de Bruijn level before the closure conversion pass, thus degenerating to a first-order representation. His closure conversion is hence similar to the one of Guillemette and Monnier [2007]. As in our work, hoisting is done at the same time as closure conversion, because his target language does not capture that functions are closed.

In both implementations, infrastructural lemmas dealing with binders constitute a large part of the development. Moreover, additional information in types is necessary to ensure the program type-checks, but is irrelevant at a computational level. In contrast, we rely on the rich type system and abstraction mechanisms of Beluga to avoid all infrastructural lemmas.

The closure conversion algorithm has also served as a key benchmark for systems supporting first-class nominal abstraction such as FreshML [Pottier, 2007] and αProlog [Cheney and Urban, 2004]. Both languages provide facilities for generating names and reasoning about their freshness, which proves to be useful when computing the free variables in a term. However, capture-avoiding substitution still needs to be implemented separately. Since these languages lack dependent types, implementing a certified compiler is out of their reach.

6 Conclusion

In addition to closure conversion and hoisting, we also have implemented the translation to continuation-passing style. Our compiler not only type checks, but also coverage checks. Termination can be verified straightforwardly by the programmer, as every recursive call is made on a structurally smaller argument, such that all our functions are total. The fact that we are not only preserving types but also the scope of terms guarantees that our implementation is *essentially* correct by construction.

Although HOAS is one of the most sophisticated encoding techniques for structures with binders and offers significant benefits, problems such as closure conversion, where reasoning about the identity of free variables is needed, have been difficult to implement using an HOAS encoding. In BELUGA, contexts are first-class; we can manipulate them, and indeed recover the identity of free variables by observing the context of the term. This is unlike other system supporting HOAS such as Twelf [Pfenning and Schürmann, 1999] or Delphin [Poswolsky and Schürmann, 2008]; in Abella [Gacek, 2008], we can test variables for identity, but users need to represent and reason about contexts explicitly. More importantly, we cannot obtain an executable program from the proof.

In addition, BELUGA's computation-level recursive datatypes provide us with an elegant tool to encode properties about contexts and contextual object. Our case study clearly demonstrates the elegance of developing certified programs in BELUGA. We rely on built-in substitutions to replace bound variables with their corresponding projections in the environment; we rely on the first-class context and recursive datatypes to define a mapping of source and target variables as well as computing a strengthened context only containing the relevant free variables in a given term.

In the future, we plan to extend our compiler to System F. While the algorithms seldom change from STLC to System F, open types pose a significant challenge. This will provide further insights into what tools and abstractions are needed to make certified programming accessible to the every day programmer.

Acknowledgements. We thank Mathieu Boespflug for his feedback and work on the implementation of BELUGA, and anonymous referees for helpful suggestions and comments on drafts of this paper.

References

Cave, A., Pientka, B.: Programming with binders and indexed data-types. In: Symposium on Principles of Programming Languages, pp. 413–424. ACM (2012)

Cheney, J., Urban, C.: αProlog: A logic programming language with names, binding and α-equivalence. In: Demoen, B., Lifschitz, V. (eds.) ICLP 2004. LNCS, vol. 3132, pp. 269–283. Springer, Heidelberg (2004)

Chlipala, A.J.: Parametric higher-order abstract syntax for mechanized semantics. In: International Conference on Functional Programming, pp. 143–156. ACM (2008)

Gacek, A.: The Abella interactive theorem prover (System description). In: Armando, A., Baumgartner, P., Dowek, G. (eds.) IJCAR 2008. LNCS (LNAI), vol. 5195, pp. 154–161. Springer, Heidelberg (2008)

Guillemette, L.-J., Monnier, S.: A type-preserving closure conversion in Haskell. In: Workshop on Haskell, pp. 83–92. ACM (2007)

Guillemette, L.-J., Monnier, S.: A type-preserving compiler in Haskell. In: International Conference on Functional Programming, pp. 75–86. ACM (2008)

Hannan, J.: Type systems for closure conversions. In: Workshop on Types for Program Analysis, pp. 48–62 (1995)

Harper, R., Honsell, F., Plotkin, G.: A framework for defining logics. Journal of the ACM 40(1), 143–184 (1993)

Nanevski, A., Pfenning, F., Pientka, B.: Contextual modal type theory. Transactions on Computational Logic 9(3), 1–49 (2008)

Pfenning, F.: Elf: A language for logic definition and verified meta-programming. In: Symposium on Logic in Computer Science, pp. 313–322. IEEE (1989)

Pfenning, F., Schürmann, C.: System description: Twelf - A meta-logical framework for deductive systems. In: Ganzinger, H. (ed.) CADE 1999. LNCS (LNAI), vol. 1632, pp. 202–206. Springer, Heidelberg (1999)

Pientka, B., Dunfield, J.: Beluga: A framework for programming and reasoning with deductive systems (System description). In: Giesl, J., Hähnle, R. (eds.) IJCAR 2010. LNCS, vol. 6173, pp. 15–21. Springer, Heidelberg (2010)

Poswolsky, A., Schürmann, C.: Practical programming with higher-order encodings and dependent types. In: Drossopoulou, S. (ed.) ESOP 2008. LNCS, vol. 4960, pp. 93–107. Springer, Heidelberg (2008)

Pottier, F.: Static name control for FreshML. In: Symposium on Logic in Computer Science, pp. 356–365. IEEE (July 2007)

Formalizing Probabilistic Noninterference

Andrei Popescu[1,2], Johannes Hölzl[1], and Tobias Nipkow[1]

[1] Technische Universität München
[2] Institute of Mathematics Simion Stoilow of the Romanian Academy

Abstract. We present an Isabelle formalization of probabilistic noninterference for a multi-threaded language with uniform scheduling. Unlike in previous settings from the literature, here probabilistic behavior comes from both the scheduler and the individual threads, making the language more realistic and the mathematics more challenging. We study resumption-based and trace-based notions of probabilistic noninterference and their relationship, and also discuss compositionality w.r.t. the language constructs and type-system-like syntactic criteria. The formalization uses recent development in the Isabelle probability theory library.

1 Introduction

Language-based noninterference [26] is a major topic in computer security. To state noninterference, one typically assumes the program memory is separated into a *low*, or public, part, which an attacker is able to observe, and a *high*, or private, part, hidden to the attacker. A program satisfies noninterference if, upon running it, the high part of the initial memory does not affect the low part of the resulting memory. In other words, the program has *no information leaks* from the private part of the memory into the public one, so that a potential attacker should not be able to obtain information about private data by inspecting public data.

While research on language-based noninterference has been thriving in recent years, only little effort has been put in the mechanical verification of results in this area. In a previous paper [23], we presented a formalization of possibilistic noninterference, with a focus on compositionality and type-system-like syntactic criteria. Here, we continue this research agenda with noninterference for a probabilistic language. The general motivation for our formalization efforts is the belief, shared by more and more researchers lately, that the development of programming language metatheory should be pursued with the help (and confidence) offered by a proof assistant [1]. But there is also a more specific motivation. Previous work on probabilistic language-based noninterference is presented in a very informal fashion, even by the standards of a "pen-and-paper" mathematician. While justified by the complexity of the involved concepts, this situation is certainly not satisfactory. The work reported here tries to alleviate this problem, taking advantage of the recent development of a rich Isabelle/HOL library for probability theory [11, 12].

We start by formalizing a probabilistic multi-threaded language and its small-step operational semantics under a uniform scheduler (§2). Then we proceed with the formalization of noninterference properties (§3). At the heart of the formalization is an abstract equivalence $\sim$ on the memory states, called *indistinguishability*, where $s \sim t$

G. Gonthier and M. Norrish (Eds.): CPP 2013, LNCS 8307, pp. 259–275, 2013.
© Springer International Publishing Switzerland 2013

intuitively means that an attacker cannot distinguish between s and t. (For a concrete notion of state that assigns values to variables and a classification of variables as high or low, $\sim$ becomes the standard low equality, i.e., identity on the low variables.) In this context, noninterference of a program roughly means that selected parts of its execution are compatible with $\sim$, in that if starting in indistinguishable states they yield indistinguishable results. We consider two flavors of noninterference.

Resumption-based (or *bisimulation-based*) noninterference (§3.1), amply represented in the literature [3, 6–8, 27–32] requires that each execution chunk (where the chunk may be one single step or several steps, depending on the specific notion) is compatible with $\sim$ on states, *and that this property is also resumed in the matching continuations.* For a probabilistic language, it does not suffice to speak of solitary matching continuations; instead, one partitions the continuations and matches the sets of the partition in a probability-preserving way. A main advantage of resumption-based notions is usually compositionality w.r.t. the language constructs; as we argue in [23] for possibilistic noninterference, this can form the basis of the *automatic* inference of type-system-like criteria—and indeed, this also applies here (§3.3). Therefore, we take compositionality as a major test for a newly introduced resumption-based notion (§3.2). A first notion we consider is a variation of standard probabilistic bisimilarity, which is mostly compositional but does not interact well with thread-termination sensitive parallel composition. To cope with this problem, we define a weaker notion, 01-bisimilarity, relaxing the requirement to match continuation steps by allowing stutter moves.

Trace-based noninterference (§3.4), rather scarce in language-based settings [18, 34] but pervasive in system-based settings (overviewed in [17]), requires that the whole set of execution traces is compatible with $\sim$. In a possibilistic framework, this would mean that, given two indistinguishable states $s \sim t$, for any execution starting in s and ending in s' there exists an execution starting in t and ending in some t' such that $s' \sim t'$. In a probabilistic framework however, one needs to take a global view and equate, for each possible indistinguishability class S of the result, the (cumulated) measure of all traces starting in s and ending in S with the measure of all traces starting in t and ending in S. We formalize two natural trace-based notions representing end-to-end security guarantees of our two main resumption-based notions.

The results of our formal development [22] can be summarized as follows:

$$\text{syntactic criteria} \xrightarrow{\text{compositionality}} \text{resumption noninterference} \implies \text{trace noninterference}$$

Besides the certification aspect, our formalization makes new contributions to the state of the art in language-based noninterference:

– it considers for the first time a fully probabilistic language, where probabilistic behavior comes not only from the scheduler, but also from the individual threads (through probabilistic choice);
– it performs a comprehensive study, including both trace-based and resumption-based noninterference and their comparison.

On the other hand, the formalized language has several limitations:

– it restricts thread communication to shared-state communication;
– it does not cover dynamic thread creation;
– it is confined to a uniform scheduler, assigning equal probabilities to each thread.

Throughout the paper, we employ notations close to the formalization, but we occasionally take some liberties with the Isabelle notation in order to ease the presentation.

2 The Programming Language

We formalize a programming language featuring the usual sequential commands, extended with probabilistic choice and parallel composition under a uniform scheduler.

2.1 Syntax

The language is parameterized by the following types:
- **atom**, of atoms, ranged over by *atm*;
- **test**, of tests, ranged over by *tst*;
- **choice**, of (probabilistic) choices, ranged over by *ch*.

Standard examples of atoms and tests are assignments such as $x := x + y$ and Boolean expressions such as $x < y + x$. Moreover, as we discuss in §2.2, choices are flexible enough to cover the standard "if" conditions, as well as stateless probabilistic choice.

The type **com**, of commands, ranged over by c, d, is defined as follows:

$$\texttt{datatype } \textbf{com} = \text{Atm } \textbf{atom} \mid \text{Done} \mid \text{Seq } \textbf{com com} \mid \text{While } \textbf{test com} \mid$$
$$\text{Ch } \textbf{choice com com} \mid \text{Par } (\textbf{com list}) \mid \text{ParT } (\textbf{com list})$$

For atomic commands Atm *atm* we usually omit the constructor Atm. The lists of commands passed as arguments to Par and ParT will be indicated using explicit index notation, e.g., $[c_0, \ldots, c_{n-1}]$ is a list of length n—this is a detour from the Isabelle syntax aimed at making the presentation clearer.

A command is called *finished* if it is either Done or, inductively, a Par- or ParT-composition of finished commands:
finished Done;
$(\bigwedge_{i=0}^{n-1} \text{finished } c_i) \implies \text{finished } (\text{Par } [c_0, \ldots, c_{n-1}]);$
$(\bigwedge_{i=0}^{n-1} \text{finished } c_i) \implies \text{finished } (\text{ParT } [c_0, \ldots, c_{n-1}]).$

Seq c_1 c_2 is the sequential composition of c_1 and c_2, written in concrete syntax[1] as $c_1 ; c_2$. While *tst* c is the usual while loop, in concrete syntax, while *tst* do c. Ch *ch* c_1 c_2 is a choice command. Par $[c_0, \ldots, c_{n-1}]$ and ParT $[c_0, \ldots, c_{n-1}]$ are two variants of parallel composition of the thread pool $[c_0, \ldots, c_{n-1}]$, written in concrete syntax as $c_1 \parallel \ldots \parallel c_{n-1}$ and $c_1 \parallel_T \ldots \parallel_T c_{n-1}$, respectively. They differ in that the latter is termination-sensitive, removing finished threads from the thread pool.

2.2 Semantics

The semantics of the language indicates the immediate steps available to a command in a given state, where the steps are assigned weights that sum up to 1. It is parameterized by the following data:

---

[1] We use abstract syntax in theoretical results and concrete syntax in examples.

c	wt c s i	cont c s i	eff c s i
atm	1	Done	aexec c s
Done	1	Done	s
Seq c_1 c_2	wt c_1 s i	c_2, if finished c_1 Seq (cont c_1 s i) c_2, otherwise	eff c_1 s i
Ch ch c_1 c_2	cval ch s, if $i = 0$ $1 -$ cval ch s, if $i = 1$	c_1, if $i = 0$ c_2, if $i = 1$	s
While tst d	1	Seq d (While tst d), if tval tst s Done, otherwise	s
Par $[c_0, \ldots, c_{n-1}]$	$\dfrac{1}{n} *$ wt c_k s j	Par $[c_0, \ldots, \text{cont } c_k\ s\ j, \ldots, c_{n-1}]$	eff c_k s j
ParT $[c_0, \ldots, c_{n-1}]$	$\dfrac{1}{m} *$ wt c_k s j, if $\neg$ finished c_k 0, otherwise	ParT $[c_0, \ldots, \text{cont } c_k\ s\ j, \ldots, c_{n-1}]$	eff c_k s j

$(k, j) \equiv$ the unique pair in $\textbf{nat} \times \textbf{nat}$ such that $0 \le k < n \land 0 \le j < \text{brn } c_k \land i = (\sum_{l=0}^{k-1} \text{brn } c_l) + j$

$m \equiv$ the number of indexes $l \in \{0, \ldots, n-1\}$ such that $\neg$ finished c_l

Fig. 1. Probabilistic Small-Step Semantics

– a type of (memory) states, **state**, ranged over by s, t;
– an execution function for the atoms, aexec : **atom** $\rightarrow$ **state** $\rightarrow$ **state**;
– an evaluation function for the tests, tval : **test** $\rightarrow$ **state** $\rightarrow$ **bool**;
– an evaluation function for the choices, cval : **choice** $\rightarrow$ **state** $\rightarrow$ $[0, 1]$, where $[0, 1]$ is the real unit interval.

cval ch s expresses the probability with which the left branch, c_1, will be picked when executing the command Ch ch c_1 c_2 (while the right branch, c_2, will be picked with probability $1 -$ cval ch s).

For every command c, we define its *branching number* (*branching* for short), brn c:
brn $atm =$ brn Done $=$ brn (While tst d) $= 1$; brn (Seq c_1 c_2) $=$ brn c_1;
brn (Ch ch c_1 c_2) $= 2$; brn (Par $[c_0, \ldots, c_{n-1}]$) $=$ brn (ParT $[c_0, \ldots, c_{n-1}]$) $= \sum_{l=0}^{n-1}$ brn c_l.

Indexes ranging from 0 to brn $c - 1$ are used to label the single-step transitions available from a command c. Then, given current states s, these transitions are assigned *weights*, *continuation commands* (*continuations* for short) and *state effects* (*effects* for short) by the functions wt c s : $\{0, \ldots, \text{brn } c - 1\} \rightarrow [0, 1]$, cont c s : $\{0, \ldots, \text{brn } c - 1\} \rightarrow$ **com**, and eff c s : $\{0, \ldots, \text{brn } c - 1\} \rightarrow$ **state**, respectively. All these are defined in Fig. 1's table. The first column lists all possible forms of c, and the other columns show, for the three operators, the defining recursive clauses for each form (with $i \in \{0, \ldots, \text{brn } c - 1\}$).

The semantics of atm and Done are straightforward: there is one single available step (hence brn is 1) with weight 1, transiting to the terminating continuation Done. For Done, there is no effect on the state (i.e., the state s remains unchanged), and for atm, the effect is given by aexec. For technical reasons, Done has a dummy transition to itself.

A Seq c_1 c_2 command obtains its branching, weight and effect from c_1, and its continuation is the continuation of c_1 if unfinished and is c_2 otherwise. While tst d performs

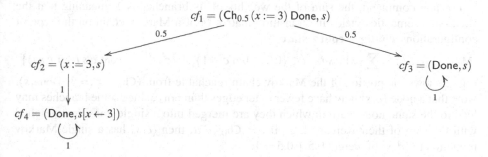

Fig. 2. Markov Chain

an unfolding step to the continuation Seq d (While tst d) if the test is True and to Done otherwise, in both cases with weight 1 and no effect.

A choice command Ch ch c_1 c_2 is assumed to perform an effectless branching according to ch. It has 2 branches, labeled 0 and 1: the left one with weight cval ch s and continuation c_1, and the right one with complementary weight $1 -$ cval ch s and continuation c_2.

If c has the form Par $[c_0, \ldots, c_{n-1}]$, then it consists of n threads, $c_0, \ldots, c_{n-1}$, running in parallel under a uniform scheduler assigning them equal probabilities. The branching of c is thus the sum of the branchings of c_l for $l \in \{0, \ldots, n-1\}$. A branch label i of c determines uniquely the numbers k and j so that i corresponds to the j's branch of c_k, via the equation $i = (\sum_{l=0}^{k-1} \text{brn } c_l) + j$. The weight of i in c is [the probability of picking thread c_k (out of n possibilities)] times [the weight of j in c_k], i.e., $(1/n) * \text{wt } c_k \ s \ j$. The i-continuation from c is obtained by replacing, in the thread pool, c_k with its j-continuation cont $c_k \ s \ j$. The i-effect of c is the j-effect of c_k, eff $c_k \ s \ j$.

ParT behaves like Par, except that it is termination-sensitive, in that finished threads are not taken into consideration (being assigned weight 0), and the choice is made among the m unfinished threads—consequently, the weight of an unfinished thread is $1/m$. In case all threads are finished, we simply add a single idle transition, with probability 1—this trivial case is not shown in the figure.

Example 1. Here is a simple standard instantiation of the generic notions of state, atomic statement and test. **state** consists of assignments of values to variables, **var** $\rightarrow$ **val**, where **val** is a type of values (e.g., integers) and **var** a countable type of variables. The atomic statements and the tests are built by means of arithmetic and boolean expressions applied to variables. The atom and test valuation functions are as expected.

Since the choice is allowed to depend on the state, it can capture not only standard probabilistic choice, but also the "if" statement [13]. We define **choice** to be $[0, 1] +$ **test**, i.e., a choice is either Inl x, the embedding of a unital real number, or Inr tst, the embedding of a test tst. Then cval (Inl x) $s = x$ for all s, making Ch (Inl x) c_1 c_2, simply written Ch$_x$ c_1 c_2, the stateless probabilistic choice. Moreover, cval (Inr tst) $s = 1$ if tval tst s $-$ True and $= 0$ otherwise, making Ch (Inr tst) c_1 c_2, simply written if tst then c_1 else c_2, the standard conditional statement.

For any command, the sum of the weights of its branches is 1, meaning that the small-step semantics yields the transition matrix M of a Markov chain on the type of configurations, **config** = **com** × **state**:

$$M (c,s) (c',s') = \sum \{\text{wt } c \, s \, i \mid i \in \{0,\ldots,\text{brn } c - 1\} \wedge (c',s') = (\text{cont } c \, s \, i, \text{eff } c \, s \, i)\}$$

Fig. 2 shows the portion of the Markov chain reachable from $(\text{Ch}_{0.5} (x := 3) \text{ Done}, s)$. Note that a node (c,s) may have fewer outer edges than brn c, since some branches may lead to the same node, case in which they are merged into a single transition weighed with the sum of their weights. E.g., if $c = \text{Ch}_{0.5} d \, d$, then (c,s) has a single Markov transition to (d,s) of weight $0.5 + 0.5 = 1$.

Let $\text{Trace}_{(c,s)} = \{(c_i,s_i)_{i \in \text{nat}} \mid (c_0,s_0) = (c,s) \wedge \forall i \in \text{nat}. \, M (c_i,s_i) (c_{i+1},s_{i+1}) > 0\}$, the set of *traces starting at* (c,s) ((c,s)-*traces* for short). A basic event for (c,s) is the set of all (c,s)-traces of a given finite prefix $(c_i,s_i)_{i=0}^{n}$; the measure of such a basic event is the product of transition weights $\prod_{i=0}^{n} M (c_i,s_i) (c_{i+1},s_{i+1})$. Let $\text{Alg}_{(c,s)}$ be the σ-algebra generated by the basic events, i.e., the smallest collection of subsets of $\text{Trace}_{(c,s)}$ that is closed under countable union and complement and contains every basic event. By standard probability theory [15], there is a unique probability measure $\text{Pr}_{(c,s)} : \text{Alg}_{(c,s)} \rightarrow [0,1]$ extending the measure of basic events. ([11] describes in detail the formalization of the involved standard constructions.) For example, in Fig. 2's Markov chain, we have only two (c_0,s)-traces, $cf_1 \, cf_2 \, cf_4^\omega$ and $cf_1 \, cf_3^\omega$. In general, the set of traces may be infinite, even uncountable. We have $\text{Pr}_{(c_0,s)} \{cf_1 \, cf_2 \, cf_4^\omega\} = \text{Pr}_{(c_0,s)} \{cf_1 \, cf_3^\omega\} = 0.5$.

3   Noninterference

We fix a relation $\sim$ on states, called *indistinguishability*, where $s \sim t$ is meant to say "s and t are indistinguishable by the attacker."

Example 2. In the context of Example 1, $\sim$ is often defined as follows. Variables are classified as either low (lo) or high (hi) by a given security level function sec : **var** $\rightarrow$ $\{\text{lo}, \text{hi}\}$. Then $\sim$ is defined as coincidence on the low variables, with the intuition that the attacker can only observe these: $s \sim t \equiv \forall x \in \textbf{var}. \, \text{sec } x = \text{lo} \Longrightarrow s \, x = t \, x$.

Noninterference of a program states that its execution is compatible with the indistinguishability relation: given two indistinguishable states s and t, (partially) executing the program once starting from s and once starting from t yield indistinguishable states.

There are two main types of formulation of noninterference: as an indefinite indistinguishability resumption property (bisimilulation) and as a property of alternative execution traces. The seminal paper [33] proposes an end-to-end noninterference property using big-step semantics, which can be seen as a property of traces. Much of subsequent work [3, 6–8, 27–32] prefers small-step semantics and resumption-based notions, although trace-based notions are also considered [18, 34].

In the presence of concurrency, resumption-based notions have been shown to be more compositional (which was no surprise, since this phenomenon is known from process algebra). Sabelfeld and Sands [27] were the first to observe the tight connection between the compositionality of resumption-based notions and sufficient type-system criteria—in a previous paper [23], we used this idea to devise a uniform methodology for extracting syntactic criteria from compositionality.

On the other hand, trace-based notions are often more intuitive to grasp, as they do not involve the alternation complexity of bisimulations. Also, trace-based notions can benefit from other kinds of static analyses, such as data-race analysis [34].

Next, we study and relate the two flavors of noninterference, including compositionality and syntactic criteria, for the introduced probabilistic language.

3.1 Resumption-Based Noninterference

We define the following notions of self isomorphism, siso, and discreetness, discr, *coinductively as greatest fixed points*, i.e., as the *weakest* predicates satisfying certain equations. (They are probabilistic counterparts of possibilistic notions introduced in [23].)

For siso, one requires that, if started in indistinguishable states, executions take the same branches *with the same probabilities*. For discr, one requires that, during the computation, the states stay indistinguishable from the initial state.

siso $c \equiv (\forall s\, t\, i.\ s \sim t\ \wedge\ i < \text{brn } c \implies \text{wt } c\, s\, i = \text{wt } c\, t\, i\ \wedge\ \text{cont } c\, s\, i = \text{cont } c\, t\, i\ \wedge$
eff $c\, s\, i \sim$ eff $c\, t\, i)\ \wedge\ (\forall s\, i.\ i < \text{brn } c \implies \text{siso } (\text{cont } c\, s\, i))$
discr $c \equiv \forall s\, i.\ i < \text{brn } c \implies s \sim \text{eff } c\, s\, i\ \wedge\ \text{discr } (\text{cont } c\, s\, i)$

siso and cont are very demanding notions of security. To define weaker notions, we need to allow alternative executions to take different branches, while also allowing execution to change the indistinguishability class of the state. It is easy to notice that these two relaxations lead us to the consideration of (binary) relations rather than unary predicates. Indeed, if the command c branches according to a high test in two continuations d_1 and d_2, then the notion of security of c is conditioned by the notion of "equivalence" of d_1 and d_2. This equivalence will be defined as bisimilarity, while security of c will be defined as self bisimilarity (c bisimilar to itself).

To introduce probabilistic bisimulation, we need a few preparations. Given $I \subseteq \{0, \dots, \text{brn } c - 1\}$, we write Wt $c\, s\, I$ for the cumulated weights from (c, s) of the labels in I, namely, $\sum_{i \in I} \text{wt } c\, s\, i$. Given sets A and P, we say P is a *partition* of A, written part $A\, P$, if P consists of mutually disjoint nonempty sets whose union is A.

The following predicate match$^c_{\xi}$ (read "match continuation against continuation") shows, for a relation on commands θ and two commands c and d, how the steps taken by c and d are matched unambiguously and weight-exhaustively, so that their effects are indistinguishable and their continuations are in θ:
match$^c_{\xi}$ $\theta\, c\, d \equiv$
$\forall s\, t.\ s \sim t \implies \exists P\, Q\, F.$
part $\{0, \dots, \text{brn } c - 1\}\, P\ \wedge\ $ part $\{0, \dots, \text{brn } d - 1\}\, Q\ \wedge\ [F : P \to Q\ \text{bijection}]\ \wedge$
$(\forall I \in P.\ \text{Wt } c\, s\, I = \text{Wt } d\, t\, (F\, I)\ \wedge$
$\qquad (\forall i \in I.\ \forall j \in F\, I.\ \text{eff } c\, s\, i \sim \text{eff } d\, t\, j\ \wedge\ \theta\ (\text{cont } c\, s\, i)\ (\text{cont } d\, t\, j)))$

Thus, match$^c_{\xi}$ $\theta\, c\, d$ states that there exist partitions P and Q of the branches of c and d and a bijective correspondence $F : P \to Q$ so that, for any corresponding sets of branches I and $F\, I$:

The cumulated weights are the same.
– For any pair (i, j) of branches in these sets, the effects are indistinguishable and the continuations are in θ.

Strong bisimilarity, $\approx_s$, is defined coinductively as the largest (i.e., weakest) relation satisfying $\forall c, d.\ c \approx_s d \iff \text{match}^C_C (\approx_s)\ c\ d$, or, equivalently, the largest relation satisfying $\forall c, d.\ c \approx_s d \implies \text{match}^C_C (\approx_s)\ c\ d$. If we ignore preservation of the state indistinguishability, this boils down to a well known property of Markov chains called probabilistic bisimulation or lumpability [15, 16].

According to the insight obtained in [23], a good during-execution noninterference candidate should be compositional with the language constructs and weaker than both siso and discr. It turns out that $\approx_s$ has many of these characteristics, in particular, it will be shown to commute with all the constructs except for While and ParT.

To compensate for the lack of ParT-compositionality of $\approx_s$, we introduce a weaker relation, $\approx_{01}$, that we call 01-*bisimilarity* because it requires a step to be matched by either no step (a stutter move) or one step. Its characteristic matcher is

$\text{match}^C_{01C}\ \theta\ c\ d \equiv$

$\forall s\ t.\ s \sim t \implies \exists P\ Q\ I_0\ F.$

$\text{part}\ \{0, \ldots, \text{brn}\ c - 1\}\ P \wedge \text{part}\ \{0, \ldots, \text{brn}\ d - 1\}\ Q \wedge I_0 \in P \wedge [F : P \to Q\ \text{bijection}]$

$$(\forall I \in P - \{I_0\}.\ \frac{\text{Wt}\ c\ s\ I}{1 - \text{Wt}\ c\ s\ I_0} = \frac{\text{Wt}\ d\ t\ (F\ I)}{1 - \text{Wt}\ d\ t\ (F\ I_0)} \wedge$$

$$(\forall i \in I.\ \forall j \in F\ I.\ \text{eff}\ c\ s\ i \sim \text{eff}\ d\ t\ j \wedge \theta\ (\text{cont}\ c\ s\ i)\ (\text{cont}\ d\ t\ j))) \wedge$$

$$(\forall i \in I_0.\ s \sim \text{eff}\ c\ s\ i \wedge \theta\ (\text{cont}\ c\ s\ i)\ d) \wedge$$

$$(\forall j \in F\ I_0.\ t \sim \text{eff}\ d\ t\ j \wedge \theta\ c\ (\text{cont}\ d\ t\ j))$$

$\text{match}^C_{01C}\ \theta\ c\ d$ relaxes match^C_C to allow matching continuation steps not only by continuation steps, but also by stutter moves. Thus, in the partitions P and Q, one singles out sets of stutter branches I_0 and $F\ I_0$ whose effects are required to preserve indistinguishability and whose continuations are required to be in relation θ with the other party's source command, d or c. Moreover, the cumulated weights in corresponding non-stutter sets of branches are no longer required to be equal, but only equal relatively to the cumulated weights of all non-stutter branches (i.e., to 1 minus the cumulated weights of the stutter ones).

01-*bisimilarity*, $\approx_{01}$, is now defined analogously to $\approx_s$, as the largest relation satisfying $\forall c\ d.\ c \approx_{01} d \iff \text{match}^C_{01C} (\approx_{01})\ c\ d$. The notion of security associated to a bisimilarity is its diagonal version, which we call self bisimilarity. Thus, c is called *self strongly-bisimilar* if $c \approx_s c$ and *self* 01-*bisimilar* if $c \approx_{01} c$.

The 01-steps relaxation scheme is well-known in language-based possibilistic bisimulations [6–8, 23], and corresponds to the triangle unwinding scheme in system-based security [17]—our relation $\approx_{01}$ seems to be its first probabilistic adaptation.

Note that all the above four notions of security are defined employing universal quantification over the relevant current states (either s alone for discr or the indistinguishable states s and t for the other notions), which are therefore "refreshed" at each resumption point. This means that security is defined *interactively*, guaranteeing correct behavior under the assumption that the environment (consisting perhaps of the attacker or of the other threads) may change the state at any point. This is crucial for compositionality [23, 25, 27]. It is immediate to prove by coinduction that $\approx_{01}$ is weaker than $\approx_s$, which in turn is weaker than siso and discr:

Prop 1. *The implications shown in the left of Fig. 3 hold.*

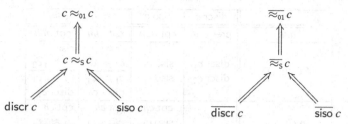

Fig. 3. Resumption-Based Notions and Syntactic Criteria

3.2 Compositionality

Here we establish the compositionality properties of the two resumption-based notions w.r.t. the language constructs and discuss their relative strengths and weaknesses.

First, we need atomic properties of preservation and compatibility adapted to the abstract notions of state and indistinguishability relation. An atom *atm* is called $\sim$-*preserving*, written pres *atm*, if $\forall s.$ aexec *atm* $s \sim s$; it is called $\sim$-*compatible*, written cpt *atm*, if $\forall s\, t.\ s \sim t \implies$ aexec *atm* $s \sim$ aexec *atm* t. A test *tst* is called $\sim$-*compatible*, written cpt *tst*, if $\forall s\, t.\ s \sim t \implies$ tval *tst* $s =$ tval *tst* t. A choice *ch* is called $\sim$-*compatible*, written cpt *ch*, if $\forall s\, t.\ s \sim t \implies$ cval *ch* $s =$ cval *ch* t.

In the setting of Example 2, for atoms, $\sim$-preservation means no assignment to low variables and $\sim$-compatibility means no direct leaks, i.e., no assignment to low variables of expressions depending on high variables. Moreover, for tests, $\sim$-compatibility means no dependence on high variables. A stateless choice is always compatible and an "if" choice is compatible if it is so as a test.

The next proposition states various compositionality results, schematically represented in Fig. 4 as follows. The first column lists the possible forms of a command c (c may be an atom *atm*, or have the form Seq $c_1\, c_2$, etc.). The next columns list conditions under which the predicates stated on the first row hold for c. Thus, e.g., row 4 column 3 says: if cpt *ch*, siso c_1 and siso c_2, then siso (Ch *ch* $c_1\, c_2$). The horizontal line in row 3 columns 4 and 5 represents an "or"—thus, e.g., row 3 column 4 says: if either [siso c_1 and $c_2 \approx_s c_2$] or [$c_1 \approx_s c_1$ and discr c_2] then Seq $c_1\, c_2 \approx_s$ Seq $c_1\, c_2$.

Prop 2. *The compositionality facts stated in Fig. 4 hold.*

As expected, compatibility is a minimal security requirement for atoms, with discreetness requiring even preservation. Sequential composition behaves perfectly w.r.t. siso and discr, but for $\approx_s$ and $\approx_{01}$ it requires strengthening either to siso on the left component or to discr on the right component. For choice, compatibility is required for all notions except for discr; in the particular case of "if" tests, this becomes the well-known "no high test" condition; for stateless choices, the condition is vacuously true.

The compositionality w.r.t. Par and ParT reveal some interesting phenomena. While in the possibilistic case we have shown that, in the presence of the aforementioned interactivity proviso, parallel composition is unconditionally compositional, here the situation is less convenient. Possibilistically, it makes no difference whether or not a finished thread is removed from the pool. Indeed, scheduling it to take a stutter move has no possibilistic effect. However, it does have the effect of delaying the steps taken by other threads. This is already discussed in [29, 30] for sequential threads, and is

c	discr c	siso c	$c \approx_s c$	$c \approx_{01} c$
atm	pres *atm*	cpt *atm*	cpt *atm*	cpt *atm*
Seq c_1 c_2	discr c_1 discr c_2	siso c_1 siso c_2	siso c_1 $c_2 \approx_s c_2$ $c_1 \approx_s c_1$ discr c_2	siso c_1 $c_2 \approx_{01} c_2$ $c_1 \approx_{01} c_1$ discr c_2
Ch *ch* c_1 c_2	discr c_1 discr c_2	cpt *ch* siso c_1 siso c_2	cpt *ch* $c_1 \approx_s c_1$ $c_2 \approx_s c_2$	cpt *ch* $c_1 \approx_{01} c_1$ $c_2 \approx_{01} c_2$
While *tst* d	discr d	cpt *tst* siso d	False	False
Par $[c_0,\dots,c_{n-1}]$	discr c_l $0 \le l < n$	siso c_l $0 \le l < n$	$c_l \approx_s c_l$ $0 \le l < n$	False
ParT $[c_0,\dots,c_{n-1}]$	discr c_l $0 \le l < n$	False	False	$c_l \approx_s c_l$ $0 \le l < n$

Fig. 4. Compositionality of Resumption-Based Noninterference

reflected in our case by $\approx_s$ not being compositional w.r.t. ParT. Also, $\approx_{01}$ is not ParT-compositional either. Fortunately, ParT composition of $\approx_s$-related threads yields $\approx_{01}$-related results, which saves the day—this is the main reason for introducing $\approx_{01}$.

Another problem that seems specific to probabilistic semantics is that $\approx_s$ and $\approx_{01}$ are not compositional w.r.t. While. The main reason is that both $\approx_s$ and $\approx_{01}$ are termination-insensitive, and hence they do not detect, in a while loop, when the body command has finished executing, which makes synchronization impossible in the bisimilarity game.

These compositionality problems are actually not as bad as they may seem: as we discuss in §3.3, when proving noninterference of a command c, if a notion fails to be compositional w.r.t. the construct from the top of c, one can fall back on a stronger notion, for which the proof can progress.

3.3 Syntactic Criteria

With the implications between bisimilarities and their compositionality facts in place, we can automatically infer type-system criteria. This was described in [23, §6] as a "table-and-graph" method for a possibilistic programming language. Since the analysis from there is semantics-independent, it also applies here. For each security notion $\chi \in \{\text{discr}, \text{siso}, \approx_s, \approx_{01}\}$, we define a function $\overline{\chi} : \textbf{com} \to \textbf{bool}$ following a potential attempt to prove $\chi\, c$, first using the corresponding compositionality fact from Fig. 4's table, and, if this fails, falling back on stronger notions given by the predecessors of χ from Fig. 3's left graph. Thus, the cell corresponding to the form of c and the notion χ contains some properties of the components of c and possibly a side condition. Then:
– if the side condition holds, then $\overline{\chi}\, c$ is defined recursively as the conjunction of all $\overline{\chi'}\, c'$, where $\chi'\, c'$ are the listed conditions for the components c' of c;
– otherwise, $\overline{\chi}\, c$ is defined as the disjunction of all $\overline{\chi'}\, c$, where χ' are the immediate predecessors of χ in Fig. 3's left graph.

Concretely:

$\overline{\text{discr}}\ atm \Longleftrightarrow \text{pres}\ atm;\ \overline{\text{discr}}\ (\text{Seq}\ c_1\ c_2) \Longleftrightarrow \overline{\text{discr}}\ (\text{Ch}\ ch\ c_1\ c_2) \Longleftrightarrow \overline{\text{discr}}\ c_1 \wedge \overline{\text{discr}}\ c_2;$
$\overline{\text{discr}}\ (\text{While}\ tst\ d) \Longleftrightarrow \overline{\text{discr}}\ d;$
$\overline{\text{discr}}\ (\text{Par}\ [c_0,\dots,c_{n-1}]) \Longleftrightarrow \overline{\text{discr}}\ (\text{ParT}\ [c_0,\dots,c_{n-1}]) \Longleftrightarrow \bigwedge_{i=0}^{n-1} \overline{\text{discr}}\ c_i;$

$\overline{\text{siso}}\ atm \Longleftrightarrow \text{cpt}\ atm;\ \overline{\text{siso}}\ (\text{Seq}\ c_1\ c_2) \Longleftrightarrow \overline{\text{siso}}\ c_1 \wedge \overline{\text{siso}}\ c_2;$
$\overline{\text{siso}}\ (\text{Ch}\ ch\ c_1\ c_2) \Longleftrightarrow \text{cpt}\ ch \wedge \overline{\text{siso}}\ c_1 \wedge \overline{\text{siso}}\ c_2;\ \overline{\text{siso}}\ (\text{While}\ tst\ d) \Longleftrightarrow \text{cpt}\ tst \wedge \overline{\text{siso}}\ d;$
$\overline{\text{siso}}\ (\text{Par}\ [c_0,\dots,c_{n-1}]) \Longleftrightarrow \bigwedge_{i=0}^{n-1} \overline{\text{siso}}\ c_i;\ \overline{\text{siso}}\ (\text{ParT}\ [c_0,\dots,c_{n-1}]) \Longleftrightarrow \text{False};$

$\overline{\approx_s}\ atm \Longleftrightarrow \text{cpt}\ atm;\ \overline{\approx_s}\ (\text{Seq}\ c_1\ c_2) \Longleftrightarrow (\overline{\text{siso}}\ c_1 \wedge \overline{\approx_s}\ c_2) \vee (\overline{\approx_s}\ c_1 \wedge \overline{\text{discr}}\ c_2);$
$\overline{\approx_s}\ (\text{Ch}\ ch\ c_1\ c_2) \Longleftrightarrow \text{cpt}\ ch \wedge \overline{\approx_s}\ c_1 \wedge \overline{\approx_s}\ c_2;$
$\overline{\approx_s}\ (\text{While}\ tst\ d) \Longleftrightarrow \overline{\text{siso}}\ (\text{While}\ tst\ d) \vee \overline{\text{discr}}\ (\text{While}\ tst\ d);$
$\overline{\approx_s}\ (\text{Par}\ [c_0,\dots,c_{n-1}]) \Longleftrightarrow \bigwedge_{i=0}^{n-1} \overline{\approx_s}\ c_i;$
$\overline{\approx_s}\ (\text{ParT}\ [c_0,\dots,c_{n-1}]) \Longleftrightarrow \overline{\text{siso}}\ (\text{ParT}\ [c_0,\dots,c_{n-1}]) \vee \overline{\text{discr}}\ (\text{ParT}\ [c_0,\dots,c_{n-1}]);$

$\overline{\approx_{01}}\ atm \Longleftrightarrow \text{cpt}\ atm;\ \overline{\approx_{01}}\ (\text{Seq}\ c_1\ c_2) \Longleftrightarrow (\overline{\text{siso}}\ c_1 \wedge \overline{\approx_{01}}\ c_2) \vee (\overline{\approx_{01}}\ c_1 \wedge \overline{\text{discr}}\ c_2);$
$\overline{\approx_{01}}\ (\text{Ch}\ ch\ c_1\ c_2) \Longleftrightarrow \text{cpt}\ ch \wedge \overline{\approx_{01}}\ c_1 \wedge \overline{\approx_{01}}\ c_2;\ \overline{\approx_{01}}\ (\text{While}\ tst\ d) \Longleftrightarrow \overline{\approx_s}\ (\text{While}\ tst\ d);$
$\overline{\approx_{01}}\ (\text{Par}\ [c_0,\dots,c_{n-1}]) \Longleftrightarrow \overline{\approx_s}\ (\text{Par}\ [c_0,\dots,c_{n-1}]);$
$\overline{\approx_{01}}\ (\text{ParT}\ [c_0,\dots,c_{n-1}]) \Longleftrightarrow \bigwedge_{i=0}^{n-1} \overline{\approx_s}\ c_i.$

The above are valid recursive definitions: each operator $\overline{\chi}$ is defined recursively in terms of itself and/or in terms of previously defined operators. Note how, when compositionality for a notion fails, stronger notions are invoked, e.g.:
$\overline{\approx_s}\ (\text{While}\ tst\ d) \Longleftrightarrow \overline{\text{siso}}\ (\text{While}\ tst\ d) \vee \overline{\text{discr}}\ (\text{While}\ tst\ d) \Longleftrightarrow (\text{cpt}\ tst \wedge \overline{\text{siso}}\ d) \vee \overline{\text{discr}}\ d$

We can prove that the syntactic notions are indeed sufficient criteria for their semantic counterparts, and that the former inherit the hierarchy of the latter.

Prop 3. *(1) For any χ in Fig. 3 left, it holds that $\forall c.\ \overline{\chi}\ c \Longrightarrow \chi\ c$.*
(2) The implications shown in the right of Fig. 4 hold.

The next example illustrates the semantic notions and their syntactic criteria:

Example 3. Consider the following commands, with h, h' high and l, l' low variables.
$- d_0$: $h' := 0$; while $h > 0$ do $\text{Ch}_{0.5}\ (h := 0)\ (h' := h' + 1)$
$- d_1$:   while $h > 0$ do $\text{Ch}_{0.5}\ (h := h - 1)\ (h := h + 1)$
$- d_2$:   if $l = 0$ then $l' := 1$ else d_0
$- d_3$: $h := 5$; $(d_0 \parallel_T l := 1)$
$- d_4$:   (if $h = 0$ then $h := 1$; $h := 2$ else $h := 3$) ; $l := 4$
$- d_5$: $d_4 \parallel_T l := 5$

Provided initially $h > 0$, d_0 has the effect of assigning a random geometrically distributed integer value to h', and d_1 performs a one-dimensional random walk with the value of h (the so-called gambler's ruin), resulting invariably in exit (at $h = 0$) with probability 1. d_3 illustrates one advantage of being able to nest parallel composition inside sequential composition—the possibility to make some global initializations (here, $h := 5$) before starting up the thread pool (here, containing two threads, d_0 and $l := 1$).

d_0 and d_1 are discreet, d_2 is self strongly bisimilar, and d_3 is self 01-bisimilar, but not self strongly bisimilar due to the presence of ParT whose compositionality requires the shift from $\approx_s$ to $\approx_{01}$. (d_3 would become self strongly bisimilar had we replaced ParT with Par.) Indeed, d_0–d_3 are deemed secure by the syntactic criteria from Fig. 3, e.g.:

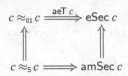

$$c \approx_{01} c \xRightarrow{\text{aeT } c} \text{eSec } c$$

$$c \approx_s c \Longrightarrow \text{amSec } c$$

Fig. 5. Resumption-Based and Trace-Based Notions of Security

$$\overline{\approx_{01}} \, d_3$$

$\overline{\text{siso}} \, (h := 5) \wedge \overline{\approx_{01}} \, (d_0 \,\|_T\, l := 1)$	$\vee$	$\overline{\approx_{01}} \, (h := 5) \wedge \overline{\text{discr}} \, (d_0 \,\|_T\, l := 1)$	$\Longleftrightarrow$
True $\wedge \overline{\approx_s} \, d_0 \wedge \overline{\approx_s} \, (l := 1)$	$\vee$	True $\wedge \overline{\text{discr}} \, d_0 \wedge \overline{\text{discr}} \, (l := 1)$	$\Longleftrightarrow$
		...	$\Longleftrightarrow$
True $\wedge$ True $\wedge$ True	$\vee$	True $\wedge$ True $\wedge$ False	$\Longleftrightarrow$
True	$\vee$	False	$\Longleftrightarrow$
	True		

On the other hand, d_4 and d_5 are not secure, not even according to $\approx_{01}$. The problem with d_4 is that the timing of the low assignment $l := 4$ depends on the value of the high variable h, which can cause probabilistic leaks when placed in parallel with other threads that may update l. d_5 shows such a situation: the initial value of h influences the likelihood that $l := 4$ is executed before $l := 5$. And indeed, d_4 and d_5 are rejected by all the syntactic criteria, e.g.:

$$\overline{\approx_{01}} \, d_4$$

$\overline{\text{siso}} \, (\text{if } h = 0 \, ...) \wedge \overline{\approx_{01}} \, (l := 4)$	$\vee$	$\overline{\approx_{01}} \, (\text{if } h = 0 \, ...) \wedge \overline{\text{discr}} \, (l := 4)$	$\Longleftrightarrow$
		...	$\Longleftrightarrow$
False $\wedge$ True	$\vee$	True $\wedge$ False	$\Longleftrightarrow$
	False		

3.4 Trace-Based Noninterference

Both $\approx_s$ and $\approx_{01}$ protect against the following end-to-end kind of probabilistic attack: The attacker may run the program multiple times and collect statistical information about the distribution of the final state up to $\sim$ (which corresponds to the low part of the memory); however, this data will never allow the attacker to infer anything about the initial state beyond the $\sim$-abstraction (which corresponds to the high part of the memory). Such a property is best formalized as trace-based noninterference.

For technical reasons, all our execution traces are infinite, with dummy transitions added for finished commands. We call *terminating* those traces reaching a configuration whose command is finished: $\text{termin} \, (c_i, s_i)_{i \in \mathbf{nat}} \equiv \exists i \in \mathbf{nat}. \text{finished } c_i$. Since dummy transitions do not affect the state, *the final state* of a terminating trace, $\text{fstate} \, (c_i, s_i)_{i \in \mathbf{nat}}$, is well defined as the unique s such that $\exists i. s = s_i \wedge \text{finished } c_i$.

We define the following sets of traces, for any (c, s), n and t:
$$T_{(c,s),n,t} \equiv \{(c_i, s_i)_{i \in \mathbf{nat}} \in \text{Trace}_{(c,s)} \mid s_n \sim t\},$$
the set of (c, s)-traces whose n-th configuration's state is indistinguishable from t;
$$T_{(c,s),t} \equiv \{(c_i, s_i)_{i \in \mathbf{nat}} \in \text{Trace}_{(c,s)} \mid \text{termin} \, (c_i, s_i)_{i \in \mathbf{nat}} \wedge \text{fstate} \, (c_i, s_i)_{i \in \mathbf{nat}} \sim t\},$$
the set of terminating (c, s)-traces whose final state is indistinguishable from t.

The set of terminating states, as well as $T_{(c,s),n,t}$ and $T_{(c,s),t}$, are all measurable sets since they can be written as countable unions of countable intersections of basic events. We say c *almost everywhere terminates*, written aeT c, if $\forall s. \text{Pr}_{(c,s)} \, \{(c_i, s_i)_{i \in \mathbf{nat}} \in \text{Trace}_{(c,s)} \mid \text{termin} \, (c_i, s_i)_{i \in \mathbf{nat}}\} = 1$, i.e., the set of terminating (c, s)-traces has measure 1.

We can now define the following trace-based notions of noninterference:
– *Any-moment security* states that, for any two executions starting in indistinguishable states and any given time, the probability of being at that time in any given indistinguishability class is the same:

$$\mathsf{amSec}\; c \equiv \forall s_1\; s_2.\; s_1 \sim s_2 \Longrightarrow \forall n\; t.\; \mathrm{Pr}_{(c,s_1)}\; T_{(c,s_1),n,t} = \mathrm{Pr}_{(c,s_2)}\; T_{(c,s_2),n,t}\; - \text{End}$$

security states that, for any two executions starting in indistinguishable states, the probability of ending up in any given indistinguishability class is the same:

$$\mathsf{eSec}\; c \equiv \forall s_1\; s_2.\; s_1 \sim s_2 \Longrightarrow \forall t.\; \mathrm{Pr}_{(c,s_1)}\; T_{(c,s_1),t} = \mathrm{Pr}_{(c,s_2)}\; T_{(c,s_2),t}$$

Any-moment security is a strong guarantee: even if one is able to observe the distribution of the low memory at any given moment, one still cannot infer anything about the initial high memory. On the other hand, end security warrants something weaker: that the final distribution of the low memory tells nothing about the initial high memory. One can prove that $\approx_s$ implies any-moment security, and that this in turn implies end security. More interestingly, $\approx_{01}$ implies end security if we also assume almost-everywhere termination; roughly, the last assumption is necessary to make sure that the "bisimulation noise" caused by stutter moves cannot delay synchronization forever, but eventually becomes negligible.

Prop 4. *The implications listed in Fig. 5 hold.*

In Example 3, d_0–d_3 are all $\approx_{01}$-secure programs and are also almost-everywhere terminating, hence they satisfy eSec by Prop. 4. Moreover, since d_2 is $\approx_s$-secure, it satisfies the stronger property amSec by Prop. 4. d_5 does not satisfy eSec, since the distribution of the final low memory reveals whether h is 0 or not: if $h = 0$, then 1 out of 4 executions yields $l = 4$; otherwise, only 1 out of 8. On the other hand, even though d_4 is not $\approx_{01}$-secure, it obviously satisfies eSec, since all its executions yield $l = 4$.

4   Overview and Statistics

Our formal development [22] amounts to about 8000 lines of scripts in Isabelle [20]. Fig. 6 shows the main theory structure, indicating for each theory the number of lines and the corresponding sections of the paper. The types and functions parameterizing the language syntax and semantics, as well as the state-indistinguishability relation $\sim$, are fixed in Isabelle locales [14]—theory Concrete instantiates these locales as discussed in Examples 1 and 2.

The language semantics was tedious to formalize due to parallel composition (especially the termination-sensitive one), which involves list-index manipulation—employing iterated binary parallel composition instead (as in the possibilistic case [23]) was not an option, since the scheduler needs to address the thread pool as a whole. On the other hand, probabilistic semantics displays a certain conceptual simplification over traditional nondeterministic semantics: for each language construct, the direct rules and the inversion rules are merged into "direct" quantitative equations (as described in Fig. 1).

We defined the probabilistic bisimulations on the concrete branches of the operational semantics, and not on the more abstract Markov-chain transitions (which may identify some of the branches); indeed, the branches provided us with a good notation

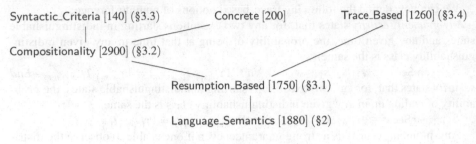

Fig. 6. Isabelle Theory Structure

for partitioning the continuations and, more importantly, with the right level of abstraction for proving compositionality facts without having to query whether some continuations happen to be equal. On the other hand, we used the general-purpose Markov-chain construction of the traces and their probability space [11] as opposed to building them from branches, which of course saved us much background work but complicated a little the proofs relating resumption notions with trace notions.

The largest and most laborious part (roughly 36% of the whole development) deals with the compositionality results listed in Prop. 2—the bisimulation relations provided as witnesses in coinductive proofs involved tedious constructions of partitions and sums over sets. Isabelle's Sledgehammer tool for deploying external automatic theorem provers [21], very helpful in discharging goals on possibilistic bisimulations, was less helpful here, where the $\forall\exists$ scheme of traditional bisimulations gives way to a quantitative $\forall\Sigma$ scheme. Another laborious task was establishing the connection between trace-based and resumption-based notions, which also involved heavy sum reasoning.

Having the compositionality preparations, the inference of syntactic criteria (Prop. 3) was immediate, with the induction goals discharged automatically by the simplifier and the classical reasoner. The route through compositionality facts localized at each language construct does better justice to noninterference results than previous formulations from the literature [18, 28–30, 32], which rely on complex monolithic proofs.

5   Conclusions and Related Work

We have formalized noninterference properties for a multi-threaded language with probabilistic choice and uniform scheduler. Distinguishing features of our approach are the comprehensive study, covering both resumption-based and trace-based notions, and the automatic extraction of syntactic criteria from compositionality. Moreover, all previous workconsiders systems of sequential deterministic threads run in parallel by a probabilistic scheduler. Our language is more powerful, allowing pervasive probabilistic behavior, including probabilistic threads. This makes the mathematical analysis more challenging, since each thread yields a Markov chain, which needs to be combined by parallel composition in the larger Markov chain of the thread pool. In fact, the very notions of thread and thread pool are relative here, since the language allows nesting parallel composition into other constructs (e.g., having Seq on top of Par or ParT), although, as we have seen, security requirements restrict some of this expressiveness.

If we are to identify a "pen-and-paper" reference for our formalized resumption-based notions, the closest is a series of papers by Smith and others [28–30, 32], which progressively introduces notions analogous to ours. Specifically, [32] introduces self isomorphism, [28] strong bisimilarity, and [29, 30] a notion weaker than our 01-bisimilarity called weak bisimilarity. In each case, the type system proved sound in there is equivalent to our syntactic criterion uniformly extracted from compositionality (Prop. 3).

A further point of convergence with the above works is the consideration of various flavors of parallel composition. [32] and [28] consider the termination-insensitive Par, while later work [29, 30] focuses on the termination-sensitive ParT. Retrospectively, in the light of the compositionality facts of Prop. 2, this is not surprising, since siso and $\approx_s$ are both compositional w.r.t. Par, and $\approx_{01}$ is quasi-compositional w.r.t. ParT.

Interestingly, all of the above works prove that the type system implies the corresponding resumption-based version, but they allude informally to a trace-based notion as the ultimately targeted security guarantee. E.g., [28, page 10] reads: "the probability that the low variables have certain values after k steps is the same when starting from (O, μ) as where starting from (O, v)" (where (O, μ) and (O, v) are bisimilar configurations)—we have formalized this as any-moment security. Also, [29, page 8] reads: "the probability that the low variables end up with some values from (O, μ) is the same as the probability that they end up with those values from (O, v)"—we have formalized this as end security. Establishing formally the relationship between a resumption-based notion and a trace-based notion can range from routine (as in $\approx_s$ versus any-moment security) to highly nontrivial (as in $\approx_{01}$ versus end security).

There are some extensions and generalizations of probabilistic semantics and noninterference not covered by our formalization. Smith [29] also considers a protect command enforcing atomicity of execution. (In principle, our formalization can handle this language construct by "instantiating" the **atom** parameter to a type mutually recursive with **com**.) He also sketches an extension to dynamic thread creation. Sabelfeld and Sands [27] show that a type system corresponding to our syntactic criterion $\overline{\text{siso}}$ for self isomorphism is strong enough to ensure noninterference for any scheduler, not only the uniform one. Mantel and Sudbrock [18] relax the $\overline{\text{siso}}$ requirement, while still covering relevant schedulers such as uniform and round robin. Our own "pen-and-paper" work [24] generalizes the results of Smith based on weak bisimilarity [29, 30] to a different class of schedulers than Mantel and Sudbrock's, providing an arguably more manageable criterion for schedulers and a stronger security guarantee.

Our end security is a generalization of the one from [18], where one defines the property only for globally terminating thread pools—this simplifying assumption allows an elementary treatment of the relevant probabilities, not requiring measure theory. We relax the termination requirement to almost-everywhere termination. This relaxation is relevant for probabilistic languages, where many interesting programs terminate only almost everywhere—this also happens to be the case for d_0–d_3 in Example 3.

Probabilistic, but single-threaded languages in the style of pGCL [19] have been formalized before in HOL4 [13], Coq [2, 4, 5] and Isabelle [9]. In very recent work [10], Cock verifies in Isabelle a lattice scheduler (a uniform scheduler that distinguishes between high and low processes) aimed at closing covert channels such as cash channels.

While the work does not target a programming language, the scheduler itself is specified as a program in pGCL and shown probabilistically noninterfering w.r.t. a version of lumpability.

Acknowledgments. We thank the reviewers for comments that helped improve the presentation. This work was supported by the DFG project Ni 491/13–2 (part of the DFG priority program Reliably Secure Software Systems–RS3) and the DFG RTG 1480.

References

1. The POPLmark challenge (2009), http://www.seas.upenn.edu/~plclub/poplmark/
2. Audebaud, P., Paulin-Mohring, C.: Proofs of randomized algorithms in Coq. S. of Comp. Prog. 74(8), 568–589 (2009)
3. Barthe, G., D'Argenio, P.R., Rezk, T.: Secure information flow by self-composition. In: CSFW, pp. 100–114 (2004)
4. Barthe, G., Daubignard, M., Kapron, B.M., Lakhnech, Y.: Computational indistinguishability logic. In: CCS, pp. 375–386 (2010)
5. Barthe, G., Grégoire, B., Béguelin, S.Z.: Formal certification of code-based cryptographic proofs. In: POPL, pp. 90–101 (2009)
6. Barthe, G., Nieto, L.P.: Formally verifying information flow type systems for concurrent and thread systems. In: FMSE, pp. 13–22 (2004)
7. Boudol, G.: On typing information flow. In: Van Hung, D., Wirsing, M. (eds.) ICTAC 2005. LNCS, vol. 3722, pp. 366–380. Springer, Heidelberg (2005)
8. Boudol, G., Castellani, I.: Noninterference for concurrent programs and thread systems. Theoretical Computer Science 281(1-2), 109–130 (2002)
9. Cock, D.: Verifying probabilistic correctness in Isabelle with pGCL. In: SSV, pp. 167–178 (2012)
10. Cock, D.: Practical probability: Applying pGCL to lattice scheduling. In: Blazy, S., Paulin-Mohring, C., Pichardie, D. (eds.) ITP 2013. LNCS, vol. 7998, pp. 311–327. Springer, Heidelberg (2013)
11. Hölzl, J.: Analyzing discrete-time Markov chains with countable state space in Isabelle/HOL. Draft, http://home.in.tum.de/~hoelzl/classifying
12. Hölzl, J., Nipkow, T.: Verifying pCTL model checking. In: Flanagan, C., König, B. (eds.) TACAS 2012. LNCS, vol. 7214, pp. 347–361. Springer, Heidelberg (2012)
13. Hurd, J., McIver, A., Morgan, C.: Probabilistic guarded commands mechanized in HOL. Theor. Comput. Sci. 346(1) (2005)
14. Kammüller, F., Wenzel, M., Paulson, L.C.: Locales - a sectioning concept for Isabelle. In: Bertot, Y., Dowek, G., Hirschowitz, A., Paulin, C., Théry, L. (eds.) TPHOLs 1999. LNCS, vol. 1690, pp. 149–166. Springer, Heidelberg (1999)
15. Kemeny, J.G., Snell, J.L., Knapp, A.W.: Denumerable Markov chains, 2nd edn. Springer (1976)
16. Larsen, K.G., Skou, A.: Bisimulation through probabilistic testing. Information and Computation 94(1), 1–28 (1991)
17. Mantel, H.: A uniform framework for the specification and verification of security properties. Ph.D. thesis, Univ. of Saarbrücken (2003)
18. Mantel, H., Sudbrock, H.: Flexible scheduler-independent security. In: Gritzalis, D., Preneel, B., Theoharidou, M. (eds.) ESORICS 2010. LNCS, vol. 6345, pp. 116–133. Springer, Heidelberg (2010)

19. McIver, A., Morgan, C.: Abstraction, Refinement and Proof for Probabilistic Systems. Springer (2005)
20. Nipkow, T., Paulson, L.C., Wenzel, M.: Isabelle/HOL: A Proof Assistant for Higher-order Logic. LNCS, vol. 2283. Springer, Heidelberg (2002)
21. Paulson, L.C., Blanchette, J.C.: Three years of experience with Sledgehammer, a practical link between automatic and interactive theorem provers. In: IWIL (2010)
22. Popescu, A., Hölzl, J.: Formal development associated with this paper, http://www21.in.tum.de/~popescua/prob.zip (to appear in the Archive of Formal Proofs, 2013)
23. Popescu, A., Hölzl, J., Nipkow, T.: Proving concurrent noninterference. In: Hawblitzel, C., Miller, D. (eds.) CPP 2012. LNCS, vol. 7679, pp. 109–125. Springer, Heidelberg (2012)
24. Popescu, A., Hölzl, J., Nipkow, T.: Noninterfering schedulers - when possibilistic noninterference implies probabilistic noninterference. In: CALCO, pp. 236–252 (2013)
25. Sabelfeld, A.: Confidentiality for multithreaded programs via bisimulation. In: Broy, M., Zamulin, A.V. (eds.) PSI 2003. LNCS, vol. 2890, pp. 260–274. Springer, Heidelberg (2004)
26. Sabelfeld, A., Myers, A.C.: Language-based information-flow security. IEEE Journal on Selected Areas in Communications 21(1), 5–19 (2003)
27. Sabelfeld, A., Sands, D.: Probabilistic noninterference for multi-threaded programs. In: CSFW, pp. 200–214 (2000)
28. Smith, G.: A new type system for secure information flow. In: CSFW, pp. 115–125 (2001)
29. Smith, G.: Probabilistic noninterference through weak probabilistic bisimulation. In: CSFW, pp. 3–13 (2003)
30. Smith, G.: Improved typings for probabilistic noninterference in a multi-threaded language. Journal of Computer Security 14(6), 591–623 (2006)
31. Smith, G., Volpano, D.: Secure information flow in a multi-threaded imperative language. In: POPL, pp. 355–364 (1998)
32. Volpano, D., Smith, G.: Probabilistic noninterference in a concurrent language. Journal of Computer Security 7(2,3), 231–253 (1999)
33. Volpano, D., Smith, G., Irvine, C.: A sound type system for secure flow analysis. Journal of Computer Security 4(2,3), 167–187 (1996)
34. Zdancewic, S., Myers, A.C.: Observational determinism for concurrent program security. In: CSFW, pp. 29–43 (2003)

Machine Assisted Proof of ARMv7 Instruction Level Isolation Properties

Narges Khakpour[1], Oliver Schwarz[1,2], and Mads Dam[1]

[1] KTH Royal Institute of Technology, Stockholm, Sweden
[2] SICS Swedish ICT, Kista, Sweden
{nargeskh,oschwarz,mfd}@kth.se

Abstract. In this paper, we formally verify security properties of the ARMv7 Instruction Set Architecture (ISA) for user mode executions. To obtain guarantees that arbitrary (and unknown) user processes are able to run isolated from privileged software and other user processes, instruction level noninterference and integrity properties are provided, along with proofs that transitions to privileged modes can only occur in a controlled manner. This work establishes a main requirement for operating system and hypervisor verification, as demonstrated for the PROSPER separation kernel. The proof is performed in the HOL4 theorem prover, taking the Cambridge model of ARM as basis. To this end, a proof tool has been developed, which assists the verification of relational state predicates semi-automatically.

Keywords: ARM instruction set, noninterference, user mode execution, kernel security, theorem proving.

1 Introduction

The ability to execute application software in a manner which is isolated from other application software running on a shared processing platform is an essential prerequisite for security. This allows user applications or virtual machines to coexist without violating confidentiality or integrity of critical data, it allows critical system resources to be protected from user manipulation, it can help to prevent fault propagation, and it can be used to save costly hardware that might otherwise be needed to provide physical separation.

Isolation is typically provided by a mix of hardware and software. A memory management unit (MMU) may be used to provide basic memory protection, and the processor may be equipped with multiple privilege levels, running application programs as userland processes and kernel routines at privileged levels, with additional abilities to access and configure critical parts of the processor, the MMU, and various storage/display/peripheral devices attached to the processor.

In such a setting, isolation is a result of the correct interplay between hardware and kernel. It is the responsibility of the kernel to correctly manipulate the processor state to achieve the desired effects, whatever they may be (context switching, logging, fault management, device management, etc). It is the

G. Gonthier and M. Norrish (Eds.): CPP 2013, LNCS 8307, pp. 276–291, 2013.
© Springer International Publishing Switzerland 2013

responsibility of the processing hardware to correctly implement the partitioning safeguards and mode transition conventions assumed by the kernel. For security, the kernel and the processor must both be correct and agree on their mode of interaction. Most formal kernel analyses in the literature [7,12,13,15,18] address the kernel software itself, in source or binary form, and leave the properties of the instruction set architecture (ISA) to be handled by fiat. Our contribution is to suggest a possible approach, including tool support, for performing the ISA specific security analysis, specifically for user mode execution.

We have identified two main concerns.

First, an implicit contract must exist which stipulates the "region of influence/dependency" of userland processes. That is, in a given user mode processor/MMU configuration it must be determined which memory locations and (control) registers can be read or written, or, in a more fine grained analysis, how information is able to flow to or from specific parts of the processor and the memory. User processes must be constrained in accessing or otherwise being influenced by critical resources of the kernel or of other user processes. This is not trivial. For instance, as shown by Duflot et al. [9], on some x86 processors it is possible for low-privilege code to overwrite higher privilege code by writing to an address that usually refers to the video card. To enable this attack, it suffices to first flip a configuration bit usually accessible from the low privilege level.

Second, kernel code relies on a set of mode switching conventions, for instance on ARM that program status registers and relevant user registers (including the program counter) are properly banked, the program counter is updated to point at the correct location in the vector table, and so on. If these conventions are not established by the processor and adhered to by the kernel, it may be possible for userland processes to induce various sorts of malicious behavior, for instance by letting a handler's link register point to a foreign address.

Performing this analysis is not trivial, particularly not if information flow is to be taken into account, as is done in this paper. All instructions, error conditions, and user to privileged mode transitions must be considered. The number of instructions is high and in modern processors a single instruction can involve a large number (order of 20-30) of atomic register or memory accesses.

In this paper, we identify and prove several partitioning-related properties of the ARMv7 ISA specification [2,3] addressing user mode execution and mode switching. The first is an instruction level noninterference property related to the non-infiltration property in [12] stating that the behavior of an ARMv7 processor in user mode only depends on its accessible resources, mostly user registers, MMU configurations and the memory allocated to that process. The second, corresponding to the non-exfiltration property of [12], is an integrity property stating that, again while in user mode, the processor is unable to modify protected resources. A third set of properties concerns mode switching conventions. These properties have been applied in the PROSPER project [5] to verify isolation for the PROSPER separation kernel [8]. The PROSPER project aims at producing and verifying a fully functional secure hypervisor for embedded

```
arm-state = <| psrs          : PSRName -> ARMpsr;
               regs          : RName -> word32;
               memory        : word32 -> word8;
               coproc        : coprocessors;
               accesses      : memory_access list;
               misc          : Monitors # ARMinfo # bool # bool |>;
```

Fig. 1. The ARM state in HOL4

systems, providing services such as guest isolation, so that only explicitly allowed communication occurs.

Our proof uses the HOL4 [4] model of ARM, developed at Cambridge by Fox et al. [10]. We extend this model by simple memory protection. The ARMv7 ISA properties outlined above are formalized and proved. To make the quite sizable proof task feasible, we have developed a helper tool based on relational Hoare logic, that is able to automate significant parts of the proof.

To the best of our knowledge our work represents the first formalized analysis of the ARMv7 ISA. Others, specifically the Cambridge HOL4 group, have developed various helper tools for assembling, disassembling, executing, and managing ARM machine code and the HOL4 ARM ISA model [10,16]. Also, the HOL4 ARM model has been used in several verification exercises in the literature, on software fault isolation (SFI) [22] and on the extension of the seL4 verification work [13] from C to binary level [20]. However, we have not yet seen general correctness properties formalized and verified for ARM at the ISA level. In fact, we believe the type of analysis presented here can be useful beyond kernel verification. For instance, formalized security properties can be useful to both improve the usefulness and precision of ISA specifications, and to enable developers obtain a concise description of secure configurations, without manual consideration of extensive architecture specifications.

2   The Formal Specification of ARM

We use Fox et al's monadic HOL4 model [10] of the ARMv7 ISA. This model covers the ARM, Thumb and ThumbEE instruction sets, comprising 81 instructions for branching, memory access, data processing, co-processor access, status access, and miscellaneous functionality. Figure 1 shows a simplified definition of an ARM state in this model. The function psrs returns the value of a processor state register (of type ARMpsr). The processor state registers include the current program status register, CPSR, in addition to the banked psrs SPSR_m for each privileged mode m, except for system mode. Program status registers encode arithmetic flags, the processor mode M, interrupt masks (I for ordinary and F for fast interrupts) and instruction encoding. The ARMv7 core provides seven processor modes: one non-privileged user mode usr, and six privileged modes (abt,fiq,irq,svc,und,sys), activated when an exception (such as an interrupt) is invoked. Variants with the TrustZone extension [1] also have a monitor mode. However, this has to be invoked from a privileged mode and we consider its usage out of scope of this paper.

The function **regs** takes a register name and returns its value. The ARM registers include sixteen general purpose registers (**r0-r15**) that are available from all modes in addition to the banked registers of each privileged mode (except of **sys**) that are available only in that mode. Among the user registers, register **r13** functions as stack pointer **SP**, register **r14** as link register **LR** and register **r15** as program counter **PC**.

The function **memory** reads a byte (**word8**) from an address (**word32**). The field **coproc** represents those coprocessor registers in **CP14** and **CP15** that implicitly influence execution. The coprocessor registers central for this work are registers **SCTLR** , **TTBR0** and **DACR** of coprocessor 15. They, together with the page table, are used to configure the MMU. The field **misc** represents the exclusive monitors used for synchronization purposes, general information about the state, e.g. the architecture version, if the system is waiting for an interrupt etc, and **accesses** records the accesses to the memory.

A *computation* in the monadic HOL4 ARM model is a term of the following (slightly beautified) type

$$\alpha\ \texttt{M} = \texttt{arm\_state} \mapsto (\alpha, \texttt{arm\_state})\ \texttt{error\_option}.$$

where **error_option** is a datatype defined as follows:

$$(\alpha, \beta)\ \texttt{error\_option} = \texttt{ValueState of } \alpha => \beta$$
$$| \texttt{ Error of string}$$

Computations act on a state **arm_state** and return either **ValueState** a s, a new state s of type **arm_state** along with a return value a of type α, or an error e. The unpredictable computations, i.e., those that are underspecified by the ARM specification return an error. The monad unit **constT** injects a value into a computation, i.e. **constT** a s = **ValueState** a s, while binding is a sequential composition operation

$$f_1 \gg=_e f_2 = \lambda s.\texttt{case } f_1 s \texttt{ of Error } c \to \texttt{Error } c$$
$$|| \texttt{ ValueState } a\ s' \to$$
$$\texttt{if } e\ s' \texttt{ then } f_2\ a\ s' \texttt{ else } f_1\ s.$$

That is, if e holds in the final state of f_1, the return value of f_1 is passed to f_2 as the input parameter, otherwise f_2 is not executed.

In addition to unit and binding, the ARM monadic specification uses standard constructs for lambda, let, and cases, as well as the monad operations parallel composition ($f_1 |||_e f_2$), positive conditional (**condT** e f), full conditional (**if** e **then** f_1 **else** f_2), error (**errorT** a), and an iterator (**forT**$_e$ l h f), (inductively) defined in Figure 2.

3   Memory Management

The Memory Management Unit (MMU) enforces memory access policies and is therefore important for isolation. MMU configurations consist of page tables in

errorT a = Error a
condT e f = if e then f else constT ()
if e then f_1 else f_2 = λs.if e s then f_1 s else f_2 s
f_1 $|||_e$ f_2 = $f_1 \gg=_e (\lambda x. f_2 \gg=_e (\lambda y. \text{constT } (x, y)))$
forT$_e$ l h f = if $l > h$ then constT []
                    else $((f\ l) \gg=_e (\lambda r. \text{forT}_e\ (l+1)\ h\ f \gg=_e (\lambda l. \text{constT}\ r :: l)))$

Fig. 2. Auxiliary monad operations

memory and dedicated registers of CP15. Specific to ARM is the possibility of partitioning pages into collections of memory regions, so-called *domains*. The theorems in this paper are based on the concrete MMU configurations (memory ranges, the page table setup etc.) used in the PROSPER kernel. The coprocessor registers involved are SCTLR, TTBR0 and DACR. The SCTLR register determines whether the MMU is enabled, TTBR0 contains the base address of the page table, and DACR manages the ARM domains.

MMU Extension. The evaluation function permitted takes as parameters a byte address, a flag indicating whether reading or writing access is to be evaluated, the values of SCTLR, TTBR0 and DACR, a flag indicating whether permissions are to be checked against a privileged mode, and the memory containing the page tables. The pair of booleans returned by permitted states whether the access permission on the specified byte is defined in the given configuration and the outcome of that decision (true if access is granted). The PROSPER kernel uses a basic version of permitted, supporting one-level page tables without address translation, but including the interpretation of ARM domains. It is shown that permitted is defined for all addresses in all reachable states.

The history of memory accesses is tracked in the accesses field of the machine state, allowing to compute the set of memory pages accessed by an instruction. To stop computation after the first access violation, $\gg=_{nav}$ has been chosen as standard binding operator, where nav s ("no access violation") is true if and only if there is no entry in the access list of machine state s that causes permitted to return a negative answer int the current configuration of s. The recording of an access always happens before the access itself.

The instruction execution function next (see Figure 3) takes an exception/interrupt flag irpt and a state s and produces the consequent state, by either initiating the demanded exception or by fetching and executing the next instruction pointed to by the PC in s. If an access violation is recorded after instruction fetching or execution, a prefetch or data abort exception (respectively) is initiated. The access list is cleared between the single steps, preventing the execution from halting and instead proceeding with exception handling. Occasionally, the unconditional binding $\gg=_T$ is used.

MMU Configuration. Let accessible i a express that address a is readable and writable by user process i. The predicate mmu_setup i s holds if and only

```
next irpt s =
(clear_alist ≫=nav
 (λu. if irpt = NoInterrupt then
         waiting_for_interrupt ≫=nav
           (λwfi. condT (¬wfi)
                   (fetch_instruction ≫=T
                   (λ(opc, ins). is_viol ≫=T (λav. clear_alist ≫=nav
                   (λu. if av then prefetch_abort
                        else
                           (execute ins ≫=T (λu. is_viol ≫=T
                           (λav. condT av
                                   (clear_alist ≫=nav
                                   (λu. data_abort)))))))))))
      else take_exception irpt ≫=nav (λu. clear_wait_for_irpt))) s
```

Fig. 3. The **next** computation

if (i) state s implements the desired access policy for process i, (ii) no MMU configuration for any address is underspecified, and (iii) none of the active page tables in s (represented by the address set **page_table_adds** s) is accessible according to the policy.

```
mmu_setup i s = ∀add, is_write, u, p.
   (u,p) = permitted add is_write (mmu_registers s) F s.memory
⇒ u ∧ ((accessible a i) ⇔ p)
      ∧ (a ∈ (page_table_adds s) ⇒ ¬(accessible a i))
```

4   Security Properties

We next turn to formalizing the instruction level partitioning properties. For user mode execution we formulate the requirements in terms of non-infiltration and non-exfiltration properties (cf. [12]), adapted to our setting.

Our model does not include caches, timing or hardware extensions such as TrustZone or virtualization support. Devices are not part of the model either; however, interrupts and other exceptions are taken into account, apart from fast interrupts and resets. Accordingly, the **fiq** and **mon** modes are outside of our analysis. As discussed, the chosen memory configuration is specific to the PROS-PER project. Consequences of a limited coprocessor model and underspecified instructions are discussed in Section 8.

4.1   Non-infiltration

Confidentiality of the kernel and neighboring user processes is guaranteed by non-infiltration, a noninterference-like property at the user mode single instruction level. Consider two machine states in user mode that are *low equivalent* in the sense that the two states agree on the resources (registers and memory locations)

that are permitted to influence user mode execution, but do not necessarily agree
on other resources. Non-infiltration holds if the poststates, after execution of one
instruction, remain low equivalent (or produce the same error).

Theorem 1. *Non-infiltration*

```
∀s1, s2, i, irpt.  mode s1 = mode s2 = usr ∧ bisim i s1 s2
⇒ (∃t1, t2. next irpt s1 = ValueState () t1
          ∧ next irpt s2 = ValueState () t2 ∧ bisim i t1 t2)
∨ (∃e. next irpt s1 = Error e ∧ next irpt s2 = Error e)
```

The relation `bisim` is the low equivalence relation. User mode processes are
allowed to be influenced by the user mode registers, the memory assigned to
them, the `CPSR`, the coprocessors, pending access violations and the `misc` state
component. Exclusive monitors (as field of `misc`) can inherently influence and
be influenced by user mode software and need thus to be cleared by kernels on
context switches.

```
bisim i s1 s2 =
    mmu_setup i s1 ∧ mmu_setup i s2 ∧ (equal_user_regs s1 s2)
  ∧ (∀a. (accessible i a) ⇒ (s1.memory a = s2.memory a))
  ∧ (s1.psrs(CPSR)= s2.psrs(CPSR)) ∧ (s1.coproc.state = s2.coproc.state)
  ∧ (nav s1 = nav s2) ∧ (s1.misc = s2.misc)
  ∧ s1.psrs(spsr_(mode s1)) = s2.psrs(spsr_(mode s2))
  ∧ s1.regs(lr_(mode s1)) = s2.regs(lr_(mode s2))
```

The two last items have been included to assure that `SPSR` and link register
(of a possibly privileged poststate) only depend on resources allowed to influence
user mode execution as well, so that they can actually be restored later on.

4.2   Non-exfiltration

Non-exfiltration guarantees the integrity of resources foreign to the active user
process. It expresses that, given an MMU setup for user process i active, the
execution of a single instruction in user mode will not modify any other resources
but those considered to be modifiable by i.

Theorem 2. *Non-exfiltration*

```
∀s, t, i, irpt. mode s = usr ∧ mmu_setup i s
  ∧ next irpt s = ValueState () t ⇒ unmodified i s t
```

Here, `unmodified` expresses the desired relation between the prestate s and the
poststate t of an active process i. We require that coprocessors, the fast interrupt
flag and any memory not belonging to i remain unchanged. The only registers
allowed to change are the `CPSR`, the user mode registers, and the `PSR` and the
link register of the mode in t. The interrupt flag of the `CPSR` is not modified
when staying in user mode.

```
unmodified i s t =
    (s.coproc = t.coproc) ∧ (s.psrs(CPSR).F = t.psrs(CPSR).F)
 ∧ (∀a. ¬(accessible i a) ⇒ (s.memory a = t.memory a))
 ∧ ((mode s ∈ {usr, mode t} ∧ mode t ∈ {usr, fiq, irq, svc, abt, und})
    ⇒( (∀reg. reg ∉ accessible_regs(mode t) ⇒ s.regs(reg) = t.regs(reg))
     ∧ (∀psr. psr ∉ {CPSR, spsr_(mode t)} ⇒ s.psrs(psr) = t.psrs(psr))
     ∧ (mode t = usr ⇒((s.psrs(CPSR)).I = (t.psrs(CPSR)).I))))
```

4.3 Switching to Privileged Modes

Secure user mode execution is not by itself sufficient. It is also necessary to consider transitions to privileged modes to prevent user processes from privileged execution rights. No user process should be able to effect a mode change with the PC set to a memory location of his choice. Instead, all entry points into privileged modes should be in the exception vector table. Similarly, even though user processes are allowed to choose a different endianness for their own execution, that should not influence the interpretation of the system handlers when switching back to privileged mode. Theorem 3 covers those additional constraints.

Theorem 3. *Privileged Constraints*

```
∀s, t, i, irpt. mode s = usr ∧ mmu_setup i s
  ∧ next irpt s = ValueState () t ⇒ priv_const s t
```

Besides the above properties, the relation `priv_const` lists the reachable processor modes[1] and assures that interrupts are masked when entering a privileged mode. Also, status register flags regarded as unwritable will be copied from the CPSR in prestate s to the SPSR in poststate t. This guarantees that a kernel can restore the saved program status register without further modifications when jumping back to the user process. Otherwise, user processes would be able to make the kernel enable/disable interrupts or change their execution mode. All access violations, if there were any, will have been handled (**nav** t).

```
priv_const s t =
    mode t ∈ {usr, fiq, irq, svc, abt, und}
  ∧ (mode t ≠ usr ⇒
    (   t.regs(PC) ∈ vt_adds(vt_base s, mode t) ∧ nav t
     ∧ (t.psrs(CPSR)).(I, J, IT, E) = (T, F, 0w, endianess s)
     ∧ (t.psrs(spsr_(mode t))).(M, I, F)
         = (usr, (s.psrs(CPSR)).I, (s.psrs(CPSR)).F)))
```

4.4 Link Register Contents in Supervisor Mode

Upon reception of a software interrupt, exception handlers in the invoked supervisor mode (svc) often need to analyze the calling instruction, in order to determine the software interrupt number for example. Therefore, verification

---

[1] Monitor and system mode can only be reached from another privileged mode.

might require assertions that the memory location pointed to by the link register actually does belong to the user process which caused the switch to supervisor mode. Formally, when going from state s in user mode to state t in supervisor mode, it is required that the svc-link register of t (i) is equal to the PC of s plus an instruction set dependent offset and (ii) corrected by the offset, points to an aligned word that is readable in t (independent of the mode). Note that offset and width of the word depend on the instruction set used by the user process, not on the one used by the handler.

Theorem 4. *Link Register Constraints*

$\forall$s, t, i, irpt, lr. mode s = usr $\land$ mmu_setup i s
$\quad \land$ next irpt s = ValueState () t $\land$ mode t = svc $\land$ lr = t.regs(LR_svc)
$\Rightarrow$ lr = s.regs(PC) + offset s
$\quad \land$ ((t.psrs(SPSR_svc)).T $\Rightarrow$ aligned_word_readable t T (lr - 2w))
$\quad \land$ ($\neg$(t.psrs(SPSR_svc)).T $\land$ $\neg$(t.psrs(SPSR_svc)).J
$\qquad \Rightarrow$ aligned_word_readable t F (lr - 4w))

Here, aligned_word_readable s b add states that the aligned word referred to by add is readable in s. Dependent on whether b is true or false, word width and alignment are 16 or 32 bit.

4.5   Safe User Mode Execution

The final aim is to guarantee that as long as the machine is executing in user mode, it causes no noninterference or integrity violations. Let $s_1 \rightsquigarrow s_n$ denote a sequence of next computations $s_1 \rightarrow s_2 \rightarrow .... \rightarrow s_n$ in user mode, i.e. mode s_i = usr, $1 \leq i < n$ and mode $s_n \neq$ usr. The following theorem assures the safe execution and safe mode switching of a user process.

Theorem 5. *Let* $s_1 \rightsquigarrow s_n$ *and* mmu_setup i s_1, *(i) if* $s_1' \rightsquigarrow s_n'$ *and* bisim i s_1 s_1' *then* bisim i s_n s_n', *(ii)* unmodified i s_1 s_n, *and (iii)* priv_const s_{n-1} s_n.

The proof of (i) and (ii) is an easy induction on n using theorems 1 and 2. Item (iii) follows from Theorem 3.

5   The Logic Framework

Considering the size and complexity of the ARM model and the instruction set, to prove the properties of the previous section tool support is essential. In this section we present proof rules for relational and invariant reasoning that help to automate the proof.

Non-infiltration The proof uses a relational Hoare logic based on assertions $\{f:R \rightarrow R'\}$ defined as follows:

$$\{f:R \rightarrow R'\} = \forall s_1, s_2. \ R \ s_1 \ s_2 \Rightarrow$$
$$(\exists a, t_1, t_2. \ f \ s_1 = \text{ValueState} \ a \ t_1 \ \land$$
$$f \ s_2 = \text{ValueState} \ a \ t_2 \ \land \ R' \ t_1 \ t_2)$$
$$\lor (\exists e. f \ s_1 = \text{Error} \ e \ \land \ f \ s_2 = \text{Error} \ e)$$

$$\text{errorTR} \frac{}{\{\text{errorT } a : \text{R\_m} \to \text{R\_m}\}} \qquad \text{constTR} \frac{}{\{\text{constT } a : \text{R\_m} \to \text{R\_m}\}}$$

$$\text{condTR} \frac{\{f : \text{R\_m} \to \text{R\_m}\}}{\{\text{condT } \psi \; f : \text{R\_m} \to \text{R\_m}\}} \qquad \text{forTR} \frac{\{f : \text{R\_m} \to \text{R\_m}\}}{\{\text{forT}_{\text{nav}} \; l \; h \; f : \text{R\_m} \to \text{R\_m}\}}$$

$$\text{conR} \frac{\{f : \text{R\_m} \to \text{R\_n}\} \quad \{f' : \text{R\_m} \to \text{R\_n}\}}{\{\text{if } \psi \text{ then } f \text{ else } f' : \text{R\_m} \to \text{R\_n}\}}$$

$$\text{widenR} \frac{\{f : \text{R\_m} \to \text{R\_n}\}}{\{f : \text{R\_m} \to \text{R\_}(n,k)\}} \qquad \text{absR} \frac{\forall y.\{f \; y : \text{R\_m} \to \text{R\_n}\}}{\{\lambda y.f : \text{R\_m} \to \text{R\_n}\}}$$

$$\text{seqTR} \frac{\{f : \text{R\_m} \to \text{R\_n}\} \quad \{f' : \text{R\_n} \to \text{R\_k}\} \quad (m = n) \vee (n = k)}{\{f \gg=_{\text{nav}} f' : \text{R\_m} \to \text{R\_}(n,k)\}}$$

$$\text{parTR} \frac{\{f : \text{R\_m} \to \text{R\_n}\} \quad \{f' : \text{R\_n} \to \text{R\_k}\} \quad (m = n) \vee (n = k)}{\{f|||_{\text{nav}} f' : \text{R\_m} \to \text{R\_}(n,k)\}}$$

Fig. 4. Relational inference rules

The judgment asserts that, if started in prestates s_1, s_2 related by prerelation R, either the executions of the monadic computation f return identical values a with poststates t_1, t_2 related by postrelation R', or else they both return the same error e.

For the analysis it suffices to consider a fixed set of relations

$$\text{R\_m} = \lambda s_1.\lambda s_2.\text{bisim i } s_1 \; s_2 \; \wedge \text{ mode } s_1 = \text{m} \wedge \text{ mode } s_2 = \text{m}$$

or $\text{R\_}(n,m) = \text{R\_n} \cup \text{R\_m}$.

Figure 4 shows the relational logic inference rules. The inference system is incomplete, but sufficient for our purpose. A relation R_m is preserved by `errorT` and `constT` (rules constTR and errorTR), and if a computation preserves one of the R_m relations then that computation can be used in a conditional or a *for* loop as well (condTR, conR and forTR). The rule widenR and absR are used to weaken the postrelation and reason about lambda computations, respectively. The rule seqTR states that the postrelation of $f \gg=_{\text{nav}} f'$ is the union of the postrelations of f and f', provided that either f preserves R_n or f' preserves R_k. If there is an access violation after f, the computation stops and R_n must hold. Otherwise, f' will execute and R_k must hold. Thus, the postrelation is the union of R_n and R_k.

Theorem 6. *All assertions* $\{f : R \to R'\}$ *derivable according to the inference rules in Figure 4 are valid.*

Non-exfiltration Similar to the non-infiltration proof, the proof of non-exfiltration uses a sound but incomplete inference system, this time concerning computation invariants of the following shape:

$$\text{INV}\langle f, \text{Q}, \text{P}\rangle = \forall s, t. \text{ Q } s \; \wedge f \; s = \text{ValueState } a \; t \implies \text{P } s \; t \wedge \text{ Q } t.$$

That is, if Q holds of the prestate then P holds of the prestate-poststate pair, and Q of the poststate. We use a simple collection of inference rules to prove

$$\text{errorTI} \frac{}{\text{INV}\langle \text{errorT } a, \mathsf{Q}, \mathsf{P}\rangle} \qquad \text{constTI} \frac{\text{refl P}}{\text{INV}\langle \text{constT } c, \mathsf{Q}, \mathsf{P}\rangle}$$

$$\text{condTI} \frac{\text{refl P} \quad \text{INV}\langle f, \mathsf{Q}, \mathsf{P}\rangle}{\text{INV}\langle \text{condT } e \ f, \mathsf{Q}, \mathsf{P}\rangle} \qquad \text{forTI} \frac{\text{refl P} \quad \text{trans P} \quad \text{INV}\langle f, \mathsf{Q}, \mathsf{P}\rangle}{\text{INV}\langle \text{forT}_e \ l \ h \ f, \mathsf{Q}, \mathsf{P}\rangle}$$

$$\text{conRI} \frac{\text{INV}\langle f, \mathsf{Q}, \mathsf{P}\rangle \quad \text{INV}\langle f', \mathsf{Q}, \mathsf{P}\rangle}{\text{INV}\langle \text{if } \psi \text{ then } f \text{ else } f', \mathsf{Q}, \mathsf{P}\rangle}$$

$$\text{absI} \frac{\forall y. \text{INV}\langle f \ y, \mathsf{Q}, \mathsf{P}\rangle}{\text{INV}\langle \lambda y. f, \mathsf{Q}, \mathsf{P}\rangle} \qquad \text{seqTI} \frac{\text{INV}\langle f, \mathsf{Q}, \mathsf{P}\rangle \quad \text{INV}\langle f', \mathsf{Q}, \mathsf{P}\rangle \quad \text{trans P}}{\text{INV}\langle f \gg=_e f', \mathsf{Q}, \mathsf{P}\rangle}$$

$$\text{parTI} \frac{\text{INV}\langle f, \mathsf{Q}, \mathsf{P}\rangle \quad \text{INV}\langle f', \mathsf{Q}, \mathsf{P}\rangle \quad \text{trans P}}{\text{INV}\langle f|||_e f', \mathsf{Q}, \mathsf{P}\rangle}$$

Fig. 5. Invariant inference rules

Q and P , shown in Figure 5. In this figure, `refl P` and `trans P` respectively state that P is reflexive and transitive. For non-exfiltration we need to prove that `unmodified i` is satisfied during the execution of each instruction both when it ends in user mode and when switching to privileged mode. A prerequisite for this is that the MMU is configured correctly during computation. To prove the non-exfiltration property, we check $\text{INV}\langle \text{next}, \texttt{mmu\_setup i}, \texttt{unmodified i} \ \rangle$.

Theorem 7. *All assertions* $\text{INV}\langle f, \mathsf{Q}, \mathsf{P}\rangle$ *derivable according to the inference rules in Figure 5 are valid.*

Privileged Constraints The final goal is to prove that `next` establishes the relation `priv_const`, a conjunction of primitive constraints P. Since the primitive constraints do not always hold during computations in privileged mode, the inference rules of Figure 5 are generally not able to prove this property. To make verification tractable, we prove primitive constraints locally at the point in the monadic computation where it is established and then use a set of inference rules to infer its correctness for the entire computation. We illustrate the proof using an example. In the ARM model, all computations which lead to a privileged mode m end by a computation called `take_m_exception`. Figure 6 shows the function `take_svc_exception` for switching to supervisor mode. Let this computation start in state s1 and end in state sn. Consider the primitive constraint P_{psr} stating that SPSR_svc of the final state sn must be equal to CPSR of the initial state s1. Let t and t', respectively be the initial state and final state of `write_spsr cr` and m be the mode of t'. The computation `write_spsr cr` writes the value of free variable cr into SPSR_m and establishes the property $P'_{psr} \overset{\text{def}}{=} t'.psrs(\text{SPSR\_m}) = cr$. We call `write_spsr cr` a P'_{psr}-establisher. A computation g is P-establisher, if independently of its input state, P holds in its output state, i.e.

$$\text{P}-\text{establ}(g) = \forall s, a, t. \ g \ s = \texttt{ValueState } a \ t \ \wedge \ \text{nav } t \implies \text{P } t$$

We can prove that the block starting from `write_spsr cr` establishes P'_{psr} as well, because the rest of the computations of this block does not modify this property. Then we can prove that the free variable cr takes the value s1.psrs(CPSR),

```
take_svc_exception  = IT_advance  ≫=nav
   (λ u.(read_reg 15w |||nav exc_vector_base |||nav read_cpsr |||nav
                read_scr |||nav read_sctlr )≫=nav
   (λ(pc,ExcVectorBase,cr,scr,sctlr).
      (condT (cr.M = 0b10110w) (write_scr  (scr with NS := F))  |||nav
      write_cpsr (cr with M := 0b10011w)) ≫=nav
      (λ (u1,u2). (write_spsr cr |||nav
         write_reg 14w (if cr.T then pc - 2w else pc - 4w) |||nav
         (read_cpsr ≫=nav
         (λ cr'.write_cpsr (cr' with
                              <| I := T; IT := 0b00000000w;J := F;
                                 T := sctlr.TE; E := sctlr.EE |>))) |||nav
      branch_to (ExcVectorBase + 8w)) ≫=nav unit4)))
```

Fig. 6. The HOL4 code for switching to svc mode [4]

$$\text{seqTS1} \frac{\text{P-establ}(f) \quad \text{INV}\langle f', \text{P}, \top\rangle}{\text{P-establ}(f \gg=_{nav} f')} \quad \text{seqTS2} \frac{\text{P-establ}(f)}{\text{P-establ}(f' \gg=_{nav} f)}$$

$$\text{parTS1} \frac{\text{P-establ}(f) \quad \text{INV}\langle f', \text{P}, \top\rangle}{\text{P-establ}(f \mid\mid\mid_{nav} f')} \quad \text{parTS2} \frac{\text{P-establ}(f)}{\text{P-establ}(f' \mid\mid\mid_{nav} f)}$$

$$\text{absS} \frac{\forall y.\text{P-establ}(f\ y)}{\text{P-establ}(\lambda y.f)}$$

Fig. 7. Privileged constraints inference rules

and m is bound to svc. Thus, sn.psrs(SPSR_svc) = s1.psrs(CPSR) holds for the computation block from write_spsr cr. As this block is a P_{psr}-establisher, we conclude that the computations before write_spsr do not influence the established property and P_{psr} is satisfied by take_svc_exception.

Figure 7 shows the P-establisher inference rules. These rules along with the inference rules of Figure 5 are used to prove the privileged constraints. The rule seqTS1 states that if the monadic computation f is a P-establisher and P is an invariant of f', then the sequential composition $f \gg=_{nav} f'$ is P-establisher. The rule seqTS2 describes that if the monadic computation f is a P-establisher, then $f' \gg=_{nav} f$ is also P-establisher. Similar rules are defined for the $\mid\mid\mid_{nav}$ operator.

Theorem 8. *All assertions* P-establ(f) *derivable according to the inference rules in Figure 7 are valid.*

6   Implementation and Evaluation

Implementation We use the HOL4 theorem prover to verify our properties. The central assets of our work are available from [5]. We have developed a tool, ARM-prover, to automate the verification process based on the proof systems in Fig. 4

and 5. To avoid having to explore the instruction set more than once the prover actually combines the theorems 1, 2 and 3 into one.

The proof systems do not provide rules for `case` and `let` statements. These are easily handled using standard HOL4 simplification. Other monadic expressions are refined using the inference rules in Fig. 4 and 5 in a top down fashion. The proofs for "write" primitives as well as register and memory accesses in user mode are done manually, but the tool can handle some of the "read" computations directly, allowing to prove a large share of the workload automatically.

A particular difficulty concerns binding. When a binding expression $f_1 \gg=_{nav} f_2$ is decomposed the return value of f_1 becomes unbound in f_2. To handle this we simplify computations by embedding more information before calling the prover, using some auxiliary lemmas. For example, the following formula states that `cpsr` in computation H following `read_cpsr` can be substituted by the CPSR in prestate s with mode m.

$$(\text{mode } s = m) \Rightarrow (\texttt{read\_cpsr} \gg=_{nav} (\lambda \texttt{cpsr. H(cpsr)})) \ s =$$
$$(\texttt{read\_cpsr} \gg=_{nav} (\lambda \texttt{cpsr. H(s.psrs(CPSR) with M:=m)})) \ s$$

For the case that an instruction leads to a privileged mode, the last execution phase of the instruction, called switching phase, is in privileged mode. However, the privileged constraints first have to be established over the course of several steps and do not hold from the beginning. Since we can not use the ARM-prover tool to prove them automatically, we prove the privileged constraints for the switching phase manually.

Evaluation. The Cambridge model of ARM is 9 kLOC. In addition to the ARM model, we rely mainly on the relatively small inference kernel of the HOL4 theorem prover, our MMU extension (about 180 lines of definitions) and the formulation of the discussed properties (about 290 lines). The entire proof script has a length of about 13 kLOC and needs roughly an hour to run on an Intel(R) Xeon(R) X3470 core. We invested about one person year of effort into this work.

7   Related Work

Several recent works address kernel verification. Some target information flow properties [7,12,15,18], based on variants of noninterference [11]. Other work establishes a refinement relation between kernel code, in some representation, and an abstract specification. For the seL4 microkernel this was first performed for its C implementation [13] and is now extended to binary level [20]. As is the case with most refinement/simulation-based approaches, this work does not address information flow. In recent work on seL4 verification, Murray et al. [14,15] present an unwinding-style characterization of intransitive noninterference. They introduce a proof calculus on nondeterministic state monads that is similar to that of this work. Their assertions are more general, however our proof rules cover several monadic operators and statements. In addition, we introduce rules to prove properties about executions that relate the final state of a computation to its initial state.

Alkassar et al. [6] describe the emulation of a simplified MIPS machine in C. The emulator allows the use of VCC to automatically check that every reachable state of a guest on a hypervisor is also reachable when the guest is running on a completely isolated machine. The C emulator has been adopted to verify parts of the hypervisor that mix C and assembly [17], and allows unknown user processes to be considered. Information flow properties are not considered, however.

Wilding et al. [21] formally proved exfiltration, infiltration and mediation theorems for the partitioning system of the AAMP7G microprocessor in ACL2. The hardware architecture differs from the one of ARM in several points, such as that there are no user-visible registers or that AAMP7G itself functions as a separation kernel. Proofs were performed using abstraction/refinement techniques and address kernel microcode. The verification led to a MILS certificate on Evaluation Assurance Level 7.

The ARMor system [22] sandboxes applications on ARM and provides formal verification of memory safety and control flow integrity, using the Cambridge HOL4 ARM model. Its software fault isolation does not use hardware features such as an MMU, but uses instead rewriting and subsequent verification of the compiled programs. This implies performance overhead, limitations on supported programs and verification processes in the extend of hours for each program. Furthermore, ARMor only establishes memory write protection; neither confidentiality nor protection of privileged registers is addressed.

Most works on kernel verification address handler code only and do not consider user mode execution. In a few cases [6,19] user mode execution is considered, but without justification in terms of concrete processor access modalities. The main contribution of our work, over and beyond the above works, is that we attempt to justify the critical assumptions on processor level information flow in user mode execution through analysis at the level of a formalized ISA model.

Heitmeyer et al. [12] introduce non-exfiltration, non-infiltration, kernel integrity and data/control separation properties to verify a separation kernel. Since we focus on user-mode execution, those properties apply only partially here. Our non-infiltration property is the same as in [12], but the non-exfiltration property in our work covers both their kernel integrity and non-exfiltration.

8 Conclusion

We introduced and proved several security properties including a non-exfiltration, a non-infiltration and a safe switching property for user mode executions on the ARM architecture, using the Cambridge HOL4 ISA model. A logical framework based on (relational) Hoare logic has been developed for the analysis, supported by a tool, ARM-prover, which helps automate the proof. The ARM-prover can be used to prove general invariants about the ARM model (i.e., statements that need to hold at each execution point). We are planning to continue the development of the ARM-prover to improve automation further and cater for more general proof tasks.

Our results concerning register contents are generally valid and with small adaptations applicable in isolation verification of other hypervisors, separation kernels, and operating systems. Statements on memory safety depend on our specific setup. A reformulation that is independent of concrete MMU configurations should require a minor effort and is planned for future work.

The HOL4 model of ARM supports a partial coprocessor model. We made the assumption that the access to coprocessors via dedicated instructions is always denied in user mode. To have a more precise analysis and cover all possible side channels, a more comprehensive model of the available coprocessors involving all registers, the coprocessors' behavior and an acceptance/rejection-mechanism for register reads and writes that follows the specification is required. During context switches kernels need to mediate coprocessor registers user-accessible by dedicated coprocessor instructions. All other coprocessor registers are guaranteed to be non-modifiable in user mode. However, kernels must not introduce information flow from non-active processes to the coprocessor registers that are part of the present ARM model, since those might influence user mode execution.

Instructions that are underspecified ("unpredictable") in the ARM Architecture Reference Manual (ARMARM) are problematic. The ARM specification states that "*unpredictable* behavior must not perform any function that cannot be performed at the current or lower level of privilege using instructions that are not *unpredictable*"[3]. In one interpretation of this statement, theorems 2, 3 and 4 are valid on unpredictable instructions as well. In general, this is not true for non-infiltration. Yet, ARMARM requires further that "*unpredictable* behavior must not represent security holes" [2]. This formulation is very vague. However, we make the assumption that non-infiltration is preserved. In fact, we argue that the security properties we have presented provide manufacturers of ARM processors with a precise description of secure behavior for unpredictable cases.

Acknowledgments. Work supported by framework grant "IT 2010" from the Swedish Foundation for Strategic Research.

References

1. ARM TrustZone technology,
   http://www.arm.com/products/processors/technologies/trustzone.php
2. ARMv7-A architecture reference manual, issue B,
   http://infocenter.arm.com/help/index.jsp?topic=/com.arm.doc.ddi0406b
3. ARMv7-A architecture reference manual, issue C,
   http://infocenter.arm.com/help/index.jsp?topic=/com.arm.doc.ddi0406c
4. HOL4, http://hol.sourceforge.net/
5. PROSPER project, http://prosper.sics.se/
6. Alkassar, E., Hillebrand, M.A., Paul, W.J., Petrova, E.: Automated verification of a small hypervisor. In: Leavens, G.T., O'Hearn, P., Rajamani, S.K. (eds.) VSTTE 2010. LNCS, vol. 6217, pp. 40–54. Springer, Heidelberg (2010)
7. Barthe, G., Betarte, G., Campo, J.D., Luna, C.: Formally verifying isolation and availability in an idealized model of virtualization. In: Butler, M., Schulte, W. (eds.) FM 2011. LNCS, vol. 6664, pp. 231–245. Springer, Heidelberg (2011)

8. Dam, M., Guanciale, R., Khakpour, N., Nemati, H., Schwarz, O.: Formal verification of information flow security for a simple ARM-based separation kernel. In: Proceedings of the 2013 ACM SIGSAC Conference on Computer and Communications Security, CCS 2013 (2013)
9. Duflot, L., Etiemble, D., Grumelard, O.: Using CPU system management mode to circumvent operating system security functions. In: Proc. CanSecWest (2006)
10. Fox, A., Myreen, M.O.: A trustworthy monadic formalization of the ARMv7 instruction set architecture. In: Kaufmann, M., Paulson, L.C. (eds.) ITP 2010. LNCS, vol. 6172, pp. 243–258. Springer, Heidelberg (2010)
11. Goguen, J.A., Meseguer, J.: Security policies and security models. In: IEEE Symposium on Security and Privacy, pp. 11–20 (1982)
12. Heitmeyer, C., Archer, M., Leonard, E., McLean, J.: Applying formal methods to a certifiably secure software system. IEEE Trans. Softw. Eng. 34(1), 82–98 (2008)
13. Klein, G., Elphinstone, K., Heiser, G., Andronick, J., Cock, D., Derrin, P., Elkaduwe, D., Engelhardt, K., Kolanski, R., Norrish, M., Sewell, T., Tuch, H., Winwood, S.: seL4: formal verification of an OS kernel. In: Matthews, J.N., Anderson, T.E. (eds.) SOSP, pp. 207–220. ACM (2009)
14. Murray, T.C., Matichuk, D., Brassil, M., Gammie, P., Bourke, T., Seefried, S., Lewis, C., Gao, X., Klein, G.: seL4: From general purpose to a proof of information flow enforcement. In: IEEE Symposium on Security and Privacy, pp. 415–429. IEEE Computer Society (2013)
15. Murray, T., Matichuk, D., Brassil, M., Gammie, P., Klein, G.: Noninterference for operating system kernels. In: Hawblitzel, C., Miller, D. (eds.) CPP 2012. LNCS, vol. 7679, pp. 126–142. Springer, Heidelberg (2012)
16. Myreen, M.O., Fox, A., Gordon, M.J.C.: Hoare logic for ARM machine code. In: Arbab, F., Sirjani, M. (eds.) FSEN 2007. LNCS, vol. 4767, pp. 272–286. Springer, Heidelberg (2007)
17. Paul, W., Schmaltz, S., Shadrin, A.: Completing the automated verification of a small hypervisor – assembler code verification. In: Eleftherakis, G., Hinchey, M., Holcombe, M. (eds.) SEFM 2012. LNCS, vol. 7504, pp. 188–202. Springer, Heidelberg (2012)
18. Richards, R.J.: Modeling and security analysis of a commercial real-time operating system kernel. In: Hardin, D.S. (ed.) Design and Verification of Microprocessor Systems for High-Assurance Applications, pp. 301–322 (2010)
19. Rushby, J.: Formally verified hardware encapsulation mechanism for security, integrity, and safety. Technical report, DTIC Document (2002)
20. Sewell, T., Myreen, M.O., Klein, G.: Translation validation for a verified OS kernel. In: Proceedings of the 34th ACM SIGPLAN Conference on Programming Language Design and Implementation (PLDI), pp. 471–482 (2013)
21. Wilding, M.M., Greve, D.A., Richards, R.J., Hardin, D.S.: Formal verification of partition management for the AAMP7G microprocessor. In: Hardin, D.S. (ed.) Design and Verification of Microprocessor Systems for High-Assurance Applications, pp. 175–191. Springer US (2010)
22. Zhao, L., Li, G., De Sutter, B., Regehr, J.: ARMor: Fully verified software fault isolation. In: Proceedings of the International Conference on Embedded Software, EMSOFT 2011, pp. 289–298 (2011)

A Formal Model and Correctness Proof for an Access Control Policy Framework

Chunhan Wu[1,2], Xingyuan Zhang[1], and Christian Urban[2]

[1] PLA University of Science and Technology, China
[2] King's College London, UK

Abstract. If an access control policy promises that a resource is protected in a system, how do we know it is really protected? To give an answer we formalise in this paper the Role-Compatibility Model—a framework, introduced by Ott, in which access control policies can be expressed. We also give a dynamic model determining which security related events can happen while a system is running. We prove that if a policy in this framework ensures a resource is protected, then there is really no sequence of events that would compromise the security of this resource. We also prove the opposite: if a policy does not prevent a security compromise of a resource, then there is a sequence of events that will compromise it. Consequently, a static policy check is sufficient (sound and complete) in order to guarantee or expose the security of resources before running the system. Our formal model and correctness proof are mechanised in the Isabelle/HOL theorem prover using Paulson's inductive method for reasoning about valid sequences of events. Our results apply to the Role-Compatibility Model, but can be readily adapted to other role-based access control models.

1 Introduction

Role-based access control models are used in many operating systems for enforcing security properties. The *Role-Compatibility Model* (RC-Model), introduced by Ott [5,6], is one such role-based access control model. It defines *roles*, which are associated with processes, and defines *types*, which are associated with system resources, such as files and directories. The RC-Model also includes types for interprocess communication, that is message queues, sockets and shared memory. A policy in the RC-Model gives every user a default role, and also specifies how roles can be changed. Moreover, it specifies which types of resources a role has permission to access, and also the *mode* with which the role can access the resources, for example read, write, send, receive and so on.

The RC-Model is built on top of a collection of system calls provided by the operating system, for instance system calls for reading and writing files, cloning and killing of processes, and sending and receiving messages. The purpose of the RC-Model is to restrict access to these system calls and thereby enforce security properties of the system. A problem with the RC-Model and role-based access control models in general is that a system administrator has to specify an appropriate access control policy. The difficulty with this is that *"what you specify is what you get but not necessarily what you want"* [4, Page 242]. To overcome this difficulty, a system administrator needs some kind of sanity check for whether an access control policy is really securing resources. Existing

G. Gonthier and M. Norrish (Eds.): CPP 2013, LNCS 8307, pp. 292–307, 2013.
© Springer International Publishing Switzerland 2013

works, for example [9,10], provide sanity checks for policies by specifying properties and using model checking techniques to ensure a policy at hand satisfies these properties. However, these checks only address the problem on the level of policies—they can only check "on the surface" whether the policy reflects the intentions of the system administrator—these checks are not justified by the actual behaviour of the operating system. The main problem this paper addresses is to check when a policy matches the intentions of a system administrator *and* given such a policy, the operating system actually enforces this policy.

Our work is related to the preliminary work by Archer et al [1] about the security model of SELinux. They also give a dynamic model of system calls on which the access controls are implemented. Their dynamic model is defined in terms of IO automata and mechanised in the PVS theorem prover. For specifying and reasoning about automata they use the TAME tool in PVS. Their work checks well-formedness properties of access policies by type-checking generated definitions in PVS. They can also ensure some "*simple properties*" (their terminology), for example whether a process with a particular PID is present in every reachable state from an initial state. They also consider "*deeper properties*", for example whether only a process with root-permissions or one of its descendents ever gets permission to write to kernel log files. They write that they can state such deeper properties about access policies, but about checking such properties they write that "*the feasibility of doing so is currently an open question*" [1, Page 167]. We improve upon their results by using our sound and complete static policy check to make this feasible.

The work we report is also closely related to the work on *grsecurity*, an access control system developed as a patch on top of Linux kernel [2]. It installs a reference monitor to restrict access to system resources. They model a dynamic semantics of the operating system with four rules dealing with executing a file, setting a role and setting an UID as well as GID. These rules are parametrerised by an arbitrary but fixed access policy. Although, there are only four rules, their state-space is in general infinite, like in our work. They therfore give an abstracted semantics, which gives them a finite state-space. For example the abstracted semantics dispenses with users and roles by introducing abstract users and abstract roles. They obtain a soundness result for their abstract semantics and under some weak assumptions also a completeness result. Comparing this to our work, we will have a much more fine-grained model of the underlying operating system. We will also obtain a soundness result, but more importantly obtain also a completeness result. But since we have a much more fine-grained model, it will depend on some stronger assumptions. The abstract semantics in [2] is used for model-checking policies according to whether, for example, information flow properties are ensured. Since their formalism consists of only a few rules, they can get away with "pencil-and-paper proofs", whereas reasoning about our more detailed model containing substantially more rules really necessitates the support of a theorem prover and completely formalised models.

Our formal models and correctness proofs are mechanised in the interactive theorem prover Isabelle/HOL. The mechanisation of the models is a prerequisite for any correctness proof about the RC-Model, since it includes a large number of interdependent concepts and very complex operations that determine roles and types. In our opinion it is futile to attempt to reason about them by just using "pencil-and-paper". Following good experience in earlier mechanisation work [11], we use Paulson's inductive method for reasoning about sequences of events [8]. For example we model system calls as events and reason about an inductive definition of valid traces, that is lists of events. Central to this paper is a notion of a resource being *tainted*, which for example means it contains a virus or a back door. We use our model of system calls in order to characterise how such a tainted object can "spread" through the system. For a system administrator the important question is whether such a tainted file, possibly introduced by a user, can affect core system files and render the whole system insecure, or whether it can be contained by the access policy. Our results show that a corresponding check can be performed statically by analysing the initial state of the system and the access policy.

Contributions: We give a complete formalisation of the RC-Model in the interactive theorem prover Isabelle/HOL. We also give a dynamic model of the operating system by formalising all security related events that can happen while the system is running. As far as we are aware, we are the first ones who formally prove that if a policy in the RC-Model satisfies an access property, then there is no sequence of events (system calls) that can violate this access property. We also prove the opposite: if a policy does not meet an access property, then there is a sequence of events that will violate this property in our model of the operating system. With these two results in place we can show that a static policy check is sufficient in order to guarantee the access properties before running the system. Again as far as we know, no such check has been designed and proved correct before.

2 Preliminaries about the RC-Model

The Role-Compatibility Model (RC-Model) is a role-based access control model. It has been introduced by Ott [5] and is used in running systems for example to secure Apache servers. It provides a more fine-grained control over access permissions than simple Unix-style access control models. This more fine-grained control solves the problem of server processes running as root with too many access permissions in order to accomplish a task at hand. In the RC-Model, system administrators are able to restrict what the role of server is allowed to do and in doing so reduce the attack surface of a system.

Policies in the RC-Model talk about *users*, *roles*, *types* and *objects*. Objects are processes, files or IPCs (interprocess communication objects—such as message queues, sockets and shared memory). Objects are the resources of a system an RC-policy can restrict access to. In what follows we use the letter u to stand for users, r for roles, p for processes, f for files and i for IPCs. We also use *obj* as a generic variable for objects. The RC-Model has the following eight kinds of access modes to objects:

Read, Write, Execute, ChangeOwner, Create, Send, Receive and Delete

In the RC-Model, roles group users according to tasks they need to accomplish. Users have a default role specified by the policy, which is the role they start with whenever they log into the system. A process contains the information about its owner (a user), its role and its type, whereby a type in the RC-Model allows system administrators to group resources according to a common criteria. Such detailed information is needed in the RC-Model, for example, in order to allow a process to change its ownership. For this the RC-Model checks the role of the process and its type: if the access control policy states that the role has *ChangeOwner* access mode for processes of that type, then the process is permitted to assume a new owner.

Files in the RC-Model contain the information about their types. A policy then specifies whether a process with a given role can access a file under a certain access mode. Files, however, also include in the RC-Model information about roles. This information is used when a process is permitted to execute a file. By doing so it might change its role. This is often used in the context of web-servers when a cgi-script is uploaded and then executed by the server. The resulting process should have much more restricted access permissions. This kind of behaviour when executing a file can be specified in an RC-policy in several ways: first, the role of the process does not change when executing a file; second, the process takes on the role specified with the file; or third, use the role of the owner, who currently owns this process. The RC-Model also makes assumptions on how types can change. For example for files and IPCs the type can never change once they are created. But processes can change their types according to the roles they have.

As can be seen, the information contained in a policy in the RC-Model can be rather complex: Roles and types, for example, are policy-dependent, meaning each policy needs to define a set of roles and a set of types. Apart from recording for each role the information which type of resource it can access and under which access-mode, it also needs to include a role compatibility set. This set specifies how one role can change into another role. Moreover it needs to include default information for cases when new processes or files are created. For example, when a process clones itself, the type of the new process is determined as follows: the policy might specify a default type whenever a process with a certain role is cloned, or the policy might specify that the cloned process inherits the type of the parent process.

Ott implemented the RC-Model on top of Linux, but only specified it as a set of informal rules, partially given as logic formulas, partially given as rules in "English". Unfortunately, some presentations about the RC-Model give conflicting definitions for some concepts—for example when defining the semantics of the special role "inherit parent". In [5] it means inherit the initial role of the parent directory, but in [7] it means inherit the role of the parent process. In our formalisation we mainly follow the version given in [5]. In the next section we give a mechanised model of the system calls on which the RC-Model is implemented.

3 Dynamic Model of System Calls

Central to the RC-Model are processes, since they initiate any action involving resources and access control. We use natural numbers to stand for process IDs, but do not model the fact that the number of processes in any practical system is limited. Similarly, IPCs and users are represented by natural numbers. The thirteen actions a process can perform are represented by the following datatype of *events*

$$
\begin{aligned}
event ::= \quad &CreateFile\ p\ f & | \ &ReadFile\ p\ f & | \ &Send\ p\ i & | \ Kill\ p\ p' \\
| \ &WriteFile\ p\ f & | \ &Execute\ p\ f & | \ &Recv\ p\ i & \\
| \ &DeleteFile\ p\ f & | \ &Clone\ p\ p' & | \ &CreateIPC\ p\ i & \\
| \ &ChangeOwner\ p\ u & | \ &ChangeRole\ p\ r & | \ &DeleteIPC\ p\ i &
\end{aligned}
$$

with the idea that for example in *Clone* a process p is cloned and the new process has the ID p'; with *Kill* the intention is that the process p kills another process with ID p'. We will later give the definition what the role r can stand for in the constructor *ChangeRole* (namely *normal roles* only). As is custom in Unix, there is no difference between a directory and a file. The files f in the definition above are simply lists of strings. For example, the file */usr/bin/make* is represented by the list [*make, bin, usr*] and the *root*-directory is the *Nil*-list. Following the presentation in [5], our model of IPCs is rather simple-minded: we only have events for creation and deletion of IPCs, as well as sending and receiving messages.

Events essentially transform one state of the system into another. The system starts with an initial state determining which processes, files and IPCs are active at the start of the system. We assume the users of the system are fixed in the initial state; we also assume that the policy does not change while the system is running. We have three sets, namely *init_procs*, *init_files* and *init_ipcs* specifying the processes, files and IPCs present in the initial state. We will often use the abbreviation

$$
obj \in init \overset{def}{=} obj \in init\_files \vee obj \in init\_procs \vee obj \in init\_ipcs
$$

There are some assumptions we make about the files present in the initial state: we always require that the *root*-directory [] is part of the initial state and for every file in the initial state (excluding []) we require that its parent is also part of the initial state. A state is determined by a list of events, called the *trace*. The empty trace, or empty list, stands for the initial state. Given a trace s, we prepend an event to s to stand for the state in which the event just happened. We need to define functions that allow us to make some observations about traces. One such function is called *current_procs* and calculates the set of "alive" processes in a state:

$$
\begin{aligned}
current\_procs\ [] &\overset{def}{=} init\_procs \\
current\_procs\ (Clone\ p\ p'::s) &\overset{def}{=} \{p'\} \cup current\_procs\ s \\
current\_procs\ (Kill\ p\ p'::s) &\overset{def}{=} current\_procs\ s - \{p'\} \\
current\_procs\ (\_::s) &\overset{def}{=} current\_procs\ s
\end{aligned}
$$

The first clause states that in the empty trace the processes are given by *init_processes*. The events for cloning a process, respectively killing a process, update this set of processes appropriately. Otherwise the set of live processes is unchanged. We have similar functions for alive files and IPCs, called *current_files* and *current_ipcs*.

We can use these functions in order to formally model which events are *admissible* by the operating system in each state. We show just three rules that give the gist of this definition. First the rule for changing an owner of a process:

$$
\frac{p \in current\_procs\ s \qquad u \in init\_users}{admissible\ s\ (ChangeOwner\ p\ u)}
$$

We require that the process p is alive in the state s (first premise) and that the new owner is a user that existed in the initial state (second premise). Next the rule for creating a new file:

$$\frac{p \in current\_procs\ s \quad f \notin current\_files\ s \quad is\_parent\ f\ pf \quad pf \in current\_files\ s}{admissible\ s\ (CreateFile\ p\ f)}$$

It states that a file f can be created by a process p being alive in the state s, the new file does not exist already in this state and there exists a parent file pf for the new file. The parent file is just the tail of the list representing f. Finally, the rule for cloning a process:

$$\frac{p \in current\_procs\ s \quad p' \notin current\_procs\ s}{admissible\ s\ (Clone\ p\ p')}$$

Clearly the operating system should only allow to clone a process p if the process is currently alive. The cloned process will get the process ID generated by the operating system, but this process ID should not already exist. The admissibility rules for the other events impose similar conditions.

However, the admissibility check by the operating system is only one "side" of the constraints the RC-Model imposes. We also need to model the constraints of the access policy. For this we introduce separate *granted*-rules involving the sets *permissions* and *compatible r*: the former contains triples describing access control rules; the latter specifies for each role r which roles are compatible with r. These sets are used in the RC-Model when a process having a role r takes on a new role r'. For example, a login-process might belong to root; once the user logs in, however, the role of the process should change to the user's default role. The corresponding *granted*-rule is as follows

$$\frac{is\_current\_role\ s\ p\ r \quad r' \in compatible\ r}{granted\ s\ (ChangeRole\ p\ r')}$$

where we check whether the process p has currently role r and whether the RC-policy states that r' is in the role compatibility set of r.

The complication in the RC-Model arises from the way the current role of a process in a state s is calculated—represented by the predicate *is_current_role* in our formalisation. For defining this predicate we need to trace the role of a process from the initial state to the current state. In the initial state all processes have the role given by the function *init_current_role*. If a *Clone* event happens then the new process will inherit the role from the parent process. Similarly, if a *ChangeRole* event happens, then as seen in the rule above we just change the role accordingly. More interesting is an *Execute* event in the RC-Model. For this event we have to check the role attached to the file to be executed. There are a number of cases: If the role of the file is a *normal* role, then the process will just take on this role when executing the file (this is like the setuid mechanism in Unix). But there are also four *special* roles in the RC-Model: *Inherit-ProcessRole*, *InheritUserRole*, *InheritParentRole* and *InheritUpMixed*. For example, if a file to be executed has *InheritProcessRole* attached to it, then the process that executes this file keeps its role regardless of the information attached to the file. In this way programs can be can quarantined; *InheritUserRole* can be used for login shells to make sure they run with the user's default role. The purpose of the other special roles is to determine the role of a process according to the directory in which the files are stored.

Having the notion of current role in place, we can define the granted rule for the *Execute*-event: Suppose a process p wants to execute a file f. The RC-Model first fetches the role r of this process (in the current state s) and the type t of the file. It then checks if the tuple $(r, t, Execute)$ is part of the policy, that is in our formalisation being an element in the set *permissions*. The corresponding rule is as follows

$$\frac{is\_current\_role\ s\ p\ r \quad is\_file\_type\ s\ f\ t \quad (r,\ t,\ Execute) \in permissions}{granted\ s\ (Execute\ p\ f)}$$

The next *granted*-rule concerns the *CreateFile* event. If this event occurs, then we have two rules in our RC-Model depending on how the type of the created file is derived. If the type is inherited from the parent directory pf, then the *granted*-rule is as follows:

$$\frac{\begin{array}{c}is\_parent\ f\ pf \quad is\_file\_type\ s\ pf\ t \quad is\_current\_role\ s\ p\ r \\ default\_type\ r = InheritPatentType \quad (r,\ t,\ Write) \in permissions\end{array}}{granted\ s\ (CreateFile\ p\ f)}$$

We check whether pf is the parent file (directory) of f and check whether the type of pf is t. We also need to fetch the role r of the process that seeks to get permission for creating the file. If the default type of this role r states that the type of the newly created file will be inherited from the parent file type, then we only need to check that the policy states that r has permission to write into the directory pf.

The situation is different if the default type of role r is some *normal* type, like text-file or executable. In such cases we want that the process creates some predetermined type of files. Therefore in the rule we have to check whether the role is allowed to create a file of that type, and also check whether the role is allowed to write any new file into the parent file (directory). The corresponding rule is as follows.

$$\frac{\begin{array}{c}is\_parent\ f\ pf \\ is\_file\_type\ s\ pf\ t \quad is\_current\_role\ s\ p\ r \quad default\_type\ r = NormalFileType\ t' \\ (r,\ t,\ Write) \in permissions \quad (r,\ t',\ Create) \in permissions\end{array}}{granted\ s\ (CreateFile\ p\ f)}$$

Interestingly, the type-information in the RC-model is also used for processes, for example when they need to change their owner. For this we have the rule

$$\frac{is\_current\_role\ s\ p\ r \quad is\_process\_type\ s\ p\ t \quad (r,\ t,\ ChangeOwner) \in permissions}{granted\ s\ (ChangeOwner\ p\ u)}$$

whereby we have to obtain both the role and type of the process p, and then check whether the policy allows a *ChangeOwner*-event for that role and type.

Overall we have 13 rules for the admissibility check by the operating system and 14 rules for the granted check by the RC-Model. They are used to characterise when an event e is *valid* to occur in a state s. This can be inductively defined as the set of valid states.

$$\frac{}{valid \; []} \qquad \frac{valid \; s \quad admissible \; s \; e \quad granted \; s \; e}{valid \; (e::s)}$$

4   The Tainted Relation

The novel notion we introduce in this paper is the *tainted* relation. It characterises how a system can become infected when a file in the system contains, for example, a virus. We assume that the initial state contains some tainted objects (we call them *seeds*). Therefore in the initial state [] an object is tainted, if it is an element in *seeds*.

$$\frac{obj \in seeds}{obj \in tainted \; []}$$

Let us first assume such a tainted object is a file *f*. If a process reads or executes a tainted file, then this process becomes tainted (in the state where the corresponding event occurs).

$$\frac{f \in tainted \; s \quad valid \; (Execute \; p \; f::s)}{p \in tainted \; (Execute \; p \; f::s)} \qquad \frac{f \in tainted \; s \quad valid \; (ReadFile \; p \; f::s)}{p \in tainted \; (ReadFile \; p \; f::s)}$$

We have a similar rule for a tainted IPC, namely

$$\frac{i \in tainted \; s \quad valid \; (Recv \; p \; i::s)}{p \in tainted \; (Recv \; p \; i::s)}$$

which means if we receive anything from a tainted IPC, then the process becomes tainted. A process is also tainted when it is a produced by a *Clone*-event.

$$\frac{p \in tainted \; s \quad valid \; (Clone \; p \; p'::s)}{p' \in tainted \; (Clone \; p \; p'::s)}$$

However, the tainting relationship must also work in the "other" direction, meaning if a process is tainted, then every file that is written or created will be tainted. This is captured by the four rules:

$$\frac{p \in tainted \; s \quad valid \; (CreateFile \; p \; f::s)}{f \in tainted \; (CreateFile \; p \; f::s)} \qquad \frac{p \in tainted \; s \quad valid \; (WriteFile \; p \; f::s)}{f \in tainted \; (WriteFile \; p \; f::s)}$$

$$\frac{p \in tainted \; s \quad valid \; (CreateIPC \; p \; i::s)}{i \in tainted \; (CreateIPC \; p \; i::s)} \qquad \frac{p \subset tainted \; s \quad valid \; (Send \; p \; i::s)}{i \in tainted \; (Send \; p \; i::s)}$$

Finally, we have three rules that state whenever an object is tainted in a state *s*, then it will be still tainted in the next state *e::s*, provided the object is still *alive* in that state. We have such a rule for each kind of objects, for example for files the rule is:

$$\frac{f \in tainted \; s \quad valid \; (e::s) \quad f \in current\_files \; (e::s)}{f \in tainted \; (e::s)}$$

Similarly for alive processes and IPCs (then respectively with premises $p \in current\_procs$ $(e::s)$ and $i \in current\_ipcs$ $(e::s)$). When an object present in the initial state can be tainted in *some* state (system run), we say it is *taintable*:

$$taintable \; obj \stackrel{def}{=} obj \in init \wedge \exists s. \; obj \in tainted \; s$$

Before we can describe our static check deciding when a file is taintable, we need to describe the notions *deleted* and *undeletable* for objects. The former characterises whether there is an event that deletes these objects (files, processes or IPCs). For this we have the following four rules:

$$\frac{\overline{deleted \; p' \; (Kill \; p \; p'::s)}}{\overline{deleted \; f \; (DeleteFile \; p \; f::s)}} \qquad \frac{deleted \; obj \; s}{deleted \; obj \; (e::s)}$$

$$\overline{deleted \; i \; (DeleteIPC \; p \; i::s)}$$

Note that an object cannot be deleted in the initial state []. An object is then said to be *undeletable* provided it did exist in the initial state and there does not exists a valid state in which the object is deleted:

$$undeletable \; obj \stackrel{def}{=} obj \in init \wedge \neg \; (\exists s. \; valid \; s \wedge deleted \; obj \; s)$$

The point of this definition is that our static taintable check will only be complete for undeletable objects. But these are the ones system administrators are typically interested in (for example system files).

It should be clear that we cannot hope for a meaningful check by just trying out all possible valid states in our dynamic model. The reason is that there are potentially infinitely many of them and therefore the search space would be infinite. For example starting from an initial state containing a process p and a file pf, we can create files f_1, f_2, ... via *CreateFile*-events. This can be pictured roughly as follows:

Initial state:
$$\{p, pf\} \quad \Longrightarrow \quad \{p, pf, f_1::pf\} \quad \Longrightarrow \quad \{p, pf, f_1::pf, f_2::f_1::pf\} \; ...$$
$$\qquad CreateFile \; p \; (f_1::pf) \qquad CreateFile \; p \; (f_2::f_1::pf)$$

Instead, the idea of our static check is to use the policies of the RC-model for generating an answer, since they provide always a finite "description of the system". As we will see in the next section, this needs some care, however.

5 Our Static Check

Assume there is a tainted file in the system and suppose we face the problem of finding out whether this file can affect other files, IPCs or processes? One idea is to work on the level of policies only, and check which operations are permitted by the role and type of this file. Then one builds the "transitive closure" of this information and checks for example whether the role *root* has become affected, in which case the whole system is compromised. This is indeed the solution investigated in [3] in the context of information flow and SELinux.

Unfortunately, restricting the calculations to only use policies is too simplistic for obtaining a check that is sound and complete—it over-approximates the dynamic tainted relation defined in the previous section. To see the problem consider the case where the tainted file has, say, the type *bin*. If the RC-policy contains a role *r* that can both read and write *bin*-files, we would conclude that all *bin*-files can potentially be tainted. That is indeed the case, *if* there is a process having this role *r* running in the system. But if there is *not*, then the tainted file cannot "spread". A similar problem arises in case there are two processes having the same role *r*, and this role is restricted to read files only. Now if one of the processes is tainted, then the simple check involving only policies would incorrectly infer that all processes involving that role are tainted. But since the policy for *r* is restricted to be read-only, there is in fact no danger that both processes can become tainted.

The main idea of our sound and complete check is to find a "middle" ground between the potentially infinite dynamic model and the too coarse information contained in the RC-policies. Our solution is to define a "static" version of the tainted relation, called *tainted^s*, that records relatively precisely the information about the initial state of the system (the one in which an object might be a *seed* and therefore tainted). However, we are less precise about the objects created in every subsequent state. The result is that we can avoid the potential infinity of the dynamic model. For the *tainted^s*-relation we will consider the following three kinds of *items* recording the information we need about processes, files and IPCs, respectively:

Recorded information:

Processes:	$P(r, dr, t, u)^{po}$
Files:	$F(t, a)^{fo}$
IPCs:	$I(t)^{io}$

For a process we record its role *r*, its default role *dr* (used to determine the role when executing a file or changing the owner of a process), its type *t* and its owner *u*. For a file we record just the type *t* and its *anchor a* (we define this notion shortly). For IPCs we only record its type *t*. Note the superscripts *po*, *fo* and *io* in each item. They are optional arguments and depend on whether the corresponding object is present in the initial state or not. If it *is*, then for processes and IPCs we will record *Some id*, where *id* is the natural number that uniquely identifies a process or IPC; for files we just record their path *Some f*. If the object is *not* present in the initial state, that is newly created, then we just have *None* as superscript. Let us illustrate the different superscripts with the following example where the initial state contains a process *p* and a file (directory) *pf*. Then this process creates a file via a *CreateFile*-event and after that reads the created file via a *Read* event:

Initial state:
$$\{p, pf\} \quad \overset{}{\Longrightarrow} \quad \{p, pf, f::pf\} \quad \overset{}{\Longrightarrow} \quad \{p, pf, f::pf\}$$
$$\qquad CreateFile\ p\ (f::pf) \qquad\qquad ReadFile\ p\ (f::pf)$$

For the two objects in the initial state our static check records the information $P(r, dr, t, u)^{Some\ p}$ and $F(t', a)^{Some\ pf}$ (assuming r, t and so on are the corresponding roles, types etc). In both cases we have the superscript $Some(...)$ since they are objects present in the initial state. For the file $f::pf$ created by the *CreateFile*-event, we record $F(t', a')^{None}$, since it is a newly created file. The *ReadFile*-event does not change the set of objects, therefore no new information needs to be recorded. The problem we are avoiding with this setup of recording the precise information for the initial state is where two processes have the same role and type information, but only one is tainted in the initial state, but the other is not. The recorded unique process ID allows us to distinguish between both processes. For all newly created objects, on the other hand, we do not care. This is crucial, because otherwise exploring all possible "reachable" objects can lead to the potential infinity like in the dynamic model.

An *anchor* for a file is the "nearest" directory that is present in the initial state and has not been deleted in a state s. Its definition is the recursive function

$$anchor\ s\ [] \quad \overset{def}{=} \quad if \neg\ deleted\ []\ s\ then\ Some\ []\ else\ None$$

$$anchor\ s\ (f::pf) \quad \overset{def}{=} \quad if\ f::pf \in init\_files \land \neg\ deleted\ (f::pf)\ s$$
$$then\ Some\ (f::pf)\ else\ anchor\ s\ pf$$

generating an optional value. The first clause states that the *root*-directory is always its own anchor unless it has been deleted. If a file is present in the initial state and not deleted in s, then it is also its own anchor, otherwise the anchor will be the anchor of the parent directory. For example if we have a directory pf in the initial state, then its anchor is *Some pf* (assuming it is not deleted). If we create a new file in this directory, say $f::pf$, then its anchor will also be *Some pf*. The purpose of *anchor* is to determine the role information when a file is executed, because the role of the corresponding process, according to the RC-model, is determined by the role information of the anchor of the file to be executed.

There is one last problem we have to solve before we can give the rules of our $tainted^s$-check. Suppose an RC-policy includes the rule $(r, foo, Write) \in permissions$, that is a process of role r is allowed to write files of type *foo*. If there is a tainted process with this role, we would conclude that also every file of that type can potentially become tainted. However, that is not the case if the initial state does not contain any file with type *foo* and the RC-policy does not allow the creation of such files, that is does not contain an access rule $(r, foo, Create) \in permissions$. In a sense the original $(r, foo, Write)$ is "useless" and should not contribute to the relation characterising the objects that are tainted. To exclude such "useless" access rules, we define a relation $reachable^s$ restricting our search space to only configurations that correspond to states in our dynamic model. We first have a rule for reachable items of the form $F(t, f)^{Some\ f}$ where the file f with type t is present in the initial state.

$$\frac{f \in init\_files \quad is\_file\_type\ []\ f\ t}{F(t, f)^{Some\ f} \in reachable^s}$$

We have similar reachability rules for processes and IPCs that are part of the initial state. Next is the reachability rule in case a file is created

$$F(t, a)^{fo} \in reachable^s$$
$$P(r, dr, pt, u)^{po} \in reachable^s \quad default\_type\ r = NormalFileType\ t'$$
$$(r,\ t,\ Write) \in permissions \quad (r,\ t',\ Create) \in permissions$$
$$\overline{\qquad\qquad F(t', a)^{None} \in reachable^s \qquad\qquad}$$

where we require that we have a reachable parent directory, recorded as $F(t, a)^{fo}$, and also a process that can create the file, recorded as $P(r, dr, pt, u)^{po}$. As can be seen, we also require that we have both $(r, t, Write)$ and $(r, t', Create)$ in the *permissions* set for this rule to apply. If we did *not* impose this requirement about the RC-policy, then there would be no way to create a file with *NormalFileType* t' according to our "dynamic" model. However in case we want to create a file of type *InheritPatentType*, then we only need the access-rule $(r, t, Write)$:

$$F(t, a)^{fo} \in reachable^s \quad P(r, dr, pt, u)^{po} \in reachable^s$$
$$default\_type\ r = InheritPatentType \quad (r,\ t,\ Write) \in permissions$$
$$\overline{\qquad\qquad F(t, a)^{None} \in reachable^s \qquad\qquad}$$

We also have reachability rules for processes executing files, and for changing their roles and owners, for example

$$\frac{P(r, dr, t, u)^{po} \in reachable^s \quad r' \in compatible\ r}{P(r', dr, t, u)^{po} \in reachable^s}$$

which states that when we have a process with role r, and the role r' is in the corresponding role-compatibility set, then also a process with role r' is reachable.

The crucial difference between between the "dynamic" notion of validity and the "static" notion of *reachable*^{s} is that there can be infinitely many valid states, but assuming the initial state contains only finitely many objects, then also *reachable*^{s} will be finite. To see the difference, consider the infinite "chain" of events just cloning a process p_0:

Initial state:
$$\{p_0\} \quad \Longrightarrow \quad \{p_0, p_1\} \quad \Longrightarrow \quad \{p_0, p_1, p_2\} \ \cdots$$
$$\qquad Clone\ p_0\ p_1 \qquad\qquad Clone\ p_0\ p_2$$

The corresponding reachable objects are

$$\{P(r, dr, t, u)^{Some\ (p_0)}\} \quad \Longrightarrow \quad \{P(r, dr, t, u)^{Some\ (p_0)}, P(r, dr, t, u)^{None}\}$$

where no further progress can be made because the information recorded about p_2, p_3 and so on is just the same as for p_1, namely $P(r, dr, t, u)^{None}$. Indeed we can prove the lemma:

Lemma 1. *If finite init, then finite reachable*^{s}.

This fact of *reachable*^{s} being finite enables us to design a decidable tainted-check. For this we introduce inductive rules defining the set *tainted*^{s}. Like in the "dynamic" version of tainted, if an object is element of *seeds*, then it is *tainted*^{s}.

$$\frac{obj \in seeds}{[\![obj]\!] \in tainted^s}$$

The function $[\![\_]\!]$ extracts the static information from an object. For example for a process it extracts the role, default role, type and user; for a file the type and the anchor. If a process is tainted and creates a file with a normal type t' then also the created file is tainted. The corresponding rule is

$$\frac{\begin{array}{c} P(r, dr, pt, u)^{po} \in tainted^s \\ F(t, a)^{fo} \in reachable^s \quad default\_type \; r = NormalFileType \; t' \\ (r, \; t, \; Write) \in permissions \quad (r, \; t', \; Create) \in permissions \end{array}}{F(t', a)^{None} \in tainted^s}$$

If a tainted process creates a file that inherits the type of the directory, then the file will also be tainted:

$$\frac{\begin{array}{c} P(r, dr, pt, u)^{po} \in tainted^s \quad F(t, a)^{fo} \in reachable^s \\ default\_type \; r = InheritPatentType \quad (r, \; t, \; Write) \in permissions \end{array}}{F(t, a)^{None} \in tainted^s}$$

If a tainted process changes its role, then also with this changed role it will be tainted:

$$\frac{P(r, dr, t, u)^{po} \in tainted^s \quad r' \in compatible \; r}{P(r', dr, t, u)^{po} \in tainted^s}$$

Similarly when a process changes its owner. If a file is tainted, and a process has read-permission to that type of files, then the process becomes tainted. The corresponding rule is

$$\frac{F(t, a)^{fo} \in tainted^s \quad P(r, dr, pt, u)^{po} \in reachable^s \quad (r, \; t, \; Read) \in permissions}{P(r, dr, pt, u)^{po} \in tainted^s}$$

If a process is tainted and it has write-permission for files of type t, then these files will be tainted:

$$\frac{P(r, dr, pt, u)^{po} \in tainted^s \quad F(t, a)^{fo} \in reachable^s \quad (r, \; t, \; Write) \in permissions}{F(t, a)^{fo} \in tainted^s}$$

We omit the remaining rules for executing a file, cloning a process and rules involving IPCs, which are similar. A simple consequence of our definitions is that every tainted object is also reachable:

Lemma 2. $tainted^s \subseteq reachable^s$

which in turn means that the set of $tainted^s$ items is finite by Lemma 1.

Returning to our original question about whether tainted objects can spread in the system. To answer this question, we take these tainted objects as seeds and calculate the set of items that are *tainted*^{s}. We proved this set is finite and can be enumerated using the rules for *tainted*^{s}. However, this set is about items, not about whether objects are tainted or not. Assuming an item in *tainted*^{s} arises from an object present in the initial state, we have recorded enough information to translate items back into objects via the function $|\_|$:

$$|P(r, dr, t, u)^{po}| \stackrel{def}{=} po$$
$$|F(t, a)^{fo}| \stackrel{def}{=} fo$$
$$|I(t)^{io}| \stackrel{def}{=} io$$

Using this function, we can define when an object is *taintable*^{s} in terms of an item being *tainted*^{s}, namely

$$taintable^s \; obj \stackrel{def}{=} \exists item. \; item \in tainted^s \wedge |item| = Some \; obj$$

Note that *taintable*^{s} is only about objects that are present in the initial state, because for all other items $|\_|$ returns *None*.

With these definitions in place, we can state our theorem about the soundness of our static *taintable*^{s}-check for objects.

Theorem 1 (Soundness). *If taintable*^{s} *obj then taintable obj.*

The proof of this theorem generates for every object that is "flagged" as *taintable*^{s} by our check, a sequence of events which shows how the object can become tainted in the dynamic model. We can also state a completeness theorem for our *taintable*^{s}-check.

Theorem 2 (Completeness). *If undeletable obj and taintable obj then taintable*^{s} *obj.*

This completeness theorem however needs to be restricted to undeletebale objects. The reason is that a tainted process can be killed by another process, and after that can be "recreated" by a cloning event from an untainted process—remember we have no control over which process ID a process will be assigned with. Clearly, in this case the cloned process should be considered untainted, and indeed our dynamic tainted relation is defined in this way. The problem is that a static test cannot know about a process being killed and then recreated. Therefore the static test will not be able to "detect" the difference. Therefore we solve this problem by considering only objects that are present in the initial state and cannot be deleted. By the latter we mean that the RC-policy stipulates an object cannot be deleted (for example it has been created by *root* in single-user mode, but in the everyday running of the system the RC-policy forbids to delete an object belonging to *root*). Like *taintable*^{s}, we also have a static check for when a file is undeletable according to an RC-policy.

This restriction to undeletable objects might be seen as a great weakness of our result, but in practice this seems to cover the interesting scenarios encountered by system administrators. They want to know whether a virus-infected file introduced by a user can affect the core system files. Our test allows the system administrator to find this out provided the RC-policy makes the core system files undeletable. We assume that this proviso is already part of best practice rule for running a system.

We envisage our test to be useful in two kind of situations: First, if there was a break-in into a system, then, clearly, the system administrator can find out whether the existing access policy was strong enough to contain the break-in, or whether core system files could have been affected. In the first case, the system administrator can just plug the hole and forget about the break-in; in the other case the system administrator is wise to completely reinstall the system. Second, the system administrator can proactively check whether an RC-policy is strong enough to withstand serious break-ins. To do so one has to identify the set of "core" system files that the policy should protect and mark every possible entry point for an attacker as tainted (they are the seeds of the $tainted^s$ relation). Then the test will reveal whether the policy is strong enough or needs to be redesigned. For this redesign, the sequence of events our check generates should be informative.

6   Conclusion and Related Works

We have presented the first completely formalised dynamic model of the Role-Compatibility Model. This is a framework, introduced by Ott [5], in which role-based access control policies can be formulated and is used in practice, for example, for securing Apache servers. Previously, the RC-Model was presented as a collection of rules partly given in "English", partly given as formulas. During the formalisation we uncovered an inconsistency in the semantics of the special role *InheritParentRole* in the existing works about the RC-Model [5,7]. By proving the soundness and completeness of our static $taintable^s$-check, we have formally related the dynamic behaviour of the operating system implementing access control and the static behaviour of the access policies of the RC-Model. The crucial idea in our static check is to record precisely the information available about the initial state (in which some resources might be tainted), but be less precise about the subsequent states. The former fact essentially gives us the soundness of our check, while the latter results in a finite search space.

The two most closely related works are by Archer et al and by Guttman et al [1,3]. The first describes a formalisation of the dynamic behaviour of SELinux carried out in the theorem prover PVS. However, they cannot use their formalisation in order to prove any "deep" properties about access control rules [1, Page 167]. The second analyses access control policies in the context of information flow. Since this work is completely on the level of policies, it does not lead to a sound and complete check for files being taintable (a dynamic notion defined in terms of operations performed by the operating system). While our results concern the RC-Model, we expect that they equally apply to the access control model of SELinux. In fact, we expect that the formalisation is simpler for SELinux, since its rules governing roles are much simpler than in the RC-Model. The definition of our admissibility rules can be copied verbatim for SELinux; we would need to modify our granted rules and slightly adapt our static check. We leave this as future work. Another direction of future work could be to reason formally about confidentiality in access control models. This would, of course, need the explicit assumption about the absence of any covert channels in systems.

Our formalisation is carried out in the Isabelle/HOL theorem prover. It uses Paulson's inductive method for reasoning about sequences of events [8]. We have approximately 1000 lines of code for definitions and 6000 lines of code for proofs. Our formalisation is available from the Mercurial repository at `http://www.dcs.kcl.ac.uk/staff/urbanc/cgi-bin/repos.cgi/rc/`.

References

1. Archer, M., Leonard, E.I., Pradella, M.: Analyzing Security-Enhanced Linux Policy Specifications. In: Proc. of the 4th IEEE International Workshop on Policies for Distributed Systems and Networks (POLICY), pp. 158–169 (2003)
2. Bugliesi, M., Calzavara, S., Focardi, R., Squarcina, M.: Gran: Model Checking Grsecurity RBAC Policies. In: Proc. of the 25th IEEE Computer Security Foundations Symposium (CSF), pp. 126–138 (2012)
3. Guttman, J.D., Herzog, A.L., Ramsdell, J.D., Skorupka, C.W.: Verifying Information Flow Goals in Security-Enhanced Linux. Journal of Computer Security 13(1), 115–134 (2005)
4. Jha, S., Li, N., Tripunitara, M.V., Wang, Q., Winsborough, W.H.: Towards Formal Verification of Role-Based Access Control Policies. IEEE Transactions Dependable and Secure Computing 5(4), 242–255 (2008)
5. Ott, A.: The Role Compatibility Security Model. In: Proc. of the 7th Nordic Workshop on Secure IT Systems, NordSec (2002)
6. Ott, A.: Mandatory Rule Set Based Access Control in Linux: A Multi-Policy Security Framework and Role Model Solution for Access Control in Networked Linux Systems. PhD thesis, University of Hamburg (2007)
7. Ott, A., Fischer-Hübner, S.: A Role-Compatibility Model for Secure System Administration, http://www.rsbac.org/doc/media/rc-paper.php
8. Paulson, L.C.: The Inductive Approach to Verifying Cryptographic Protocols. Journal of Computer Security 6(1-2), 85–128 (1998)
9. Sarna-Starosta, B., Stoller, S.D.: Policy Analysis for Security-Enhanced Linux. In: Proc. of the 2004 Workshop on Issues in the Theory of Security (WITS), pp. 1–12 (2004)
10. Uzun, E., Atluri, V., Sural, S., Vaidya, J., Parlato, G., Ferrara, A.L., Madhusudan, P.: Analyzing Temporal Role Based Access Control Models. In: Proc. of the 17th ACM Symposium on Access Control Models and Technologies (SACMAT), pp. 177–186 (2012)
11. Zhang, X., Urban, C., Wu, C.: Priority Inheritance Protocol Proved Correct. In: Beringer, L., Felty, A. (eds.) ITP 2012. LNCS, vol. 7406, pp. 217–232. Springer, Heidelberg (2012)

Author Index